PHARMACEUTICAL REGULATORY INSPECTIONS

PHARMACEUTICAL REGULATORY INSPECTIONS

"EVERY INSPECTION A SUCCESS STORY – A PRACTICAL GUIDE TO MAKE IT HAPPEN"

EDITED BY

MADHU RAJU SAGHEE
QUALITY ASSURANCE, MICRO LABS, INDIA
&
DIRECTOR (PHSS) – INDIA

© 2014 Euromed Communications

Published by Euromed Communications, Passfield, England.
www.euromedcommunications.com

Designed by Sarah Beale, Get Set Go
Printed in England by 4edge Ltd, Hockley, Essex.

ISBN 978-1-899015-89-4

DEDICATED TO
JOE RIDGE
FOUNDER OF EUROMED COMMUNICATIONS
A VISIONARY WRITER, EDITOR AND PUBLISHER

AND

S.M. MUDDA,
EXECUTIVE DIRECTOR - TECHNICAL, MICRO LABS, INDIA
A QUINTESSENTIAL PHARMACEUTICAL PROFESSIONAL
WHO ESPOUSES THE IMPORTANCE OF PQS
(PHARMACEUTICAL QUALITY SYSTEM) AS THE LYNCHPIN OF
BMS (BUSINESS MANAGEMENT SYSTEM)

CONTENTS

FOREWORD

In the 35+ years that I have been involved with the US and international pharmaceutical industries both from within the FDA and from the consultant's perspective, companies and sites that are inspected by regulatory authorities struggle to fully understand the reasons and objectives for inspections, inspection approaches taken by the regulators, how to properly prepare to host regulatory inspections, and how to effectively communicate with the inspectors and their agencies.

These difficulties were apparent in the mid-1970s and continue today. As the number of regulatory authorities performing inspections, particularly at foreign sites increases, together with the number of inspections being conducted, a fuller understanding of the inspections process becomes more important to ensure a better chance of a successful inspection and consequent product approval.

Another point which must be considered and not often realized is that there are new generations of employees, and managers who continue to enter the pharmaceutical industry, as well as new inspectors in the regulatory agencies. I say this because "my generation" has lived through the generic drug scandal, the evolution specific Good Manufacturing Practice (GMP) requirements for sterile products and APIs, the movement of documentation from paper based systems to electronic formats, and enormous advances in technology and automation which has formed much of the basis for the industry as it is today. The "new generation" need the information passed along to better cope with contemporary regulatory activities.

For example, not too long ago I was conducting a GMP-related training course and casually mentioned the generic drug scandal. About half of the attendees had puzzled looks on their faces. I then realized that a large number of people working in the pharmaceutical industry today have no knowledge of those dark days for the industry and the FDA. That episode changed the way the FDA does business. This was the genesis of the FDA pre-approval inspection program, as well as requirements for handling of out-of-specification analytical results, raw data management by control of analytical worksheets and batch records, and the importance of properly documented investigations. It is extremely helpful if the "new generation" can understand the basis and reasons why we do certain things today.

Having said that, the selection of chapters in this book represents a collection of topics of great breadth and importance to the pharmaceutical industry, especially for someone new to a position that requires interfacing with regulatory authorities. The individual chapters have been contributed by an exemplary group of authors with collectively more than 225 years of experience in government and industry in the United States, Europe, Japan, and India. The authors share their knowledge and practical experiences to provide a great deal of insight to assist pharmaceutical industry professionals to prepare for and manage inspections by various international regulatory authorities.

I am not aware of any other contemporary book that provides the comprehensive information that is contained in *Pharmaceutical Regulatory Inspections*. As indicated on this book's cover, this is a **practical** guide to achieving compliant pharmaceutical manufacturing operations.

A fundamental requisite for successful regulatory inspections is preparation. Preparation has many facets that include first and foremost, establishing compliant GMP and Quality Systems.

But an important element of preparation also includes knowledge and understanding of the laws and GMP requirements - "why we must do certain things". These chapters provide the background information as well as outlines and suggested activities.

The first chapter sets the background for GMP Regulations providing an overview of GMP including a discussion of the history of GMPs and the current requirements. Global GMPs are discussed highlighting some similarities and differences between international regulations and guidelines.

Chapter 2 goes on to provide an overview of U.S. Food and Drug Administration (FDA) regulation and enforcement of drug products from the all-important legal perspective, beginning with a basic overview of the regulatory requirements, as well as information with respect to FDA's enforcement authority.

Several years ago the FDA adopted the System-Based Inspection Approach which essentially enables the FDA to assess individual or a combination of systems without using the time needed to conduct a complete "front door to back door" inspection. Use of the System-Based inspection approach does not change any GMP requirements, but does categorize each GMP element into one of the six FDA-defined systems. Understanding this approach, and that the Quality System will always be covered during inspections, is helpful to prepare and predict where the inspection will be going.

The FDA drug approval process will involve one or more Pre-Approval Inspections (PAI) and one chapter is devoted to the PAI and several other chapters contain valuable information with respect to the PAI. The Pre-Approval Inspection actually is "GMP +", which in addition to determining GMP compliance at the facility, the inspection will also verify that the data contained in the associated submission upon which FDA will base its approval decision, is accurate and credible.

Preparing for a PAI requires an understanding of the origin and reasons for the FDA's Pre-Approval Inspection program. Knowing why the program exists helps identify the issues and areas that will be the focus of an FDA inspection. Proactive preparations will lead to a successful PAI and the timely approval of the associated application.

Once a manufacturing site is found acceptable by the FDA, there will be subsequent routine inspections. Although the frequency of the routine follow-up inspections is quite variable, reinspections are performed based on risk. Sterile products, being more risky than non-sterile Active Pharmaceutical Ingredients, are likely to be re-inspected at a greater frequency. If a routine re-inspection fails the agency has several options including issuing a Warning Letter, or pursuing a Consent Decree, or product seizures. Foreign sites following a failed inspection can be placed on "Automatic Detention" in a relatively short time, which effectively prohibits a company from importing any or all products into the U.S.

When it comes to drug product approval and marketing in the European Union (EU), there are a number of changes to the process. With most companies striving to achieve a presence in the global marketplace, adherence to the European regulatory requirements as well as the US FDA, is extremely important. Unlike the U.S., Europe is an economic and political partnership between member states (27 European countries with 44 National Competent Authorities), who are linked through common regulations and whose governments are at liberty to interpret European law when implementing it into their own legislation. The result is a much more varied and individual approach to inspections than experienced elsewhere. Chapter 6 covers the key legislation in the EU, the GMP Guidance, the major agencies and the specifics of an inspection by EU inspectors.

Also important in the global marketplace is Japan, with its own marketing authorization license and approval system. Pharmaceutical products are regulated with rules established by the Pharmaceutical Affairs Law (PAL) that provides the legal basis for the GMP inspection activity implemented by the Pharmaceuticals and Medical Devices Agency (PMDA). Details with regard to the PMDA requirements are covered in Chapter 7.

Sterile pharmaceutical products carry the highest risk due to their route of administration which could lead to patient harm or even death if they are not sterile. The complexities of sterile drug production operations, especially aseptic processes, while maintaining sterility is an onerous task. These products are subject to the most frequent and stringent inspections by regulatory authorities. Therefore this book has devoted an entire chapter to the topic of sterile product manufacturing.

Another important facet to regulatory processes is the inspection of Active Pharmaceutical Ingredients (APIs). Since the vast majority of APIs are produced in foreign countries, the FDA plans many foreign inspections to focus on APIs. It should be noted that many FDA investigators performing foreign inspections of API manufacturers will have gained little experience in the home district and therefore during inspections hosts should be aware of this situation and develop plans to facilitate the inspection.

Chapters 8 and 11 are particularly useful to the understanding of FDA inspections at foreign API sites, with chapter 8 addressing international inspections and chapter 11 devoted to API inspections. Chapter 11 also includes the FDA Compliance Program (their "SOP") for conducting API inspections and a handy checklist to aid the manufacturer in preparing for the inspection of API facilities.

Once an inspection is completed there may be follow-up activities required by the company, especially with respect to responding to an FDA-483, List of Observations and verbal observations made by the inspection team. At this point an adequate response is of the utmost importance to avoid further action by the agency. The response must be well written, concise but provide sufficient details of corrections, address actions to the specific observation as well as system-wide global actions, without being defensive and without re-stating arguments that were already lost during the inspection. There is a 15 business day time frame for submission of the response following conclusion of the inspection. Chapter 9 is dedicated to providing a resource to post inspection activities.

The final chapter is a proactive model for achieving compliance. The purpose of this chapter is to describe an approach for a continuous improvement program and an intention that every site will be ready to host a regulatory inspection at any time. The chapter outlines the usefulness of the ICH Guidance documents (ICH Q8, Pharmaceutical Development, ICH Q9, Quality Risk Management, and ICH Q10, the Pharmaceutical Quality System) to achieve and maintain compliance, stating the importance of company top management being involved in and supporting Quality and compliance.

Understanding background information with respect to laws and regulations, how agencies conduct inspections, why the regulators do what they do, what they require, and what they expect, are very important elements of proper preparation. Differences in the conduct, requirements and expectations may be subtle but very important to know. How to manage an inspection and its aftermath can be crucial to the final outcome of inspections and subsequent product approvals. These chapters provide a great deal of information in an organized and easy to understand format. This will be a very useful resource for anyone involved in regulatory inspections.

Best wishes for a positive result with regard to your inspection.

Peter D. Smith, Vice President, Strategic Compliance
PAREXEL Consulting

PREFACE

It is of the utmost importance that pharmaceuticals are manufactured in a way which ensures that medicines are effective in terms of their label claim and are safe to use. The means by which safety of medicines is ensured is through regulations and the work of regulatory agencies, particularly compliance inspections.

Regulations are defined in a series of national and international standards which collectively form the body of guidance termed Good Manufacturing Practice (GMP). Whilst GMP is periodically defined through updates to published standards, pharmaceutical organizations are expected to be conversant with the current practices, technologies, tests and ways of working which will safeguard the product and the patient. This is the body of work and practices which form "current" GMP. This is drawn from inspectorate findings, published papers, and guidance documents.

In terms of guidance documents the twenty-first century has seen a plethora of documents relating to new technologies, risk management and quality systems, which the pharmaceutical manufacturer is expected to absorb and to incorporate into their operational systems. This has made the regulatory compliance requirements sometimes difficult to identify and sometimes difficult to meet.

The burden is upon the pharmaceutical organization to maintain knowledge of cGMP and of inspectorate trends. This is not always a straightforward task. One of the objectives of this book was to provide a framework, coupled with in-depth analysis, of current issues pertaining to pharmaceutical regulatory inspections. To achieve this, a group of leading international experts were approached to write chapters on various topics which relate to regulations, standards and inspection approaches. The outcome of this approach which provides a detailed overview of the important legislative and practical aspects of regulatory inspections is contained within this book

In presenting this material, the other objectives of the book were realized. These were, firstly, to provide detail on various major regulatory agency inspectional models, emphasizing the similarities and differences in the approaches taken by international regulators (particularly between the US Food and Drug Administration and the European Medicines Agency). The second objective was to provide the reader with pragmatic, detailed and meaningful advice in relation to preparing for a regulatory inspection. The third objective focused on highlighting the main areas of inspectorate concern in relation to different types of pharmaceutical manufacturing, from sterile products to Active Pharmaceutical

Ingredients. The key focal points are listed in the relevant chapters which are supported by detailed checklists to aid readers in assessing their own facilities.

The chapters in this book together cover all aspects of regulatory inspections. By way of an introduction, in chapter one, Tim Sandle and I provide an overview of global GMPs, where the similarities and differences between regulatory agencies are highlighted. The chapter also examines a range of different guidance bodies from organizations like ISO and ICH. In the second chapter, the focus moves to FDA specific issues. Here Seth Mailhot examines how the FDA undertakes drug enforcement and regulation.

In the third chapter, David Barr and Tim Sandle consider the 'systems approach' for regulatory inspections, with a focus upon how agencies tend to approach the process of inspections and how the organization should prepare and respond to such inspections.

The fourth chapter looks at a certain type of inspection – the pre-approval inspection – and considers what information is required to be collated by the manufacturer and the types of topics which inspectors focus on. This chapter is written by Ron Johnson.

The fifth chapter turns attention back to the FDA and here John Avellanet considers the best practices for managing and surviving FDA inspections. In a similar way, Siegfried Schmitt and Nabila Nazir, address the approaches and focal points required for EMA inspections in chapter six. To complete the tripartite overview of the major regulatory agencies, Yoshikazu Hayashi, in chapter seven, considers the approaches undertaken by Japanese inspectors.

Where the preceding three chapters address national and supranational inspections, in chapter eight, Andreas Brutsche and Tim Sandle consider the important aspects of international inspections. This advice is particularly useful for small and medium sized companies receiving inspections outside of their own territories for the first time.

Attention then moves to the actions required post-inspection, particularly on responding to inspection points and citations. This is set out in chapter nine by Tim Sandle, David Barr and myself.

The next two chapters examine inspection focal points relating to specific types of pharmaceutical manufacturing. In chapter ten, Tim Sandle examines the important regulatory issues pertaining to sterile manufacturing and in chapter eleven Siegfried Schmitt and Richard Einig consider the applicable regulatory issues of relevance to API manufacturing.

The book closes with a chapter by Mark Tucker which examines ways of optimizing regulatory compliance. Here the importance of an organization remaining up-to-date with compliance issues is addressed.

The compilation of these chapters has resulted in the creation of a unique book, where past agency inspectors and industry experts have expressed their views and provided tips for succeeding in regulatory inspections and streamlining regulatory compliance. No other single volume captures such a diverse range of experts to provide such a focused account of regulatory issues.

The editor wishes to express his thanks to each of the contributors for taking the time and trouble, and drawing upon many years of experience, to produce unique and detailed chapters. The editor would also like to thank the publisher Joe Ridge and typesetter Sarah Beale for their patience and diligence in the production of this work. The final product is a book which will aid different types of pharmaceutical organizations in preparing for regulatory inspections, understanding key regulatory issues and for reviewing inspectorate trends and findings.

Madhu Raju Saghee
Bangalore, India

Madhu Raju Saghee is currently working as AGM, Quality Assurance at Micro Labs Sterile Facility, India. Prior to this, he worked at Novartis and Gland Pharma in corporate quality and compliance functions leading and supporting both global and site compliance and validation efforts. Madhu has wide ranging expertise in the implementation of best-practice Pharmaceutical Quality System (PQS) and espouses the need for proactive attitudes and quality culture. He has immense experience in handling regulatory inspections and has helped firms to successfully pass regulatory inspections. Madhu is a frequent speaker at major industry conferences and often leads workshops on regulatory compliance.

His areas of expertise include all aspects of quality and compliance for systems, processes, facilities and operations for drug products, particularly for sterile products. Madhu is also a volunteer for Pharmaceutical and Healthcare Sciences Society (PHSS), acting as the Director for India. He is a member of the editorial board of GMP Review and the international review board of EJPPS. He is the recipient of the coveted 2013 "*Young Pharmaceutical Analyst Award*" from IDMA (Indian Drug Manufacturers Association). Madhu has edited/co-edited several books including *Achieving Quality and Compliance Excellence in Pharmaceuticals* and *Microbiology and Sterility Assurance in Pharmaceuticals and Medical Devices* and has authored many technical articles. Madhu has a Master of Science in microbiology from Andhra University. He is an active member of various industry associations including the PDA, ISPE and PHSS. He can be reached at madhu.saghee@gmail.com

John Avellanet

John Avellanet, author of "Get to Market Now! Turn FDA Compliance into a Competitive Edge in the Era of Personalized Medicine" (Logos Press), helps clients solve compliance problems. He is the founder and managing director of Cerulean Associates LLC, a lean compliance consulting firm. An internationally acknowledged expert who has trained FDA, NIH and Health Canada officials on supplier management issues, his clients include Fortune 50 pharmaceutical firms, biotechnology startups, and everything in between. He is an internationally renowned speaker, expert and authority on lean compliance practices and FDA cGMP, GLP, QSR, 21 CFR Part 11, ICH and GHTF.

Trusted by officials at FDA, NIH and the OIG, as well as by clients around the world, Mr Avellanet provides practical, business-savvy solutions to strengthen compliance while lowering costs and reducing risk. He speaks frequently at industry conferences and organizations around the world.

David Barr

David Barr is currently working as Director, Compliance at Alkermes Inc. He worked for more than 20 years in the FDA. After doing graduate work in zoology at California State University, Mr Barr began working at the FDA in 1974 as an investigator in Los Angeles. He began taking on more responsibility in such positions as Supervisory Investigator for Pharmaceutical and Device inspections, and became the Deputy Director in the Office of Compliance at the Center for Drug Evaluation and Research. In his time at the FDA, he participated in the development and writing of parts of 21 CFR 211, as well as various guidelines, such as the "Draft Guideline on Supplements to NDAs and ANDAs for Sterile Products"

For the last 10 years, Mr Barr has been sharing his expertise with Kendle clients, beginning as a Senior Consultant, and in 2001 becoming Vice President of Regulatory Compliance Consulting. He has had multiple publications in journals such as the Food and Drug Law Journal and Pharmaceutical Engineering, is a co-author of Application of Pharmaceutical CGMPs published by the Food and Drug Law Institute and he has also featured as a speaker for organizations such as the Parenteral Drug Association and the American Association of Pharmaceutical Scientists. Mr Barr also serves as an expert witness in private litigation.

Andreas Brutsche

Dr Andreas Brutsche is working at Novartis Pharma Stein AG, Stein, Switzerland. An internationally acknowledged expert, renowned speaker and authority on regulatory compliance practices, he has a wealth of experience in managing and hosting regulatory inspections. He has written several articles and book chapters in the areas of cGMP compliance and quality management systems.

Richard Einig

Richard G. Einig, PhD, RAC, CQA is Vice-President and managing partner in API Consulting and Management (APICM). He is also a consultant specializing in the pharmaceutical industry. His employment experience spans over twenty years in senior management of quality, regulatory, and development units of large international companies and start-up "biotechs". Since leaving corporate management, he has consulted in the U. S. and internationally with innovator and generic dosage form companies, API manufacturers, and medical device manufacturers. He advises clients that are operating under adverse regulatory findings on remediation of quality systems.

Dr Einig participated in developing the PhRMA Bulk Pharmaceutical Committee's Guidance on Production of Drug Substance, and is an invited speaker at domestic and international meetings on quality and processing of pharmaceutical products. He teaches seminar courses on compliant API manufacturing for the Center for Professional Advancement in the US and Europe.

Dr Einig is a member of the American Chemical Society as well as a member and carries certifications from the American Society for Quality, the Regulatory Affairs Professional Society, and the Institute for Independent Business. He received undergraduate and graduate degrees in Chemistry from St. Louis University, MBA from Webster University, and PhD from Missouri University. He can be reached at einig@APICM.com.

Yoshikazu Hayashi

Yoshikazu Hayashi is the international liaison officer in the Pharmaceuticals and Medical Devices Agency (PMDA), Japan. Mr Hayashi holds a BS in Pharmacy from Toyama Medical and Pharmaceutical University, Japan. His background is Pharmaceutical Science and Social Pharmacy. He worked for the Ministry of Health, Labour and Welfare (MHLW) for more than 25 years where he held various positions responsible for pharmacovigilance and other regulations on medicinal products for human use. He also experienced a position in the Environment Agency and was dispatched to the WHO/ International Programme on Chemical Safety (IPCS) in Geneva, Switzerland, during 1998-2001, where he took charge of the risk assessment of chemicals. When he returned to the MHLW in 2001, he was engaged in coordinating the assessment and marketing authorisation process of new medicinal products for human use between MHLW and PMDA. He also served as an ICH coordinator for MHLW from 2002 to 2003.

Mr Hayashi held prominent positions in the Ministry such as Planning Director for Research Promotion, Director of Office of Clinical Trial Promotion, and Planning Director of Blood and Blood Products Division. He moved to a new posting at the European Medicines Agency (EMA) between Nov 2009 and May 2012. He worked there as an MHLW/PMDA liaison to explore various areas of mutual interest and to coordinate/facilitate them based on the Confidentiality Arrangements between the MHLW/PMDA and the EC/EMA.

Ron Johnson

Ron Johnson, President of Becker & Associates Consulting, Inc. (Becker Consulting), has more than 30 years of senior FDA leadership experience, as well as 14 years of applied consulting practice assisting FDA-regulated companies with complex regulatory and compliance matters.

Mr Johnson's prominent FDA leadership positions include Director of Compliance at CDRH and Director of the Pacific Region. In the latter position, he managed the execution of FDA's consumer and enforcement programs in the nine western states (including Hawaii and Alaska) and the Pacific Trust Territories, including the activities of 600 administrative, inspectional, laboratory, and compliance personnel stationed in three of FDA's largest field installations. In addition, Mr Johnson has played an active role in the development of FDA policies and training of personnel as Chair of FDA's Public Affairs Field Committee, Chair of FDA's Human Resources Field Committee, a member of FDA's Enforcement Policy Council, and Course Leader for FDA's Evidence Development Training Program for New Investigators and Analysts.

He has received numerous awards in recognition of his government service and leadership, including multiple FDA Commissioner Special Citations, three FDA Awards of Merit, ten "Hammer" Awards from Vice President Al Gore and the National Performance Review, a US Public Health Service Superior Service Award, and a citation for exemplary leadership in FDA by the US House of Representatives Oversight Committee.

Seth Mailhot

Seth Mailhot leads the FDA Regulatory practice at Michael Best and Friedrich LLP and is a member of the firm's Transactional Practice Group at the Washington D.C. office.

Prior to entering private practice, Mr Mailhot spent 14 years working in the US Food and Drug Administration (FDA). While at FDA, Mr Mailhot oversaw activities of pharmaceutical, biologic, and medical device companies. Mr Mailhot's FDA regulatory experience includes enforcement and recall matters, preparation and prosecution of FDA premarket submissions (such as NDAs, ANDAs, INDs, and pre-INDs), product promotion and labeling issues, pharmaceutical exclusivity matters, and compliance with quality, regulatory, and manufacturing requirements. Mr Mailhot's background is as a chemical engineer, and he coauthored numerous technical articles while at FDA.

Mr Mailhot's understanding of manufacturing regulations and current good manufacturing practices spans all industries, with a focus on regulatory issues involving foreign manufacturing. Mr Mailhot is also a patent attorney who has worked on major intellectual property litigation cases involving FDA regulated products. Mr Mailhot also worked closely with the District of Massachusetts US Attorney's Office Health Care Fraud Unit on matters involving civil and criminal charges under the Federal Food, Drug and Cosmetic Act.

Nabila Nazir

Nabila Nazir, Senior Consultant joined PAREXEL Consulting in 2011. Nabila has over 12 years of industry experience and provides consulting advice on topics including emerging markets, labelling and medical devices. Previously held positions include working with the Medical Devices Agency and the Medicines Control Agency, as well as working as a Manager within Biogen Idec both in the UK and in the US.

Madhu Raju Saghee

Madhu Raju Saghee is currently working as AGM, Quality Assurance at Micro Labs Sterile Facility, India. Prior to this, he worked at Novartis and Gland Pharma in corporate quality and compliance functions leading and supporting both global and site compliance and validation efforts. Madhu has wide ranging expertise in the implementation of best-practice Pharmaceutical Quality System (PQS) and espouses the need for proactive attitudes and quality culture. He has immense experience in handling regulatory inspections and has helped firms to successfully pass regulatory inspections. Madhu is a frequent speaker at major industry conferences and often leads workshops on regulatory compliance.

His areas of expertise include all aspects of quality and compliance for systems, processes, facilities and operations for drug products, particularly for sterile products. Madhu is also a volunteer for Pharmaceutical and Healthcare Sciences Society (PHSS), acting as the Director for India. He is a member of the editorial board of GMP Review and the international review board of EJPPS. He is the recipient of the coveted 2013 "*Young Pharmaceutical Analyst Award*" from IDMA (Indian Drug Manufacturers Association). Madhu has edited/co-edited several books including *Achieving Quality and Compliance Excellence in Pharmaceuticals* and *Microbiology and Sterility Assurance in Pharmaceuticals and Medical Devices* and has authored many technical articles. Madhu has a Master of Science in microbiology from Andhra University. He is an active member of various industry associations including the PDA, ISPE and PHSS. He can be reached at madhu.saghee@gmail.com

Tim Sandle

Dr Tim Sandle is the Head of Microbiology at the UK Bio Products Laboratory. His role involves overseeing a range of microbiological tests, batch review, microbiological investigation and policy development. In addition, Tim is an honorary consultant with the School of Pharmacy and Pharmaceutical Sciences, University of Manchester and is a tutor for the university's pharmaceutical microbiology MSc course. Tim is a chartered biologist and holds a first class honors degree in Applied Biology; a Masters degree in education; and a PhD in the safety testing of blood products.

Tim serves on several national and international committees relating to pharmaceutical microbiology and cleanroom contamination control (including the ISO cleanroom standards), and he has acted as a spokesperson for several microbiological societies. He is a committee member of the UK and Irish microbiology society Pharmig and editor of its newsletter. Tim has written over one hundred and fifty book chapters, peer reviewed papers and technical articles relating to microbiology, this includes co-editor of the books *Microbiology and Sterility Assurance in Pharmaceuticals and Medical Devices* and *Cleanroom Management in Pharmaceuticals and Healthcare*. In addition, Tim runs an on-line microbiology blog (http://www.pharmamicroresources.com)

Siegfried Schmitt

Siegfried Schmitt, Principal Consultant, joined PAREXEL Consulting in 2007. He provides consulting services to the medical device and pharmaceutical industry on all aspects of regulatory compliance, particularly the design and implementation of Quality Management Systems and Competitive Compliance. He is the PAREXEL practice lead for Quality by Design.

Dr Schmitt's areas of expertise include all aspects of quality and compliance for systems, processes, facilities and operations for drug substances and drug products, particularly for sterile cytostatic or cytotoxic biotech products. He has previously held positions in industry as Senior Production Chemist with Roche and global Quality Director with GE Healthcare and as Validation Manager with Raytheon and Senior Lead Consultant with ABB.

Dr Schmitt is an active member of various industry associations, including DIA, PDA and ISPE, conference presenter and organiser of international events. He is also an accomplished author and editor.

Mark Tucker

Mark Tucker PhD is the founder and President of Mark Tucker, LLC, a consulting firm specializing in Inspection Management and GMP training. Mark has over 15 years experience in the area of Pharmaceuticals, including experience at FDA as an Investigator, Analyst, Compliance Officer and finally Investigations Branch Director. Mark also worked at Genentech, a leading biotech manufacturer, progressing from Associate Director to Senior Director while leading both Inspection Management and GMP Compliance audit functions.

Mark is an active member of various industry associations, including ECA, PDA, RAPS and ISPE, and is a member of the FDA Alumni Association. He has presented on FDA Inspections and Inspection Management in the US, Europe and China. Mark holds a Life Sciences BS from the University of Nebraska – Lincoln and a PhD in Biochemistry from William Marsh Rice University in Houston, Texas.

CHAPTER 1

Basic Concepts of Global GMP Requirements

Tim Sandle, Bio Products Laboratory, UK
Madhu Raju Saghee, Micro Labs, India

1.1 Introduction

This chapter is concerned with Good Manufacturing Practice (GMP) as applied to pharmaceutical processing. It provides an overview of GMP in relation to pharmaceutical manufacturing and then proceeds to discuss the historical evolution of GMP and the current requirements of GMP in general. It also looks at the 'global' GMPs, examining similarities and differences between international regulations and guidelines. Due to the international dominance of FDA, EU and WHO GMPs, the discussion is orientated towards the requirements of these important guidelines. The substantive section of the chapter outlines the main GMP requirements common across all of the major GMPs and as applicable to sterile and non-sterile pharmaceutical manufacturing. Prior to discussing GMP, the chapter outlines the concept of quality and its history in relation to pharmaceutical processing. This helps to connect GMP to the overall quality paradigm.

GMPs are concerned with the quality and safety of pharmaceutical products. People prescribing or being prescribed a medicine have little chance of detecting if it is faulty or not. People who take a medicine trust the doctor who wrote the prescription and the pharmacist who dispensed it. The doctor and pharmacist in turn put their trust in the manufacturer who has a fundamental role in ensuring that the medicine is fit for its purpose and is safe to use. Underlying this level of trust is the manufacture of medicines to an appropriate standard. This standard is described as "Good Manufacturing Practice".

At the simplest level Good Manufacturing Practices (GMPs) are a legal codification of quality principles applied in the manufacturing and testing of pharmaceutical products ("medicinal products" in European terminology, "drug products" in United States terminology)[1]. These products are administered through different delivery systems, which include oral, by injection (parenteral), via the skin (such as transdermal patch), or across the nasal membrane (such as by a specially formulated nasal spray). GMP is about ensuring that quality is built into the organisation and the processes involved in manufacturing such medicinal products.

GMPs are, ultimately, set by national governments or by supranational agencies. Thus GMPs cannot be totally separated from political or economic considerations (the objective of supranational bodies, like the European Union, establishing GMPs is connected with mutual recognition and trade between different countries). Beyond the politico-legal description, GMP is essentially about risk reduction in pharmaceutical processing based on scientific principles. The primary risks are either related to cross contamination (such a microbial contamination) and mix-ups (such as incorrect labelling)[2]. A pharmaceutical product refers to any material or product intended for human or veterinary use, either presented in its finished dosage form or as a starting material for eventual use in a final dosage form.

1.2 Quality as applied to pharmaceutical manufacturing

Before examining GMP, as a set of guidelines, it is important to place GMP in context. GMP is part of the 'quality system' and, in turn, is part of the wider concept of 'quality assurance' (and a major part of that). As a subset of Quality Assurance, GMP is one of several 'good practice' guidelines pertaining to drug manufacture and distribution (other 'good practice' guidelines are outlined below and include Good Laboratory Practice and Good Distribution Practice).

1.2.1 Quality Assurance and Quality Management

Quality Assurance can be defined as an integrated management system that provides an assurance that the contractual and legal obligations of the company, to its customer (ultimately the patient) and the community are being efficaciously fulfilled. Quality Assurance therefore incorporates GMP and other factors. The reader should note, in the pharmaceutical context, Quality Control is part of GMP (albeit an important part). Quality Assurance forms part of an organisation's Quality Management System[3].

Quality Management is a set of principles, many of which are captured in the ISO 9000[4] series (ISO 9000 captures definitions and terminology relating to quality systems and also links to ISO 9001, which details the requirements of a quality system and this standard is often referred to by auditors who are undertaking quality certifications; and to ISO 9004, which presents a set of guidelines and is orientated towards maintaining standards and seeking improvements). The ISO standards apply across a range of industries and, whilst useful as a supplier audit tool, they are not sufficiently process or product specific in themselves to replace GMPs.

Drawing upon both ISO and GMP definitions, Quality Management can be considered as a wide-ranging set of objectives. Within GMP these typically include:

- The manufacturer must produce products that are fit for their intended use and do not place patients at risk[5]

- Senior management is responsible for product quality and safety but the participation and commitment of staff is vital, as is the support of suppliers

- There must be a comprehensive and effectively implemented Quality Assurance system, incorporating Good Manufacturing Practice and Quality Control

- The system should be fully documented and its effectiveness monitored

- All parts of the Quality Assurance system should be adequately resourced with competent personnel and proper premises, equipment and facilities

ISO 9000 has a similar structure to these general principles, which embraces[6]:

- Scope of activities

- Quality Management System

- Management responsibility

- Resource management

- Product realisation

- Measurement, analysis and improvement

- Compulsory areas:

    o   Control of Documents,

    o   Control of Records,

    o   Internal Audits,

    o   Control of Nonconforming Product/Service,

    o   Corrective Action,

    o   Preventive Action.

The adoption of a Quality Management System (QMS) by a pharmaceutical manufacturer needs to be a strategic decision of an organisation, and is influenced by varying needs, objectives, the products/services provided, the processes employed and the size and structure of the organisation. A QMS must ensure that the products and services conform to customer needs and expectations.

As part of Quality Management, a satisfactory system of Quality Assurance depends on people. For this there must be sufficient qualified personnel to carry out the necessary tasks and the responsibilities of individuals should be clearly understood and recorded. In addition, all personnel should receive initial and continuing training, relevant to their needs. The adoption of these principles, the establishment of a system of Quality Assurance (QA), and the creation of management structures necessary to cope with the application of QA has happened and in different ways. Thus each organisation's Quality Management Systems and Quality Assurance functions will appear somewhat different. Nonetheless, they will include a set of 'good practice guidelines', which include:

- *Good Manufacturing Practice (GMP)*, which includes Good Control Laboratory Practice (GCLP) for Quality Control laboratories. GMP embraces each manufacturing step, from purchasing raw materials to the finished product[7].

  Also associated with GMP are the requirements for automated systems. These fall under the Good Automated Manufacturing Practice (GAMP) Guide for Validation of Automated Systems in Pharmaceutical Manufacture[8]. Within EU GMP, computerised systems come under Annex 11 and Chapter 4 – documentation. With the FDA, separate guidance is available.

- *Good Laboratory Practice (GLP)*, for laboratories conducting non-clinical studies (toxicology and pharmacology studies in animals). GLP provides a framework within which laboratory studies are planned, performed, monitored, recorded, reported and archived. The reader should note that GLP, in this context, is different from Good Control Laboratory Practice (GCLP), which is a subset of GMP and refers to quality control testing in relation to product manufacture.

- *Good Clinical Practice (GCP)*, for hospitals and clinicians conducting clinical studies on new drugs in humans and pharmacokinetics (often in conjunction with International Standard ISO 14155[9]). GCP is an international ethical and scientific quality standard for designing, conducting, recording and reporting trials that involve the participation of human subjects. Compliance with this standard is intended to provide public assurance that the rights, safety and well-being of trial subjects are protected consistent with the principles of the Declaration of Helsinki[10], and to ensure that the clinical trial data are credible. A clinical trial describes testing in which preventive, diagnostic, or therapeutic agents are given to a human population under controlled conditions to determine the agents' safety and effectiveness. The trial is, therefore a systematic investigation that tests the effects of materials or methods, according to a formal study plan (that is, a protocol), and usually in subjects having a particular disease or class of diseases.

- *Good Regulatory Practice (GRP)*, for the management of regulatory commitments, procedures and documentation.

- *Good Distribution Practice (GDP)* deals with the guidelines for the proper distribution of medicinal products for human use.

- *Good Transportation Practice (GTP)* deals with the guidelines for the proper domestic and international transportation of medicinal products for human use

Collectively, these 'good practice' quality requirements are referred to as "GxP" requirements. They are similar insofar as they contain quality centred philosophies. Medical device manufacture is not directly covered by the above regulations, although the principles which apply to medical device manufacture are generally similar in terms of the quality centred philosophy.

1.2.2   Quality as applied to pharmaceutical manufacturing

Before examining GMP specifically, it is important to discuss 'quality' in relation to pharmaceutical processing. Understanding the overall approach to quality is important because the way in which a company manages quality impacts significantly upon the way in which GMP standards are implemented and maintained.

Quality means different things to different people and there are a wide range of contrasting views and opinions concerning what 'quality' means and often terms are ill-defined, used interchangeably and inconsistently[11]. In everyday conversation, and as reflected by the advertising of consumer products, it reflects the subjective terms 'luxurious' or 'expensive'. In manufacturing quality has traditionally meant 'quality control', which implies the inspection or measurement of the work of others or simple conformance to a standard[12]. Whilst quality control plays an important part in any quality system it is often indicative that quality is only the concern of the QC department rather than a philosophy which cascades throughout the organisation[13]. Quality in relation to an organisation, like a pharmaceutical company, is about meeting agreed requirements.

Versions of the old UK GMP 'Orange Guide' provide a useful definition of 'quality', as an organisational wide requirement:

> "The essential nature of a thing and the totality of its attributes and properties which bear upon its fitness for its intended purpose[14]."

Quality in relation to medicines has a long history, dating back to the pioneering work of Lister and Pasteur, who attempted to reduce the risks associated with surgery from bacterial contamination[15]. As scientific thinking advanced, so too did the idea of quality.

It is not only through scientific progress and developments with contamination control that pharmaceutical processing has benefited from quality, for like most mass production industries the pharmaceutical industry adopted management practices and production methods like Taylorism, and through to the so-termed post-1945 'quality revolution'[16]. Along with many industries in the post-war era pharmaceutical processing has adopted the principles of quality management associated with paradigms like Total Quality Management (TQM)[17] and Kaizen (the Japanese continuous improvement philosophy)[18]. TQM was based on the quality management teachings of WE Deming and J Juran, whose ideas were utilised by the Japanese government post-1945 to rebuild its war-damaged economy[19]. TQM's underlying philosophy is that imperfect quality leads to wasted resources[20].

Joseph Juran's ideas, although not aimed at the pharmaceutical industry, nevertheless established some important principles of quality which are readily transferrable to pharmaceutical production[21]. Juran's key points in relation to organisational quality were:

- Quality, Safety and Effectiveness must be designed and built into the product

- Quality cannot be tested into the product

- Each step in a manufacturing process must be controlled, to maximise the probability that the finished product meets all its quality and design specifications

- The end result is 'fitness for purpose'

Describing quality as an organisational wide measure and not simply 'quality control', Juran adopted a wide definition of quality, which he saw as "the totality of activities which must be carried out to achieve the quality objectives of the company."

Whilst the approaches to quality have varied over decades, some driven by economic changes within capitalism (notably from mass production to niche markets), whereas other approaches are perhaps no more than re-packaging established ideas, what is important are the essential principles[22]. The principles of quality can be surmised as:

- The commitment of the whole business to the idea of quality

- The extension of quality to all aspects of the organisation: work, process, product, etc

- Utilisation of technology to support business objectives

- Avoiding process and manufacturing 'defects'

- Joining different aspects of quality in different areas of the organisation together in a systematic manner

- The long-term commitment of the workforce to the ideas of quality (the so-termed 'human factor'). Here quality is an attitude of mind, and is closely associated to how people behave, how they perform, and what training they have received[23]

- Investing in people to raise workplace skills

- Instilling quality to the leadership of the organisation and ensuring that it is a priority. This is normally through adopting a Quality Management System. Quality management is a permanent strategy involving every person in every department (research, quality control, production, marketing, purchasing, etc.)

- Developing a culture of continuous improvement and having a focus on preventative actions; thus quality represents a continuous cycle

Some general aspects of quality are examined below.

Quality Standards

The principles of quality have been developed through international standards. The key documents for Quality Standards are primarily the International Organisation for Standards (ISO) 9001 "Quality Systems – Model for Quality Assurance in Design, Development, Production, Installation, and Servicing," and the ISO/CD 13485 "Quality Systems – Medical Devices – Supplementary Requirements to ISO 9001." Although these are documents which cut across a range of industries and are not specifically GMP documents, they were adopted by the US FDA in 1994 as part of the agency's definition of quality standards.

The ISO 9001 document in particular allows one company to audit another company and thereby make an assessment of another organisation's quality systems.

Quality Management

The scope of Quality Management has been outlined above. The objective of Quality Management within a manufacturing operation is to improve performance by eliminating wasteful practices, increasing productivity, and providing a higher level of assurance of quality. In doing so, operational efficiencies, lower costs, and improved reputation and competitiveness, can be enhanced.

The application of Quality Management within a pharmaceutical manufacturing operation often includes the following:

- Improved testing and inspection of incoming materials

- Better process understanding by using validation data

- Extending of the concept of parametric release

- Improved in-process controls

- Representative of QC inspection

- Accurate and end-product testing

The above 'goals' are most effectively delivered when the organisation adopts a multifunctional approach to Quality Management, whereby each department within the organisation (not just those directly involved with pharmaceutical processing) becomes focused upon complying with quality rules and delivering a safe and efficacious product.

Extrinsic and intrinsic quality

Quality can be separated into intrinsic and extrinsic factors (or 'quality attributes'). Intrinsic cues are part of the physical product[24]. Extrinsic cues, although related to the product, are not physically part of it.

a)   Intrinsic quality

Intrinsic quality attributes are closely related to the product and cannot be changed without also changing the physical characteristics of the product. The intrinsic quality of a pharmaceutical product is what the guidelines (e.g. CFR Part 11 or the EU GMP Guide) describe. It is its aptness for use, its efficacy, its purity, its exact dosage and its stability.

b)   Extrinsic quality

If a tablet has a slight black spot on it, which may come from the chemically detectable impurity in one of its starting materials such as starch, it will in no way affect the safety or efficacy of the tablet. This is an extrinsic quality (or non-quality) feature. If the date of expiry on a carton is not as clear as it should be, or the colour of a carton is not quite as green as it should be, these are extrinsic quality problems. They will not affect the safety of the patient. They may harm the image or reputation of the firm who makes it, but the patient will still get full benefit from the product.

Quality and pharmaceutical regulations

It is apparent from the discussion thus far in this chapter that there is an historical connectedness between quality in pharmaceuticals and standards and regulations (and this is described below in relation to the history of GMP). Another important driver of quality, aside from the scientific and the economic, has come from government (latterly in the form of regulatory agencies). Governments have a dual role in desiring a safe medicinal product to be available to the populace and, as major purchasers of medicines, value of money for patients. Here regulatory authorities, as government agencies, require medicines which provide high standards of Quality Assurance. The purchasing authorities, through government health plans or larger health providers like the UK NHS, require medicines which will provide both high standards of quality and value for money when purchased. There is a careful balance to be struck between 'Quality' and 'Value' (not least because matters of economic gain could potentially clash with the objective of manufacturing a safe and efficacious product). Nonetheless, the FDA considers that a focus on quality itself delivers good value:

> "Quality and productivity improvement share a common element — reduction in variability through process understanding (e.g., application of knowledge throughout the product lifecycle). Reducing variability provides a win-win opportunity from both public health and industry perspectives."[25]

The development and manufacture of pharmaceutical products is performed in a highly regulated environment. The patients and the regulatory authorities carefully monitor the activities of the pharmaceutical industry. Very often, incidents in the market lead to new regulations because authorities were forced to take appropriate actions.

The quality standards of today and requirements of regulatory authorities have arisen through events which have occurred during the past century. Over this period of time several events represented milestones in the development of quality standards.

Some examples are listed below:

- A governmental report on the atrocious conditions in the Chicago meat-packing industry led to the original Food and Drug Act in the US (1906)

- In the late 1930s, sulfanilamide elixir ('Elixir of Sulfanilamide') contaminated with diethylene glycol (the same chemical used in antifreeze) killed a large number of patients. This incident led to the Food and Cosmetic Act (1938)

- The thalidomide tragedy in Europe led to a tightening of the testing of pharmaceutical products prior to their marketing[a]. This incident prompted the Kefauver-Harris Amendments (enacted as 21 CFR and the first GMPs) (1962)[26]

- The Davenport incident in the UK in 1972. This incident is described fully in a UK Government enquiry, the Clothier Report[27]. In March 1972 a series of untoward reactions (rapid increase in core body temperature, shock, and imbalances in body fluids) were seen amongst post-operative patients in the Davenport Section of Plymouth General Hospital[28]. The common factor was that all had received intravenous administration of a 5% Dextrose Infusion Fluid. A batch of bottles of Dextrose Infusion Fluid manufactured by a major UK pharmaceutical company was found to be contaminated by *Klebsiella aerogenes* and other Gram-negative coliform bacteria. The concentration of bacteria in the bottles was sufficiently high to be visibly perceptible (in excess of 106 bacteria per mL). This led to a review of, and a considerable improvement to, GMP standards in relation to sterilisation[b]

- An incident in Nigeria, involving ethylene glycol and killing more than 100 children, really began the push towards international harmonisation (1991)

- Other incidents included the dioxin case in Belgium (1999) and contaminated blood products (involving transmissible spongiform encephalopathies (TSEs)

These are important historical examples of violations which lead to new enforcement activities by regulatory authorities.

At present the biggest obstacle is the definition that constitutes cGMPs. There is no harmonised interpretation of cGMPs between the different authorities, e.g. EMEA, FDA, Japanese Authorities, WHO, PIC, ICH.

International quality standards

Although ISO 9001 and 'universal' GMPs like ICH quality standards (which are discussed below) provide an outline for the international community there is no single 'universal' quality standard applicable to pharmaceutical manufacturing. For example, comparing the two major international quality and GMP standards – that of the European Community and USA (through the FDA) – reveal the following differences:

- **Legal differences**. In essence, the legal bases are rather similar from the inspection point of view. However, one major difference is the US Freedom of Information Act, which allows the public access to the FDA inspection reports. This is not the case in most EU member states[29]

- **Philosophical differences**. The US approach is centred on compliance and enforcement. In contrast, the European Union tends to operate on the basis of trust and cooperation

- **Operational differences**. The FDA is a much bigger organisation to the EU equivalent, with very wide-ranging activities covering food and cosmetics as well as drugs. The FDA is much more involved in biologics and bulk chemicals. The FDA inspectors are much more likely to carry out unannounced inspections in the United States and will also put more emphasis on preapproval inspection

- **Technical differences**. In general terms, there are few technical differences, although some specific aspects, such as investigations and handling, are given more emphasis by the FDA, as is the whole subject of validation

1.3   Good Manufacturing Practice

The primary requirement for the manufacture of medicinal products is Good Manufacturing Practice. (GMP). As the introduction to this chapter has set out, GMP is part of Quality Assurance, and thus a part of the Quality System. GMP is applicable to the manufacture and testing of pharmaceuticals or drugs including active pharmaceutical ingredients, excipients (chemical components other than the active pharmaceutical ingredient such as binders, fillers, diluents, disintegrants, lubricants, flavours, colours, and sweeteners). Excipients are classified as per the source of origin, which are: animal origin, plant origin, mineral origin and synthetic origin[30]. Also included are diagnostics, foods, pharmaceutical products, and medical devices. In this chapter the focus of GMP is narrowed to medicinal products.

GMP is open to differing interpretations. The original definition of GMP from the 'Orange Guide' remains a useful one:

"That part of quality assurance aimed at ensuring that products are consistently manufactured to a quality appropriate to their intended use. It is thus concerned with both manufacturing and quality control procedures."[31]

The scope of GMP within the pharmaceutical organisation embraces the production, packaging, repackaging, labelling and re-labelling of pharmaceuticals[32], and to:

- Raw materials used in the manufacture of drugs. These must be of a known, and of possibly standardised, quality and be free from contamination

- The manufacturing process must be able to produce a pharmaceutical product meeting its quality attributes

- Adequate quality control testing measures must be employed to assure that the product meets its quality specifications at time of release to market, and at the end of its shelf life

GMP is an agreed system for ensuring that products are consistently produced and controlled according to quality standards. It is designed to minimise the risks involved in any pharmaceutical production and sets the standards for the testing and release of the final product. Importantly, GMPs are *guidelines* and are not prescriptive instructions on how to manufacture products. GMPs are often presented as a series of general principles that must be observed during manufacturing. A manufacturer's quality program and manufacturing process will interpret and apply GMPs in line with the product and process, and thus there are sometimes different ways to meet GMP requirements. The way this is done rests with the manufacturer (and this often begins with a Quality Manual, from which site policies follow).

The Quality Manual normally covers the following areas:

- Table of Contents
- Introduction (brief description of the company, the company's purpose for writing a Quality Manual, and a brief description of the scope of the Quality Manual)
- Quality Policies and Objectives
- Organisation and Structure of Documentation
- Company's Products
- References
- Quality Policies for Specific Regulation Elements

Although GMP applies to manufacturing and laboratory testing, the required balance should always be towards control of manufacturing. Indeed, in the US a drug may be deemed adulterated even if it passes all of the specified release tests but is found to have been manufactured in a condition which violates current good manufacturing guidelines. Thus, good quality must always be built in during the manufacturing process; it cannot be tested into the product afterwards. Without GMP it is impossible to be sure that every unit of a medicine is of the same quality as the units of medicine tested in the laboratory (here, with sterile products the sterility test is a good case in point: it is impossible to test a sufficiently large enough proportion of the batch to have any degree of confidence that all items in the batch are sterile)[33][34].

GMP is not simply about the detection or avoidance of errors; it is about the requirement of a drug manufacturer to consistently produce medicinal products to a high standard. GMP is applicable to all aspects of drug manufacture. This begins with the purchase and receipt of the starting materials, through to the manufacturing process (premises and equipment), and to the way personnel behave (including aspects of their training and personal hygiene).

Important points shared by all GMP systems are:

- Manufacturing processes must be clearly defined and controlled. All critical processes must be validated to ensure consistency and compliance with specifications
- Quality by design: to ensure that the facility or the equipment is designed for the correct purpose from the start
- Ensuring that processes and equipment are validated
- Manufacturing processes must be controlled, and any changes to the process are evaluated. Changes that have an impact on the quality of the drug need to be validated as necessary
- Ensuring that written procedures are in place and are followed
- Ensuring that personnel are trained and understand what they are doing
- Each process step and the personnel involved in completing each step must be recorded
- Documented records must be kept
- The importance of hygiene must be understood and practiced at all times
- Premises and equipment must be maintained
- Quality must be built into the product lifecycle (this refers to all phases in the life of a product from the initial development through marketing until the product's discontinuation)[35]
- Quality must be assessed by regular audits

GMP is controlled and proven through documentation. When undertaking processes there must be detailed, written procedures in place, and for each activity, from production to laboratory testing, each step must be written down.

GMP is important from both a health protection perspective and from an economic perspective. From the health perspective, GMP failures can result in medicines being manufactured which may contain toxic substances or micro-organisms that have been unintentionally added and which may cause harm to the patient. It may also be that a medicine contains little or none of the claimed active ingredient so that it will not have the intended therapeutic effect. From the economic perspective, GMP failures are a cost to organisations in terms of time, production expenses and lost revenue.

Another facet of GMP is in terms of trade. Countries are more willing to trade with one another, and purchase medicines, if the two countries practice the same or equivalent level of GMP. Most countries will only accept import and sale of medicines that have been manufactured to internationally recognised GMP. The need to trade adds a socio-economic dimension to GMP and, arguably, a political one too. The head of the US Food and Drug Administration (FDA) (the Commissioner of the FDA), for example, is a political appointment and the bequest of the President of the USA and at times the policy of the FDA reflects the political direction of the ruling administration.

Although the acceptance of GMP guidelines for pharmaceutical products is almost universal, and to a large extent companies comply with the principles expressed by such guides, the mechanisms for achieving compliance can vary considerably from company to company. Much will depend upon how various functions (e.g. production, QC, warehousing, engineering, distribution, and marketing) involved in making medicines view GMP, how they interact and what priority they award to it. In the production and laboratory departments, GMP may have a high priority, whereas finance and personnel may have difficulty in recognising how GMP regulations should be applied to their own activities. Furthermore, each year there are several news reports which indicate that GMP standards have broken down within one manufacturer or another.

1.3.1 cGMP (current Good Manufacturing Practice)

Whilst GMP is codified through various published regulatory documents and standards (which are discussed below), regulatory agencies expect the philosophy of "cGMP" to be followed. The 'c' stands for 'current' and is placed in front of GMP in order to convey the fact that GMPs are not static but instead they evolve through technological developments and revised guidelines. The manufacturer is expected to remain up-to-date with cGMP.

Many of the current GMP approaches are based on the identification and assessment of risks. Here manufacturers are expected to use the technology and system available to minimise process risks. Some of the risks to pharmaceutical manufacturing include:

- The contamination of products, both chemical and microbial

- Incorrect packaging or labels on containers, which could mean that patients receive the wrong medicine

- Insufficient or too much active ingredient added to the formulation, resulting in ineffective treatment or adverse effects[36]

In addition to the above global concerns, GMPs will also relate to the product itself and what had been agreed between the manufacturer and the regulator. That is, GMP is required so that the manufacturer produces the product as intended and as contained in a marketing authorisation and to a product specification. Marketing authorisations are sometimes referred to as product licences and are legal documents issued by a regulatory authority which establish the detailed composition and formulation of the product and the specifications of the ingredients and of the final product. Authorisations also include details of packaging, labelling and shelf-life. Product specifications are the requirements to which the products or materials used or obtained during manufacture must conform. Specifications are the basis for quality evaluation (through Quality Control).

1.3.2   A brief history of GMP

The first country to develop standards akin to Good Manufacturing Practices was the USA. The first sets of standards to be issued were those contained within the Biologics Control Act of 1902. The Act was introduced in response to the deaths of twelve children from diphtheria antitoxin contaminated with live tetanus bacteria. Shortly afterwards, in 1906 the Pure Food and Drug Act was passed which led to the creation of the Food and Drug Administration (FDA[c]). The primary feature of this act was that all food and medicines required labelling, that the label needed to state what the active ingredient was and that the label had to be truthful. From this point it became illegal to sell contaminated food or drug products[37]. The term used to describe contaminated food and drugs in the act was 'adulterated'. The word "adulterated" has now come to be associated with products manufactured without following GMP[38].

The next developments were in the 1930s, when standards were extended to medical devices. Towards the end of the decade, in 1938, the Food, Drug and Cosmetic Act was passed. This Act placed a requirement on companies to demonstrate that their product was safe prior to the product being marketed. The Act also allowed the FDA to conduct factory inspections and allowed the FDA to take companies which were not complying with standards to court and for the courts to be able to issues penalties[39].

The basis of Good Manufacturing Practice came about with the 1944 Public Services Act. The Act included the regulation of biological products and control of communicable diseases[40]. This led to the FDA revising the manufacturing and quality control requirements for drug manufacturers. The FDA also began testing samples for each drug or active ingredient manufactured (a practice which continued until 1983). In the 1960s, due to some misapplication of drug products (including the response to the notorious thalidomide incident), the FDA gained powers to be able to regulate the marketing and advertising of medicines. From this, the first GMPs for pharmaceutical products were issued by the FDA in 1963[41]. It was here that the phrase "good manufacturing practice" first appeared in a governmental document.

As a global agency focused on health prevention, the World Health Organisation produced its first GMP text. This was published in 1967, following a resolution passed at the Twentieth World Health Assembly (resolution WHA20.34). This initial guidance was substantially revised in 1975. Contemporaneously, in the UK the first key piece of legislation orientated towards GMP was the Medicines Act of 1968. This Act set out to regulate the pharmaceutical industry to protect patients and to ensure quality and safety. The Act was subsequently amended through various 'Statutory Instruments'.

The formation of the GMPs in a form more recognisable today came about in the USA with the codification of 21CFR parts 210 and 211, which were finalised in the 1970s[42] (the last part of this process was completed in 1978 with the CFR applicable to medical devices[43]). The Code of Federal Regulations (CFR) applies to a range of services across the US and not only to medicines. The CFRs are the codification of the general and permanent rules of the federal government and contain the complete and official text of the regulations that are enforced by federal agencies (of which the FDA is one such agency).

With respect to part 210 and 211, which are applicable to medicines, these CFRs set out the:

- Minimum current methods to be used in and the facilities or controls to be used for the manufacture, processing, packaging and holding of a drug and device

- A requirement that the drug or device meets the requirements of the Act as to safety

- The requirement that the drug or device is identifiable and has the required strength, and that it meets the quality and purity characteristics that it purports or is represented to possess

In the late 1970s further revisions placed a stronger emphasis on documentation and record keeping.

In the UK, the first 'Guide to Good Pharmaceutical Manufacturing Practice' appeared in 1971, following the 1968 Medicines Act[44]. This became, due to its cover jacket, colloquially known as the "Orange Guide". It had no statutory force and described "measures for control of quality during manufacture and assembly, with particular reference to those aspects that are associated with safety". The guide was written with the "conviction that the control of quality begins before any materials are purchased, continues throughout manufacture, assembly and distribution, and cannot be 'inspected into' a product at the end of its processing". It contained a number of principles followed by guidance. The aspects covered were: buildings and surrounding areas; equipment; cleanliness and hygiene; production procedures and documentation; Quality Control; record keeping; transportation; the verification of procedures[45]. This list, when compared to current EU GMP, is very similar and the basic contents have not altered very much. GMP has therefore stood the test of time and represents a fundamentally sound set of rules and guidance. The UK "Orange Guide" went on to become the basis of much of what is now EU GMP in 1991, as a result of the passing of Directive 91/356/EEC (which formed the basis of EU-wide GMP at the time of the establishment of the single market for goods and services as part of the 1992 Maastricht Treaty)[46]. In turn, EU GMP considerably influenced what is current WHO GMP.

In 1979, the FDA finalised Good Laboratory Practices in relation to animal testing[47]. In 1983, measures to strengthen the packaging of medicines were put in place and an anti-tampering act was passed. During the 1980s the FDA began to issue the first of its guidance documents about GMPs and the production of medicines. The first of these, issued in 1983, related to computer validation; this was followed by a series of guidelines from aseptic filling to Quality Control Laboratories[48]. In 1990, the FDA undertook the start of the revision of the cGMP regulation to add the design controls authorised by the Safe Medical Devices Act. Provisions relating to quality inspection, electronic records and cleaning validation[49] were passed during the 1990s. In 1997, the FDA underwent a major reform with the FDA Modernisation Act, which granted the Agency greater control over approval of new devices and the regulation of advertising pharmaceuticals and medical devices.

The basis of international collaboration over GMP issues began with the first of the series of International Conference on Harmonisation (ICH) documents (such as the 1996 ICH E6 guidance on good clinical practices relating to human clinical trials[50]). The 2000s saw the harmonisation process extended although national GMPs remained considerably influential (as the section below acknowledges). However, progress has been made with regards to regulatory agencies co-operating and the FDA and European Medicines Agency (EMA) announced a joint inspection strategy in 2011[51]. During the first two decades of the 21st century the emphasis has been upon quality risk management and the move towards risk-based inspections[52]. An example of this was with the 2004 FDA initiative "Pharmaceutical Current Good Manufacturing Practices (cGMPs) for the 21st Century" (a document which is discussed later in this chapter).

1.3.3   Global GMPs

There is no universally recognised system of GMP accepted by all countries in the world. GMP is either applied nationally (such as the Code of Federal Regulations and the operation of the FDA in the USA), at the supranational level (such as by the European Union across its member states or the Association of South-East Asian Nations (ASEAN), or trans-nationally by those countries who elect to adopt World Health Organisation GMP.

Although national GMP theoretically applies only to the particular nation state it can have a far wider reaching influence than national borders. With the FDA, for instance, the GMP is applicable not only to medicines manufactured within the USA; it also extends outwards to any medicine to be imported into the USA. Given the size of the US market, this affords FDA GMP a considerable influence.

Whilst most countries have made it a legal requirement that manufacturers follow GMPs, the GMPs themselves are normally presented as guidelines and are, at times, open to interpretation by both the manufacturer and by the regulator. It is noticeable that GMPs are not updated all that often and that there is an onus on the pharmaceutical manufacturer to stay up-to-date with technological and quality changes. This is in keeping with CFR 210.1(a), which stated: "The regulations set forth in this part and in parts 211 through 226 of this chapter contain the minimum current good manufacturing practices for methods to be used in .....the manufacture, processing, packing, or holding of a drug". This important proviso links back to the discussion above about 'cGMP'.

1.3.3.1   National and international GMP

World Health Organization

The influence of the World Health Organization (WHO) in the setting of global standards is very strong and one of the reasons for the formation of the WHO was to "develop, establish and promote international standards with respect to food, biological, pharmaceutical and similar products". This mission is enshrined in Article 2 of the WHO constitution.

WHO GMP appears in the two volumes of "Quality assurance of pharmaceuticals: a compendium of guidelines and related materials"[53], with the first volume issued in 1997. Volume 2 focuses most strongly on GMP guidance (the first edition was issued in 1999 and several updates have followed). The GMP guidance is developed by the WHO Expert Committee on Biological Standardisation and the WHO Expert Committee on Specifications for Pharmaceutical Preparations.

The WHO version of GMP is used by pharmaceutical regulators and the pharmaceutical industry in over 100 countries worldwide, primarily in the developing world where no history of established GMP exists. There are a number of similarities between WHO GMP and global GMPs and the WHO guidance is most closely aligned to EU GMP (for example, the classification of cleanrooms in WHO GMP follows the EU GMP grade notations A, B, C and D).

Federal Drug Administration

GMPs in the US are enforced by the US Federal Drug Administration (FDA), under Section 501(B) of the 1938 Food, Drug, and Cosmetic Act (21USC351). The FDA GMPs are described as "current good manufacturing practices" (cGMP). This allows the FDA, or the US courts, to determine that a drug product is adulterated even if there is no specific regulatory requirement that was violated if the process was not performed according to what the agency interprets as industry standards.

The FDA GMPs are codified as legally enforceable regulations (the Code of Federal Regulations). The CFR is divided into 50 titles. The title that pertains to the FDA is Title 21 – Food and Drugs. Title 21 of the CFR is divided into Parts 1 to 1499. Notable parts of 21 CFR are:

- CFR Part 58: is for GLP (Good Laboratory Practices)

- CFR Part 202: Prescription Drug Advertising

- CFR Part 203: Prescription Drug Marketing

- CFR Part 210: Current Good Manufacturing Practice in Manufacturing, Processing, Packaging, or Holding of Drugs: General

- CFR Part 211: Current Good Manufacturing Practice for Finished Pharmaceuticals

- CFR Part 312: Investigational New Drug Application

- CFR Part 510: New Animal Drugs

- CFR Part 600: Biological Products: General

- CFR Part 820: Quality System Regulation

The CFRs most applicable to medicinal products are:

- Human pharmaceutical products and veterinary products (21 CFR 210-211)

- Biologically derived products (21 CFR 600 and 21 CFR 620) (part 606 deals with blood products)

- Medical devices (21 CFR 820)

These CFRs contain the minimum current standards applicable to facilities used to manufacture medicines and the controls to be adopted by the manufacturer, for the manufacture, processing, packaging and holding of a drug. Changes to CFRs are first published in the Federal Register (FR) by the appropriate executive departments and agencies of the Federal Government.

The FDA enforces GMPs through inspections, which are covered by clause 704(A) of the FD&C Act (21USC374) (regulatory inspections are discussed below).

The FDA is divided into a number of centres. Those which will be interest to readers of this chapter are:

- The centre responsible for overseeing biological products such as blood, vaccines, allergenics, and therapeutics is FDA CBER (Center for Biologics Evaluation and Research). CBER ensure that biologic products, medical devices used for collecting and processing blood products, and drugs closely related to biologics are safe and effective. CBER analyses data submitted by the manufacturer in the form of an investigational new drug application (IND) to determine if the product is safe and effective for its intended use. If CBER determines that the product is safe and effective they will approve the licence application

- The centre responsible for ensuring that prescription, over-the-counter, and generic drugs are safe and effective and work correctly is FDA CDER (Center for Drug Evaluation and Research). CDER regulates a wide range of products from fluoride toothpaste to aspirin, from sunscreens to cancer treatments. It oversees new drug development, manufacturing, approval, post market performance, drug labelling, and drug promotional information.

- The centre responsible for ensuring that medical devices are safe and effective and that radiating-emitting devices are safe is FDA CDRH (Center for Devices and Radiological Health). Medical devices are classified into three classes; Class I, Class II, Class III. The requirements of the regulations varies depending on the class level; with Class III medical devices being the most highly regulated.

- The centre responsible for ensuring food and cosmetic safety is FDA CFSAN (Center for Food Safety and Applied Nutrition). CFSAN oversees US food manufacturers, processors, and warehouses to ensure that the food supply is safe, sanitary, wholesome and properly labelled.

European Union

The European Union's GMP (EU GMP) enforces similar requirements to both WHO and FDA GMP. In 1993, all member states of the European Union adopted a common GMP framework (under Directive 93/41/EEC, which established the European Agency for the Evaluation of Medicinal Products[d]). EU GMP is collectively known as 'EudraLex', the set of rules and regulations governing medicinal products in the European Union.

EU GMP consists of a number of parts. These are divided across a series of volumes:

- Volume 1 – Pharmaceutical Legislation
- Volume 2 – Notice to Applicants
- Volume 2A deals with procedures for marketing authorisation
- Volume 2B deals with the presentation and content of the application dossier.
- Volume 2C deals with Guidelines
- Volume 3 – Guidelines
- Concerning Medicinal Products for human use in clinical trials (investigational medicinal products)
- Volume 10 – Clinical trials
- Concerning Veterinary Medicinal Products
- Volume 5 – Pharmaceutical Legislation
- Volume 6 – Notice to Applicants
- Volume 7 – Guidelines
- Volume 8 – Maximum residue limits
- Concerning Medicinal Products for Human and Veterinary use
- Volume 4 – Good Manufacturing Practices
- Volume 9 – Pharmacovigilance
- Miscellaneous:
  1. Guidelines on Good Distribution Practice of Medicinal Products for Human Use (94/C 63/03)

Volume 4 – Good Manufacturing Practices contains a number of chapters and annexes. The chapters are:

1.   Quality Management

2.   Personnel

3.   Premises and Equipment

4.   Documentation

5.   Production

6.   Quality Control

7.   Contract Manufacture and Analysis

8.   Complaints and Product Recall

9.   Self Inspection

The guidelines of European GMP are overseen by the European Medicines Agency (EMA) (although the conduction of inspections is carried out by national GMP agencies in European member countries). The EMA is divided into five directorates and its work is conducted through six committees, which are: the Committee for Medicinal Products for Human Use (CHMP), the Committee for Medicinal Products for Veterinary Use (CVMP), the Committee for Orphan Medicinal Products (COMP), and the Committee on Herbal Medicinal Products (HMPC), the Paediatric Committee (PDCO) and the Committee for Advanced Therapies (CAT).

United Kingdom

GMP is overseen by the Medicines and Healthcare products Regulatory Agency (MHRA). The MHRA follows EU GMP and, given the UK's influence in shaping EU GMP, it remains a pivotal player in shaping current practices.

Japan

GMP is overseen by the Pharmaceuticals and Medicines Devices Agency (PMDA). The primary GMP document is "Ministerial Ordinance on Standards for Manufacturing Control and Quality Control for Drugs and Quasi-drugs", Ordinance of Ministry of Health, Labour and Welfare, No. 179, 2004), which the PMDA enforces.

Japanese GMP is outlined in a different way to the GMPs of most other countries (where other nations tend to model GMPs on FDA, EU or WHO practices). The Japanese style of GMPs resembles detailed job descriptions describing the duties and responsibilities of various members of key management staff in the pharmaceutical company (such as the Japanese Product Security Pharmacist, a role similar to that of a Qualified Person in Europe).

Australia

GMP is codified as the Australian Code of Good Manufacturing Practice for Medicinal Products, which is overseen by The Therapeutic Goods Administration (TGA), as part of the Australian Government Department of Health and Ageing. The TGA's overall purpose is to protect public health and safety by regulating therapeutic goods that are supplied either imported or manufactured, or exported from Australia. Given Australia's links with the PIC/S scheme there are similarities between the Australian GMP and EU GMP.

Canada

GMP is overseen by Health Canada. The primary GMP document is "Good Manufacturing Practices (GMP) Guidelines (GUI-0001)."[54] The document draws upon aspects of GMP covered by the World Health Organisation (WHO), the Pharmaceutical Inspection Cooperation/Scheme (PIC/S), and the International Conference on Harmonisation (ICH). Specific reference is made to: premises, equipment, personnel, sanitation raw material testing, manufacturing controls, quality control, and packaging material testing finished product testing, records, samples and sterile products.

India

GMP is overseen by the Central Drugs Standard Control Organisation (under the Directorate General of Health Services, Ministry of Health and Family Welfare, Government of India). This is under Schedule M: GMP and Requirements of premises, plant and equipment for pharmaceutical products. Unlike some other national GMPs, the Indian document is relatively comprehensive in its size and scope.

China

Article 9 of the Drug Administration Law of the People's Republic of China calls for the application of Good Manufacturing Practice in the manufacture of pharmaceuticals. The law comprises a total of 10 chapters and has been in force since 2001[55].

Others

There are other national and supranational bodies which issue GMP guidance, such as: Medsafe (New Zealand), Gulf Cooperation Council (Arab nations), and Pan American Network for Drug Regulatory Harmonisation (a WHO subgroup for principally South American nations). These are not discussed in this chapter.

1.3.3.2 GMP organisations and related bodies

PIC/S

The Pharmaceutical Inspection Convention (PIC) and the Pharmaceutical Inspection Cooperation Scheme (PIC Scheme), known together as PIC/S, is a global scheme whereby different national and international regulators work together and share information in relation to regulatory inspections. PIC was established in 1970 and the Scheme was launched in 1995. The organisation, which has over 40 national members, functions as a cooperative arrangement among global regulatory and health authorities. Its objectives are:

- Mutual Recognition Of Inspections
- Harmonisation Of GMP Requirements
- Uniform Inspection Systems
- Training Of Inspectors
- Exchange Of Information
- Mutual Confidence

For many years the PIC/S was largely a European-Oceanic initiative. The international scope was widened in 2011 when the FDA became Scheme members.

The PIC/S produces a number of inspectorate guidance documents. Current PIC/S guidance documents include: blood establishments (PE 005-3), preparation of medicinal products (PE 010-3), parametric release (PI 005-3), media filling trials (PI 006-3), aseptic processing (PI 007-6), computerised systems (PI 011-3), sterility testing (PI 012-3) and isolators (PI 014-3). These documents are used by regulatory inspectors as training materials and as an aide-memoire during inspections.

ICH

The International Conference on Harmonisation (ICH) publishes a series of GMPs which apply to those countries and trade groupings that are signatories to ICH (the EU, Japan and the US), and also applies in other countries (e.g., Australia, Canada, Singapore) which adopt ICH guidelines for the manufacture and testing of active raw materials.

IPEC

The International Pharmaceutical Excipients Council (IPEC) is a trade body that serves the interests of producers, distributors and users of pharmaceutical excipients. Excipients are not covered by GMP within some territories. IPEC issue a Good Manufacturing Practices Audit Guideline for Pharmaceutical Excipients[56] and publish audit guidelines. The guidance is based on WHO best practices[57].

1.3.3.3 Similarities and differences between global GMPs

GMP regulations for drugs contain minimum requirements for the methods, facilities, and controls used in manufacturing, processing, and packing of a drug product. At a high level, GMPs of various nations are very similar; most require aspects including:

- Equipment and facilities are to be properly designed, maintained, and cleaned

- Standard Operating Procedures (SOPs) must be written and approved

- An independent Quality unit (such as Quality Control and/or Quality Assurance) must be in place

- Well trained personnel and management are required to be employed[58]

Global GMPs, specifically those produced by the FDA, WHO and the EU, share several commonalities, as well as some notable differences[59]. These are examined below.

a) **Legal standing**: Japan, Korea and the United States are the only countries where the GMPs have been enshrined in law as legal regulations. For example, with the US, GMPs are enshrined in law through the Code of Federal Regulations (as outlined earlier). In other nations, GMPs are simply best practice guidelines. In Europe, the key document controlling GMP inspections is Commission Directive 2003/94/EC. Subordinate to the above directive is the "Rules Governing Medicinal Products in the European Union", a Guide to GMP, which is then adopted by member states.

b) **Inspection approaches**. Although there has been a greater harmonisation of inspectorate approaches by regulatory agencies, differences continue to exist between the approach of FDA and EU GMP inspectors. An FDA inspection focuses on compliance, with an emphasis on documentation review: has the manufacturer done, with batch X, what the product licence Y requires? In contrast, most European inspectors look for adequate science, with a greater emphasis upon going through the process and interviewing staff and management.

c) **Inspector training**: there is a difference in the types of personnel hired to act as inspectors. FDA inspectors typically do not have any background in the healthcare manufacturing and are instead entirely trained on the job. In contrast, most EU GMP inspectors must have significant management experience within the pharmaceutical industry.

Identifying the areas of similarity and difference with the GMPs would be a complex task and would fill the entire contents of this book. **Table 1** below selects some of the major areas of GMP and shows that, for the most part, they cut across the three major GMP standards: EU GMP, the FDA (through the CFRs) and WHO. Therefore global GMPs share more *similarities* than differences.

Table 1: Table displaying similarities and differences between the scopes of the major global GMPs.

Subject	EU GMP	US CFR	WHO GMP
Scope	Introduction	21CFR210.1	Part 1
Glossary / Definitions	Introduction	21CFR210.3	Introduction
Quality Management	Part 1	21CFR211.B	Part 1
Premises	Part 3	21CFR211.C	Part 12
Equipment	Part 3	21CFR211.D	Part 13
Personnel	Part 2	21CFR211.B	Parts 9, 10 and 11
Cleaning and disinfection	Part 3	21CFR211.D	Part 12
Raw Material Testing	Part 6	21CFR211.E	Part 17
Manufacturing Control	Part 5	21CFR211.F	Part 16
Quality Control	Part 6	21CFR211.I	Part 17
Packaging Material Testing	Part 6	21CFR211.G	Part 17
Finished Product Testing	Part 6	21CFR211.I	Part 17
Records	Part 4	21CFR211.J	Part 15
Samples	Part 6	21CFR211.I	Part 15
Stability	Part 6	21CFR211.I	Part 17
Contract Manufacture	Part 7	Not covered	Part 7
Complaints and Recalls	Part 8	21CFR211.J	Part 5
Internal audits	Part 9	Not covered	Part 8
Training	Part 2	211.25(a)	Part 1
Validation policy	Part 1	Not addressed	Part 1

The primary differences, in terms of subject content, relate to self-inspection (auditing) and the requirements that the manufacturer should ensure are in place for any contract facilities used. These are two areas that are not specifically covered in the CFRs but which have a prominent place with EU and WHO GMP. A further difference arises with the role of the "Qualified Person", who is unique to EU GMP. The primary legal responsibility of the Qualified Person (QP) is to certify batches of medicinal products prior to use in a clinical trial (human medicines products only) or prior to release for sale and placing on the market (human and veterinary medicinal products). In contrast, the FDA holds the most senior member of management directly, and personally, responsible for any infractions of, or breaches in compliance with the GMPs.

1.3.4   GMP inspections

Regulatory agencies and GMP inspections

Manufacturers of medicinal products are subject to inspections by GMP regulatory authorities. Within the European Union, GMP inspections are performed by National Regulatory Agencies (e.g., GMP inspections are performed in the United Kingdom by the Medicines and Healthcare products Regulatory Agency (MHRA)). In other territories, inspections are performed by the respective national bodies, for example, in Australia by the Therapeutical Goods Administration (TGA). A regulatory agency (or body) is a legislatively empowered organisation, group, or individual charged with evaluating the compliance profile of a site and/or product as measured against applicable standards. The work of a regulatory agency is undertaken through a team of regulatory inspectors.

Each inspectorate undertakes a routine GMP inspection to ensure that drug products are produced safely and correctly. Many inspections are risk based, that is the time between inspections and the length of inspection are based on the regulatory agency's opinion of the overall quality of the manufacturer. Typically, inspections are performed every two years. The frequency between inspections is shortened when the risk assessment of the manufacturer, conducted by the regulatory authority, is considered to be greater[60].

In addition to routine inspections, inspections are held for new medicinal product and where manufacturers are planning to sell a drug into a new market. These are described as pre-approval inspections (PAI) and are intended to assess the GMP compliance of a manufacturer prior to the approval of a new drug for marketing.

Furthermore, many regulatory agencies (including the FDA and the regulatory agencies in the EU) are authorised to conduct unannounced inspections. This normally occurs when the agency have a concern about the manufacturer. This may occur if there has been a major recall or in the event of a series of patient reactions to a product. For US domestic manufacturers, many FDA routine inspections are unannounced; whereas in the EU inspections are customarily notified in advance.

Most regulatory agencies expect manufacturers to produce annual reviews or to issue compliance reports. In such documents, important changes which might impact upon product quality should be reported. For example:

- Shift in performance

- Key Personnel or staff numbers

- Company ownership/ structure or status

- Changes to processes or to products

- Alterations to facilities or major new items of equipment

Scope of GMP inspections

The topics covered and scope of GMP inspections varies according to the manufacturer and to 'current' GMP topics. In general, however, many GMP inspections will include an assessment of the following areas:

1. Buildings and Facilities

2. Manufacturing

3. Packaging and Labelling

4. Product development

5. Quality Control laboratories

6. Batch Release

7. Quality Assurance

8. Receiving and Shipping

9. Training Department

10. Validation and calibration

11. Regulatory affairs

Conduct of GMP inspections

The way by which a regulatory agency will perform a regulatory inspection varies according to the agency. Some agencies place a greater emphasis upon walking the plant and looking for non-compliant activities, others are more concerned with verifying documentation to ensure that what is written in a manufacturing licence or product specification matches what is contained within the batch record. Other inspections are balanced between these two extremes. However the balance plays out in practice, a strong emphasis is always upon documentation.

A second common feature is the importance of control, either during manufacturing (in-process) or as part of finished product release. Here inspectors are concerned with a drift in standards and the batch-by-batch consistency of manufacture. A third feature is validation, in that all manufacturing processes that could somehow impact upon the variability of the finished product must be validated (for example, as referred to in FDA CFR 211.100 (a)[61]).

1.3.5 The essentials of GMP and major themes

This section, the substantive part of this chapter, examines the essential elements of GMP as well as considering some of the major topics. This section focuses on 'traditional' GMP that is the aspects of GMP first identified from the initial standards of the 1970s and which, albeit in an updated form, remain of primary focus today. The section which follows looks at more recent GMP developments and the newer topics which form part of 21st century GMP.

Manufacturing and processing

Pharmaceutical manufacturing is centred on the manufacture of a batch of a product. A batch (sometimes called a lot) is an entity, by either time or quantity or both, of a product that is intended to have a uniform character and quality. A batch must be produced within predefined and specified conditions following a defined manufacturing cycle or process. Individual batches should be assigned a batch number (a combination of numbers and/or letters which should uniquely identify a batch or lot).

GMPs require that manufacturing processes are clearly defined and controlled and that all critical processes are validated to ensure consistency and compliance with specifications. In relation to manufacturing controls, any changes to the process must be evaluated (such as through a formal change control system). Changes that have an impact on the quality of the drug must be risk assessed and validated as necessary[62]. The manufacturing of a medicinal product must be conducted under a Master Manufacturing Record. This is the comprehensive document which describes the full manufacturing process for the manufacture of a drug substance or drug product. The process starts with the starting materials, the quantities to be used, together with a description of the manufacturing operations including details of the required in-process controls.

Further in relation to manufacturing and processing:

- Production activities must follow clearly defined procedures and should be performed only by trained and competent people

- A batch or lot number must be assigned to each product to ensure traceability and to avoid mix-ups

- All materials and products should be stored under the appropriate conditions and in an orderly fashion to permit batch segregation and stock rotation

- At each stage of processing, products and materials should be protected from microbial and other contamination. This includes the use of automated washing systems, cleaning and disinfection regimes, microbial reductive filters (for which a final sterilising filter is required for aseptically filled products[63], and devices such as ultrafilter units

- Raw materials used for manufacturing should be weighed and measured, with weighing and measuring devices being of suitable accuracy for the intended use[64]

- Equipment used should be calibrated to ensure accurate results within appropriate ranges. Of greatest concern here is with what is often described as 'critical equipment'. Critical equipment and instrumentation are those used in a process that have a direct impact on the quality, safety, purity or efficacy of the final product

- Operations (and operators) should be adequately supervised

- Actual yields and percentages of theoretical yield should be determined at the conclusion of each appropriate phase of manufacturing, processing, packaging, or holding of the drug product

- All process related calculations should be performed by one person and independently verified by a second person[65]

- Any deviation from instructions or procedures should be avoided. Within a GMP context, a deviation is:

  o   A departure from written instructions

  o   An unexpected event

  o   A departure from cGMP

- Significant amendments to the manufacturing process, including any change in equipment or materials which may affect product quality and/or reproducibility of the process should be validated

- Validation studies should be conducted to demonstrate that the process equipment or the activity actually lead to the expected results[66]

- cGMP expectations are generally towards the automation of processing, where possible[67]

- The release of effluent and waste which must be controlled as part of "contained manufacture, use"[68]

Furthermore, for each product manufactured within the facility each product must have:

- A master record that outlines the specifications and manufacturing procedures.
- Individual batch or history records to document conformance to the master record.
- Written schedules and procedures for cleaning and maintaining the equipment and manufacturing areas. GMP inspections place an important emphasis upon cleaning validation to show there is no risk of cross-contamination from product-to product residues or from microbial contamination[69].

All critical processes in pharmaceutical manufacturing should be validated before implementation according to a pre-defined protocol of tests and acceptance criteria. Validation studies, including statistically based sampling where feasible, should be conducted to ensure that products are produced with consistent quality characteristics. Acceptance criteria should be based on a defined set of specifications[70].

At the end of manufacturing, a reconciliation exercise should be undertaken. Reconciliation is used for performing a comparison between the amount of product or materials theoretically produced or used and the amount actually produced or used.

Once produced, the final product should be packaged and labelled. Packaging is an important part of the manufacturing process as it protects the product and provides information about the product. Packaging includes filling, capping, visual inspection, labelling, cartoning and packing[71][72]. Of importance to sale and distribution is the package integrity. The packaging integrity is the assurance that the designed packaging component fulfils the predefined requirements in protecting the product during transportation, storage and handling during the products full shelf life.

With labelling, the label must include the batch number, product description and expiry date (relating to the the shelf life of the product which is the date beyond which the product should no longer be used). Inside the outer packaging a product information leaflet must be placed.

The packaging process creates the finished product (the final product that has been through all stages of manufacturing, including packaging, and is in its final, labelled primary and secondary packaging). The product is then subject to batch review and release (which is discussed below).

Premises and equipment

The manufacturing area is referred to as the facility, which embraces all of the rooms, suite or plant used for the manufacture and testing of the product. These should be described in a site master file. This is the comprehensive documentation describing the facilities, utilities, computer systems, organisational structure and manufacturing processes at a site. The master file should be supported by a site Validation Master Plan (VMP), which describes the assessment of the validation for the sites facilities, utilities, computer systems and manufacturing processes. The VMP captures all areas of activities within which validation is to take place and provides an overview of the status of planning. It lists the areas, systems and projects being managed, defines the status of validation for each and gives a broad indication of when validation is to be completed. It is a general plan and would normally cover all equipment and processes. It should include all systems for which validation is planned.

The layout of the production areas must be designed to suit the sequence of operations. The aim is to reduce the chances of cross-contamination and to avoid mix-ups and errors. For example, final product must not pass through or near areas that contain intermediate products or raw materials. Further, in relation to premises and equipment[73]:

- Premises and equipment must be built and maintained to suit the operations being carried out.

- The layout and design must aim to minimise the risk of errors or hazards and permit effective cleaning and maintenance[74].

- Materials, products, and their components should be segregated to minimise confusion and the potential for mix-ups and errors. A component is any ingredient to be used in the manufacture of a drug product.

- All equipment should be designed and installed to suit its intended purpose and should not present any hazard to personnel or to the product.

- Equipment must be easy to repair and maintain and the design should allow for effective cleaning.

- The air, water, lighting, ventilation, temperature and humidity within a plant should contribute to quality and be designed so that they are not contamination sources. This includes ensuring that:

- o Lighting, temperature, humidity and ventilation are appropriate (as controlled through the Heating Ventilation and Air Conditioning [HVAC] system)

- o Interior surfaces (walls, floors and ceilings) are smooth, free from cracks and do not shed particulate matter. Surfaces must be easily cleanable and compatible with disinfectants and detergents

- o Interior surfaces must be easy to clean

- o Pipe work, light fittings, and ventilation points must be easily cleanable

- o Drains must be sized adequately and have trapped gullies

For this, an appropriate standard of engineering is required[75].

- Calibration of equipment should be carried out and documented according to established standard operating procedures and national regulations. Calibration is the demonstration that a particular instrument or device produces results within specified limits by comparison with those produced by a traceable standard over an appropriate range of measurements. Regular calibration is necessary for key items of equipment, such as temperature probes (e.g. in refrigerators), pipettes, balances and timing devices.

  Calibrated equipment should have a certificate of calibration. The certificate is a document signed by qualified authorities that testify that a system's qualification, calibration, validation or revalidation has been performed appropriately and that the results are acceptable.

- The entry of unauthorised people should be prevented, and production and storage areas should not be used as a 'right of way'.

- Measuring, weighing, recording and control equipment should be calibrated regularly.

- Defective equipment should be removed if possible or labelled as defective if not.

- Maintenance, cleaning and calibration should be performed regularly and should be recorded in log books. Maintenance of equipment should be carried out at intervals according to a documented schedule. The maintenance programmes should be established on the basis of qualification activities (what is often described as planned preventative maintenance). The intervals should be defined according to the instructions of the manufacturer of the equipment or set according to risk assessment. Where intervals are not defined by the equipment manufacturer, maintenance should be carried out at least annually. All maintenance activities should be documented[76].

- Supply utilities must be appropriate for use and validated. Utilities include those designated as process systems, which refers to utilities which could affect patient safety or product quality. These utilities should be designed, constructed, commissioned and verified to provide a service that meets a defined specification (considering product quality requirements), and prevent product contamination accordingly. Utilities which should be covered by GMP include:

  o Purified Water and Water for Injections (WFI)

  o Clean Steam

  o Nitrogen and Other Process Gases

  o Instrument Air

  o Breathing Air (through Heating Ventilation and Air Conditioning Systems)

  o Heating/Cooling System P Yes Enhanced/Qualified See specific

  o Process Vacuum

  o Potable (or Mains) Water

  o Plumbing Drains

  o Mechanical Seal Fluids

  o Chilled Water

- In relation to supplies and service, particular attention is often paid in GMP inspections to sterilisation devices and the validation of such devices (using thermometric equipment and suitable biological indicators)[77].

Maintenance records should include:

- When the equipment was last used

- What is was used for

- When it was cleaned

- When it was last inspected or repaired

- When it was last calibrated

Another important aspect is cleanliness. One of the requirements of GMP is to prevent contamination of products. Dust, particles and dirt can contaminate products as can chemicals and micro-organisms.

Some specific areas relating to premises are:

Production areas

Pharmaceutical manufacturing should be carried out in adequate facilities that are suitable for the purpose. These are cleanrooms. A cleanroom or clean zone, simply is a room that is clean. A more specialised meaning is as defined in ISO 14644-1: a room with control of particulates and set environmental parameters. Construction and use of the room is in a manner to minimise the generation and retention of particles. The classification is set by the cleanliness of the air[78].

By prescribing a grade, the areas are regarded as controlled environments. A controlled environment is any area in an aseptic process system for which airborne particulate and micro-organism levels are controlled to specific levels to the activities conducted within that environment[79].

The environments in which processing occurs are either classified (to a cleanroom standard, such as ISO 14644) or controlled. Once a room has been assigned a classification, certain environmental parameters (physical and microbiological) are to be met on a routine basis. This is by collecting data and examining the results of monitoring against pre-set criteria[80]. Requirements include[81]:

o   Classification by airborne particle counts using optical particle counters[82].

o   Meeting viable microbiological limits through regular monitoring using an environmental monitoring prorgamme[83].

o   Temperature requirements (required for EU GMP Grade B cleanrooms and certain processes, such as cold stores).

o   Relative humidity.

o   Pressure differentials (measured in Pascals).

o   Placement of an airlock between cleanrooms of different grades, for the purpose of controlling the air flow and pressure differential between those rooms. An airlock allows for the movement of personnel or goods.

o   Air velocity (for uni-directional airflow devices).

o   Air changes per hour.

o   Clean-up time (cleanroom recovery tests).

o   HEPA filter integrity (local efficiency).

o   Cleanroom clothing.

Classifications depend upon the room use. The EU GMP guide provides some examples of room uses, using an alphabetical grading system (this system is also used by the WHO).

EU GMP Grade	ISO Class	Room Use
A	4.8	Aseptic preparation and filling (critical zones under unidirectional flow)
B	7	A room containing a Grade A zone (the background environment for filling) and the area demarcated as the 'Aseptic Filling Suite' (including final stage changing rooms in the 'at rest' state)
C	8	Preparation of solutions to be filtered and production processing
D	9	Handling of components after washing; plasma stripping
U*	N/A	Freezers, computer conduits, store rooms, electrical cupboards, other rooms not in use, etc.

Table 2: EU GMP cleanroom grades and typical room uses.

Other territories, and the FDA, use the ISO 14644 numerical classification system. The equivalence between the EU/WHO cleanroom grading system and the ISO 14644 classification system, for cleanrooms in the occupied or 'in use' condition, is shown in the table below.

EU GMP	ISO 14644-1
A	4.8
B	7
C	8
D	9

Table 3: EU GMP and ISO 14644 equivalence table.

Once established, cleanroom operation parameters should be assessed through routine monitoring and by annual or six-monthly formal re-certification. In addition, cleanrooms should be subject to an environmental monitoring programme[84]. This includes regular assessment of physical parameters (such as pressure differentials) and the microbiological and particulate monitoring of the process environment[85]. The level of monitoring should be commensurate with the risk, which can be based on a number of factors including the type of product manufactured. To evaluate the microbiological conditions in a clean room area, representative data should be collected. Variations associated with work shifts, level of activity, and such factors as seasonal changes should be studied.

For aseptic filling, the highest level of environmental monitoring (in terms of numbers of samples and the frequency of monitoring) is undertaken. The aseptic preparation of ready-to-use sterile products is a highly challenging process[86]. Because aseptic processing relies on the exclusion of micro-organisms from the process stream and the prevention of micro-organisms from entering containers during filling, the level of viable counts of the manufacturing environment is an important factor relating to the level of sterility assurance of these products[87].

After filling it is important that the container-closure system for vials (stopper and crimp and cap) have been demonstrated to be robust and to avoid the ingress for microbial contamination[88]. For non-sterile products the packaging must be suitable and tamper proof.

With the environmental monitoring of aseptic facilities, the GMPs require that the monitoring be continuous, due to the higher risks associated with manufacturing such products[89]. This is normally taken as continuous particle counting at representative locations and the continuous exposure of settle plates, with other types of environmental monitoring undertaken regularly[90].

Storage areas

Storage areas should provide adequate space and should be arranged in a way that allows for dry and orderly placement of stored materials. Storage conditions should be controlled, monitored and documented to show compliance with the specifications. Equal distribution of temperature throughout the storage facility should be guaranteed and documented.

Laboratories

Testing laboratories should be designed and constructed so as to minimise the risk of errors and contamination. Laboratory areas should be separated from the processing and final product storage areas.

Process maps

For GMP inspections, a plant schematic (a blueprint or layout of the facility) must be available as an authorised document. Schematics should be reviewed and updated each year, or whenever any process changes occur. A schematic can be a simple line drawing by hand or an elaborate, mechanically drawn blueprint, depending upon the complexity and nature of the operations.

An example of a schematic is illustrated in **Figure 1** below.

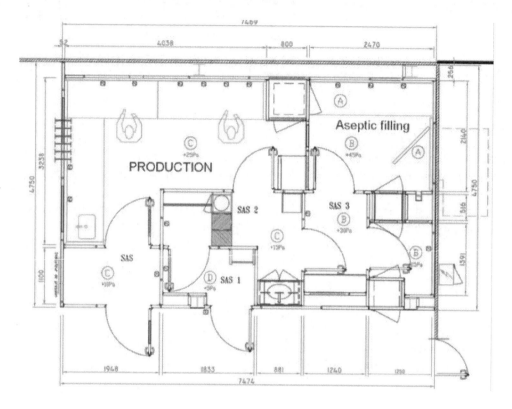

Figure 1: Example facility map schematic

In addition to a simple plant schematic, drawings that demonstrate the product or "process" flow should also be available. The process flow schematic should briefly describe the most relevant features of each processing step: time, temperature etc.

Pest control

All animals, including mammals, birds, reptiles, and insects, are potential sources of contamination in processing environments because they harbour, or could be a vector for a variety of pathogenic agents, or they could cause damage. Thus each facility should establish a pest control program to reduce the risk of contamination by rodents, insects, birds and any other pests.

An effective pest control program should include regular and frequent monitoring of affected and treated areas to accurately assess the programme's effectiveness. A staff member should be trained to implement the programme and work with outside pest control contractors as needed. Detailed pest control logs describing treatments and results should be maintained.

Third party manufacturing and suppliers

If any processing steps are subcontracted to another facility, those subcontracted operations should have GMPs of their own and should be included in any third-party audit or certification activity.

Suppliers should be audited by the manufacturer (see below) and be covered by technical agreements. A technical agreement sometimes also called quality agreement or quality technical agreement is a contract agreement which states the manufacturing and quality control provisions as well as the GMP provisions required. The rules should be stated as part of the agreement (for example, immediate information about changes in production or manufacturing failures).

Materials

Only reagents and materials from approved suppliers that meet documented requirements and specifications should be used in manufacturing or for recording laboratory tests. All materials and reagents relevant to the quality of the products should be purchased or obtained only from qualified suppliers. Materials and reagents should meet the legal requirements for medical devices. The management procedures for materials, reagents and supplies should define the specifications for acceptance of any elements that may influence the quality of the final product.

Receipt logs or records for these critical materials should indicate their acceptability on the basis of the defined specifications and should identify the person accepting them. Appropriate checks (e.g. attached certificates, expiry date, lot number, defects) should be performed on received goods in order to confirm that they correspond to the order and meet the specifications. Material should be place in quarantine and be subject to appropriate release procedures.

Once released by an authorised person, inventory records should be kept for the traceability of materials and reagents.

Documentation

Instructions and procedures must be written in clear and unambiguous language (this is part of Good Documentation Practices). Documentation is a wide term, covering those procedures, instructions, logbooks, records, raw data, manuals and policies associated with the development, manufacture, testing, marketing and distribution of a medicinal product or devices required demonstrating compliance with GMP standards and any other applicable worldwide regulatory requirements. GMP documentation includes:

- **Specifications**: These detail the requirements with which products and materials have to conform, i.e. they serve as a basis of quality evaluation.

- **Manufacturing formulae, processing, packaging and testing instructions** – 'provide detail of all starting materials, equipment and computerised systems (if any) to be used and specify all processing, packaging, sampling and testing instructions. In-process controls and process analytical technologies to be employed should be specified where relevant, together with acceptance criteria'.

- **Operating Instructions**: These detail material and equipment requirements and describe the steps to complete a task.

- **Standard Operating Procedures**: These give direction for performing certain tasks and provide higher-level instruction than operating instructions. The main components of an SOP are:

  o   Header with title

  o   Regulatory basis

  o   Reference documents

  o   Purpose

  o   Scope

  o   Responsibilities

  o   Procedure

  o   Acceptance criteria

  o   Results evaluation and review

  o   Signing, approving and reporting

- **Records**: These provide a history of each batch and provide a mechanism to check that personnel are following operating procedures and instructions.

- **Raw data** (primarily in the context of the quality control laboratory). Raw data relates to work sheets, records, memoranda, notes or exact copies thereof, that are the result of original observations and activities of a study, or a process, and are necessary for the reconstruction and evaluation of the report of that study or process. Raw data includes, for example, photographs, computer printouts, and magnetic media, including dictated observations, and recorded data from automated instruments where applicable.

- **Technical agreements** – contracts between BPL and external organisations (for example testing laboratories).

Of the records, arguably the most important is the batch manufacturing record.

Documentation is heavily featured within GMP guidelines. EU GMP describes requirements in Chapter 4 – Documentation. This chapter states that:

- 'Good documentation constitutes an essential part of the quality assurance system and is key to operating in compliance with GMP'

- 'Documents may exist in a variety of forms, including paper-based, electronic or photographic media'

- 'The main objective of the system...must be to establish, control, monitor and record all activities which directly or indirectly impact on all aspects of the quality of medicinal products'

Similarly, US 21 Code of Federal Regulations (CFR) part 211 Subpart J – Records and Reports – describe the requirements to:

- Keep batch related documentation for a period of time after batch expiry

- Have records readily available for inspection

- Review the quality standards of each product at least annually to determine whether there is a need to change product specifications, or manufacturing and/or control procedures. This includes a review of rejected batches, related investigations and any complaints, recalls, returned or reworked batches

In general, GMP matters relating to documentation include the following[91]:

o Documentation should be clearly written, since this prevents errors from spoken communication

o Specifications, instructions, procedures and records must be free from errors and available in writing

o Documents should be unambiguous, have clear and concise contents, a title and purpose

o Documents should be regularly reviewed and kept up to date

o Documents should not be hand-written (except for the entry of data)

o Any alteration to a record should be signed and dated with the original entry still visible

Documentation is necessary and allows the manufacturer to define in advance what is going to happen; to check what has been done what should have been done. Additionally, documentation is about keeping records of information, results and actions taken and to investigate problems.

There are a number of different types of document in use. Manufacturing and other instructions are contained within Standard Operating Procedures (SOPs), which describe specific tasks in a stepwise fashion, instructing staff what to do and if necessary how to do it, although the how is mainly part of training and job skills.

Record keeping

Records must be made manually or by instruments during manufacture that demonstrate that all the steps required by the defined procedures and instructions were undertaken. Records of manufacture (including distribution) should be made and these must enable the complete history of a batch to be traced. Records must be retained in a comprehensible and accessible form, for a specified time[92].

In relation to records, it is necessary to ensure that:

1. All necessary information immediately upon completion of a task is captured.

2. Records are contemporaneous.

3. Records are legible.

4. Records are signed and dated.

5. Records relate to a defined procedure.

6. Mistakes, which have been corrected, remain readable and the reason for the mistake is stated. Corrections must be signed and date. A reason for the correction must be included.

7. Any deviations from procedure must be recorded and investigated.

    Deviations need to be identified, reported, investigated, risk assessed against product/patient safety, monitored and ideally avoided in the future. The requirement for deviation investigation is detailed in the following GMP guidelines:

    *   EU GMP Chapter 1 section 1.2 – 'any significant deviations are fully recorded and investigated'

    *   US CFR 211.100(b) – 'any deviation from the written procedures shall be recorded and justified'

8. Records must only be completed by the person who actually completed the task.

9. If an activity is undocumented it must be assumed, and will certainly be assumed by regulators that it did not happen.

Records for every stage of the manufacturing process must be retained. Such records include:

- product master records
- batch or manufacturing records
- material/component control records
- personnel records
- training records
- equipment service and calibration records
- equipment logs
- Cleaning logs

Computerised systems

A computerised system may be described as a functional unit consisting of one or more computers and associated peripheral input and output devices, and associated software that uses common storage for all or part of a program and for all parts of the data necessary for the execution of the program. Hardware and software should be protected against unauthorised use or changes. Critical computerised systems should be validated before use. The system is considered critical if it has an impact on product quality, information management, storage, or tools for operational decision-making and control.

Periodic revalidation or annual checks to ensure reliability should be performed on the basis of a risk assessment. There should be procedures for each type of software and hardware, detailing the action to be taken when malfunctions or failures occur.

Batch release

Following on from records, be they paper or electronic, is the system of batch release. This is the process of reviewing and approving all pharmaceutical product manufacturing and control records and it performed by the Quality Unit to determine compliance with all established approved written procedures before a batch is released. The process of batch release, and the authority and training of the persons eligible to do so, varies according to different GMP systems. However, there should be in place a procedure describing how the batch release is performed, including how batch deviations (changes to the predefined process or condition detailed in the batch manufacturing record) are assessed and how batches are assessed for release or for rejection.

Personnel

The personnel employed within a pharmaceutical organisation must be suitably qualified and trained to carry out procedures. GMP ultimately depends on people. The individual skills and understanding of people about their work can be developed by training (see, for example EU GMP Section 2.8.12 and CFR Part 21)[93]. For example, The Code of Federal Regulations 211.25 under Personal Qualification requires that:

"Each person engaged in the manufacture, processing, packing or holding of a drug product shall have education, training, and experience, or any combination thereof, to enable that person to perform the assigned functions. Training shall be in the particular operations that the employee performs and in current good manufacturing practices (including the current good manufacturing practice regulations in this chapter and written procedures required by these regulations) as they relate to the employee's functions. Training in current good manufacturing practice shall be conducted by qualified individuals on a continuing basis and with sufficient frequency to assure that employees remain familiar with cGMP requirements applicable to them."

It is also important for product safety that staff do not carry out tasks for which they have not been trained and which either they or their supervisors think they are not competent to carry out.

Training should be given on both the theory and practice of the work being undertaken in a particular area, as well as relevant 'on -the-job' (i.e. task-based) training. Records of this training must be available. Personnel should receive initial and continuous training that is appropriate to their specific tasks. This training should be carried out by qualified personnel or trainers and should follow prearranged written programmes. Approved training programmes should be in place and should ideally include GMP and hygiene topics.

An example of a basic personnel training scheme with respect to GMP can be summarised as:

o   Make sure you have the correct written instructions before starting a task

o   Do not carry out a task for which you have not been trained or in which you do not feel competent

o   Always follow instructions precisely. Do not cut corners. If in doubt, ask

o   Check that the equipment and the materials being used are the correct type, as stated in the procedure

o   Check that the equipment you are using is clean

o   Always be on your guard for labelling errors

o   Keep everything clean and tidy (including yourself!)

o   Staff should be examining documents for any mistakes, and containers for defects and unusual events. These should be reported immediately

o   Clear and accurate records of what was done and the checks carried out must be made.

Ultimately quality depends on people. Although processes get more automated there are still many activities which require the constant care and attention of staff

Training should also focus on reducing errors (or 'error risk reduction'). Error Risk Reduction (ERR) is a process which:

•   Shows where errors might occur

•   Identifies factors that increase the risk of error

•   Reduces the risk by addressing those 'Risk Influencing Factors' (RIFs)

•   Is intended to be used pro-actively, but can also be used retrospectively to form part of investigations and CAPA (Corrective And Preventive Action)

•   Execution errors – for example, why people incorrectly perform familiar tasks, why they might fail to act when needed, or why they failed to notice something they should have

•   Activities and operational context – for example, what features of the activity or working environment might have led towards an actual error or an increased chance of an error

•   Reducing overall burden of risk, by raising staff awareness and by influencing the design of equipment, documentation, activities and processes to reduce the potential for errors to occur

One important part of training for pharmaceutical manufacturing, and which relates to personnel, is hygiene. All employees must have a good working knowledge of basic sanitation and hygiene principles. They should understand the impact of poor personal cleanliness and unsanitary practices on processing. The level of understanding needed will vary as determined by the type of operation, the task, and the assigned responsibilities. Furthermore, personnel working in cleanrooms must wear the appropriate gowns relating to the type of cleanroom in which they are working[94]. The manufacturer must control the washing, repair and laundering of cleanroom gowns.

During a GMP inspection, regulatory agencies will expect to be shown an organisation chart. An organisation chart helps clarify and document the roles of staff. This chart should identify who is responsible for the various phases of the operation. In addition to the chart, job descriptions must be available for all personnel involved with GMP activities. The job description should contain each individual's specific responsibilities relevant to each aspect of GMP. These responsibilities should be written in a manner that is clear and easy to understand to avoid confusion when describing who is responsible for making which decisions and for their consequences.

Quality Control

Quality Control is that part of GMP which is concerned with specifications, sampling and testing (and where appropriate, such as for microbiological testing, samples should be taking aseptically[95]). Quality Control is also concerned with the organisation, documentation and release procedures which ensure that the necessary and relevant tests are carried out and that neither material are released for use nor products released for supply until their quality has been judged to be satisfactory.

Quality Control is required at various stages of drug manufacturing:

- Raw material and starting material testing: This includes active pharmaceutical ingredients, excipients (including process water) and packaging materials

- Intermediate product testing: bulk tablets, capsules, granules, liquids and creams, etc

- Finished product testing

- Ongoing stability testing of products on the market

- Microbiological monitoring of the manufacturing areas

Samples for testing should be taken via an approved, and where appropriate, statistically sound sampling plan. A sampling plan describes the details of the planned sampling activity, such as the number of units or quantity of material that must be collected and the manner in which it is to be collected.

Testing includes such as areas as chemical purity or impurity, bioburden, bacterial endotoxin and, for sterile products, sterility of the finished article[96]. Testing must be performed to show that the quantity and quality of the drug is as expected. Where the result is unexpected, and falls outside the agreed specification, an out-of-specification (OOS) report must be raised and an investigation conducted. An OOS is a result that falls outside the predefined specifications or acceptance criteria. Where there is a drift from agreed parameters, or as series of OOS results have occurred, an out-of-trend (OOT) report must be raised and an investigation conducted.

Further in relation to Quality Control activities:

- Quality Control is not confined to laboratory operations, but must be involved in all decisions that may concern the quality of the product.
- The independence of Quality Control from Production is considered fundamental to the satisfactory operation of Quality Control.
- Sampling of product should take place in accordance with written procedures.
- Analytical methods should be validated.
- With contract manufacture and analysis, the manufacturer should have the same level of confidence and control as if the testing was conducted at the manufacturing site.

Quality control involves:

o Quality inspection, where outputs are assessed against inspection criteria.

o Quality sampling, through sampling schemes designed to directly measure materials or products against specifications.

o Quality testing using defined and validated methods against pre-defined acceptance criteria. This sometimes involves, for analytical testing, the use of reference standards. A reference standard is a chemical substance or mixture, or analytical standard, or material other than a test substance that is used for the purposes of establishing a basis for comparison with the test substance for known chemical or biological measurement.

o Statistical process control, to examine for variance and to identify the cause of variance.

Many quality control tests are undertaken according to pharmacopoeial monographs (books which describe drugs and drug preparations used in medicine). Although there are many pharmacopoeia, the ones which regulators expect to be followed are either:

- the International Pharmacopoeia (IP) – "official" and primarily used in the developing nations, and which fits with WHO GMP;
- the British Pharmacopoeia (BP) – "official" and primarily used in the 54 member countries of the British Commonwealth;

- the United States Pharmacopeia (USP) – "official" and primarily used in North America and in 27 other countries;

- The European Pharmacopoeia (Ph. Eur.) – "official" and primarily used in the 26 countries which are member states of the Council of Europe.

Many of the chapters within the USP are linked to the FDA CFRs and thus have a legal standing. Therefore, if a pharmaceutical product does not meet the requirements of a particular USP monograph, then it is considered to be mislabelled or adulterated.

One of the basic principles of GMP, which is echoed in ISO 9000, is for the responsibility for Quality Control to be held by somebody independent from the people responsible for production. This does not mean that the testing facilities need to be managed independently, but that the acceptance of the results of the testing must be independent, as must the initiation of any actions necessary after examining the results.

Laboratory methods should be validated. The validation should demonstrate, that the performance specifications of the system established by the kit manufacturer are met by the laboratory and those laboratory personnel are thoroughly instructed, trained and competent to operate the test system[97].

Stability programmes and stability testing are often associated with quality control activities. The term stability describes the ability of a drug product or drug substance to stay in their chemical, physical, microbiological and biopharmaceutical specified limits during its whole shelf life. The stability program is a planned and documented program assessing the stability profile of materials and products to establish their retest periods or shelf life and storage directions.

Product review

Regular periodic or rolling quality reviews should be conducted with the objective of verifying the consistency of the existing process and the appropriateness of current specifications in order to highlight trends and to identify improvements in both product and process.

Qualification and validation

Qualification and validation is undertaken in order to prove that equipment and processes consistently do what they are supposed to do. These are the actions of proving that any equipment or process works correctly and consistently and produces the expected results. It is a GMP requirement to prove control of the critical aspects of certain operations. For this testing and documentation is required. The aim is to ensure that the device meets it specification or process parameters and that it is capable of a consistent performance. Such an approach is based upon having pre-defined acceptance criteria. Acceptance criteria are necessary for making a decision to accept or reject a validation parameter and thus to determine whether a process has or has not worked as intended.

All validation activities should be well planned and clearly defined. This is usually by means of a Validation Master Plan, or VMP (this is described above in relation to the qualification of premises). Beneath the VMP are validation plans and protocols. Validation plans are summary plans which communicate management's expectations and commitments to be followed for the equipment or process to be validated. The plan describes the program to be conducted to get the items in question in a validated manner. The plan lists all of the validation activities to be completed, as well as the schedule for their completion.

From this there are a series of qualification protocols. These are procedures, stating in sufficient detail, how the qualification will be achieved. Included are specific qualification requirements for each equipment item, each system requirement, and product requirement. Each protocol should stipulate test parameters as well as decision points on what constitutes acceptable test results. The written protocols should be based on the associated qualification procedures and should be step-by-step instructions to be used in the field to qualify equipment, instruments, materials, systems and subsystems, and should include data sheets to record critical data.

In establishing a validation plan, consider all critical parameters that may be affected and impact product quality. Validation typically consists of:

- User Requirement Specification, or URS. The URS provides a clear and precise definition of what the user wants the system to do. It defines the functions to be carried out, the data on which the system will operate and the operating environment. The URS defines also any non-functional requirements, constraints such as time and costs and what deliverables are to be supplied. The emphasis should be on the required functions and not the method of implementing those functions

- Design Qualification, or DQ, the documented verification that the proposed design of the equipment and system is suitable for the intended purpose.

- Installation Qualification, or IQ, testing to verify that the equipment is installed correctly.

- Operational Qualification, or OQ, testing to verify that the equipment operates correctly.

- Performance Qualification, or PQ, testing to verify that product can be consistently be produced to specification.

Once the testing is complete, a validation report is written in which the qualification results are compared with the agreed qualification criteria.

A second concern is re-validation. Changes in production methods of a biological product may necessitate an assessment of comparability to ensure that these manufacturing changes have not affected the safety, identity, purity or efficacy of the product. This assessment could lead to re-validation of equipment or processes[98].

Equipment

All equipment should be qualified and used in accordance with validated procedures. New and repaired equipment should meet qualification requirements when installed and should be authorised before use. Qualification results should be documented. The extent of qualification depends on the critical nature and complexity of the equipment. For some equipment, installation qualification and calibration may be sufficient. More complex equipment may need a more thorough approach to qualification and validation and should include the instruments, the associated operation(s) and the software involved.

Change control

A formal change control system should be in place to plan, evaluate and document all changes that may affect the quality, traceability and manufacturing of a pharmaceutical product. This is normally by change control. Change control is a process in a formal change control system by which qualified representatives of appropriate disciplines review and approve a change request.

The change control system should guarantee a formal approval of a change before it is implemented. Furthermore, it should ensure that the impact of the proposed change is assessed and that all necessary measures — such as qualification and validation, training of personnel, adoption of working instructions, revision of contracts, definition of maintenance tasks, information for third parties and authorities — are defined and completed at the time the change is put into force. The need for additional testing and validation should be determined on a scientific basis. After the implementation of a change, a post-implementation evaluation should be carried out in order to determine whether the introduction of the change has been successful and effective. The introduction of new equipment, processes and methods should be treated as a change.

Deviations

Any deviation from standard operating procedures, validated processes, or non-conformances with specifications or other quality-related requirements should be recorded and investigated. The potential impact on the quality of the product in question, or on other products, should be evaluated. The evaluation of the cause of the deviation and of related processes that may also be implicated in the deviation should be documented.

All deviations and non-conformances should be logged in a system that allows for appropriate data review. A data review should be carried out periodically in a manner that allows for tracking and trending of data and that facilitates process improvement. Within such reviews, deviations must be trended to detect/predict reoccurrent problems and to monitor the effectiveness of the implementation of the corrective actions. The frequency of the trend analysis and reporting should be at least monthly.

Traceability

Traceback is the ability to track each raw material and component back to its source (through the supply chain) and to understand each part of the manufacturing process. The ability to identify the source of a product through traceback serves as an important component of good manufacturing practices and will prevent the occurrence of some manufacturing problems.

Such systems should be regularly tested and audited. Information gained from a traceback investigation may also be useful in identifying and eliminating a hazardous pathway. A key element of a traceback programme is batch or lot identification. Adequate lot coding and distribution records are critical. Lack of a coding system and accurate records could lead to a product recall with notification to all customers.

Distribution

The distribution of the drugs must minimise any risk to their quality (Good Distribution Practice). A system should be available for recalling any batch of drug from sale or supply.

Customer complaints and product recall

Complaints about marketed drugs must be examined and the causes of any quality defects investigated (together with any appropriate measures taken to prevent a recurrence of the defect). All complaints and other information concerning potentially defective products must be reviewed carefully according to written procedures. The person responsible for Quality Control should normally be involved in the study of product defects. There should be written recall procedures that are regularly checked and updated where necessary. Recall operations should be capable of being initiated promptly and at any time.

Recalls are actions taken by a firm to remove from the market any product that is in violation the law. Recalls of a drug may be conducted on the manufacturer's own initiative or by a regulatory agency.

There are many situations that can result in a product recall. Some are emergency situations; other situations are not. Following is a list of potential causes for recalls:

- Microbial contamination, such as contamination by spoilage organisms or harmful bacteria (non-sterile products) or an issue of non-sterility (for sterile products)

- Chemical contamination

- Foreign materials – Presence of glass, plastic or metal

- In-house sabotage

- Misbranding, such as violations of labelling laws

- Real or fraudulent customer or consumer claims

- Tampering and tampering threats

- Undeclared ingredients

US GMP, 21 CFR 211.180(e) requires that manufacturers must establish and follow written procedures for periodically reviewing complaints, recalls, returned or salvaged drug products, and investigations of product discrepancies.

The product and packaging must be clearly labelled, and include:

Product identity:

- Product name, including all brand names

- Product code numbers

- Product description

Manufacturer identity:

- Handler name and address

- Responsible individual at firm: name, title, phone, fax and email address

- Recall contact
- Contact for the public

For recalls, the following must be established:

Reason for the recall

- Explanation of cause of problem and date/ time it occurred.
- Explanation of how and when the problem was discovered.
- Determine whether the problem affects all products in the lot being recalled or only a portion of the products being recalled.
- If the clinician received a positive microbiological sample laboratory result, obtain a copy of the confirmed result.
- If the clinician received complaints associated with the problem, he must provide dates of the complaints with descriptions that include details of injury or illness, lot numbers, code dates, etc.
- If the recall is due to presence of a foreign object, describe the size composition, hardness and/or sharpness of the object.
- If the recall is due to the presence of a chemical contaminant, explain level of contamination and provide labelling, list of ingredients and material safety data sheets (MSDS) for the contaminant.
- If the recall is the result of a labelling issue, the company must provide and identify the correct and incorrect label, description and formulation.

Volume of product being recalled

- Total quantity produced.
- Date(s) produced.
- Quantity distributed.
- Quantity on hold by recalling firm and its distribution centres.
- Description of product quarantine procedures and conditions.
- Estimated amount of product remaining in the marketplace at: distributor level, retail level, and consumer level.

Distribution pattern

a. List of consignees (names, addresses, phone numbers).

b. Indicate quantity of product shipped to each consignee, including dates.

In relation to distribution:

- Control and accountability for inventory of all product involved in recall

  o   Prepare inventory and distribution status report showing where, when and to whom recalled product shipped or is currently warehoused

- Production & Quality Assurance

  o   Identify lot codes of product implicated

  o   Investigate cause of problem. Examine all production, quality and lab records

- Maintain implicated product quarantine until regulators approve product release

  o   Do not process or destroy implicated product until directed to do so by regulators through the recall coordinator

Change control

Change control is a systematic approach to managing all changes made to a product or system. The purpose is to ensure that no unnecessary changes are made, that all changes are documented, that services are not unnecessarily disrupted and that resources are used efficiently. The change control process is usually conducted as a sequence of steps proceeding from the submission of a change request.

The common steps in most change control systems are:

- **Documenting the change request**: the request is needs to be categorised and recorded. At this stage an informal assessment of the importance of that change and the risk and difficulty of implementing it is made

- **Formal assessment**: the justification for the change and risks and benefits of making/not making the change are evaluated. If the change request is accepted, a team is often assigned for its implementation. If the change request is rejected, that fact is documented and a reason given

- **Planning**: the implementation of the change is planned out

- **Designing and testing**: for major changes, the change is often tested out to determine if there would be any undesirable affects

- **Implementation and review**: the change is implemented and stakeholders review the change

- **Final assessment**: the change is evaluated to determine if it were implemented satisfactorily. In the event that it was successful then the change request is closed

Quality audits

Audits are a systematic, independent examination of the quality system, looking at the system and the results of such activities. Audits check if procedures are implemented effectively, and if the procedures are suitable to achieve quality system objectives. Audits are either conducted by the manufacturers or the key suppliers of the manufacturer or internally within the organisation. With the latter, it is important, as part of continuous improvement, that a system for self inspection (internal audit) is in place[99].

Internal audits should be conducted in order to monitor the implementation and compliance with GMP. They should be conducted in an independent and detailed way by designated competent persons. Thus a good GMP system will not function or improve without adequate audits and reviews. Audits are carried out to ensure that actual methods are adhering to the documented procedures, whilst system reviews should be carried out periodically and systematically, to ensure the system achieves the required effect.

There should be a schedule for carrying out audits, with different activities possibly requiring different frequencies. An audit should not be conducted just with the aim of revealing defects or irregularities – they are for establishing the facts rather than finding faults. Audits do indicate necessary improvement and corrective actions, but must also determine if processes are effective and that responsibilities have been correctly assigned. The emphasis on process improvement and enhancing customer satisfaction in the revised standard will require a more thoughtful approach to auditing.

The generic steps involved in an audit are:

- Initiation
- Scope
- Frequency
- Preparation

- Review of documentation
- The programme
- Working documents
- Execution
- Opening meeting
- Examination and evaluation
- Collecting evidence
- Observations
- Close the meeting with the auditee
- Report
- Preparation
- Content
- Distribution
- Completion
- Report
- Submission
- Retention

Quality Management Review

The object of a Quality Management Review is to provide an overview of the results from the Quality Control and Quality Assurance functions, and thus GMP forms part of this process. The review forms part of an organisation's Quality Management Systems (as detailed in both GMP guidance and in ISO 9001). A Quality Management System review should take place regularly and should cover:

o   Results of audits

o   Customer feedback

o   Process and product conformity

o   Status of preventative and corrective actions

o   A review of progress relating to non-compliances

o   Follow up actions from previous management reviews

o   Changes that could affect the QMS

o   Recommendations for improvements

o   Outputs should include:

    –   Improvements to the QMS and processes

    –   Improvements of a product related to customer requirements

o   Resource needs

At QMS meetings, data are often presented by participants from different departments. The data are normally measured against a target (to allow for effectiveness, efficiency and capability to be measured). This can be relatively straightforward, such as the number of internal audits completed against a target level (such as 95% of audits to be completed to schedule) or more complex when relating to statistical process control charts for laboratory test data.

Arguably, the three most important aspects of GMP considered at such meetings are audit review; a review of processing issues; and consideration of corrective and preventive actions (CAPA). CAPA are put into place to mitigate the immediate and future risks.

- Corrective Action = action taken immediately to safeguard the activity or batch in progress at the time. A corrective action may be a short-term fix until proper preventive action has been agreed.

- Preventive Action = action taking to avoid the same error or incident happening again.

Any given CAPA program must adhere to very specific standards. In regards to corrective action, organisations must firstly take actions to eliminate the causes of non-conformity in order to prevent recurrence. These actions drive the root cause analysis and the focus should be on improving the process, not just correcting a specific incident. Secondly, the corrective actions should be appropriate to the effect of the non-conformity. Thirdly, the corrective action procedure needs to be documented, which should include a review of the non-conformance including any customer complaints, a determination of the cause of non-conformance, a determination of the action needed to prevent recurrence, and an effectiveness evaluation to record the results of the action[100].

Regulators will expect to see a progression from corrective action to preventive action. As an organisation matures and grows, there should be a shift toward decreasing corrective actions and increasing preventive actions. Theoretically, an increase in preventive actions will lead to a decrease in corrective actions.

Developments in API and excipient GMPs

Active pharmaceutical ingredients (API) refer to any substances or mixture of substances intended to be used in the manufacture of a pharmaceutical dosage form and that, when added, become the active ingredient. The API thus furnishes pharmacological activity. APIs have fallen under GMP regulations in the US for many years but they were only adopted within Europe from 2005, as part of International Conference on Harmonisation, GMP standard ICH Q7A[101].

With excipients the legislation is varied, where some excipients are classed as similar to APIs. Within Europe, Directive 2004/27/EC[102] specifically mandated the implementation of GMP for "certain excipients" including:

- excipients prepared from materials derived from a TSE-relevant animal species, with the notable exception of lactose

- excipients derived from human/animal material with potential viral contamination risk

- excipients claimed to be sterile (or sold as sterile) and used without further sterilisation

- excipients with the specification or claim that they are endotoxin/pyrogen controlled; or specific excipients, namely propylene glycol and glycerol

1.3.6   Current Good Manufacturing Practice (GMP for the 21st Century)

A discussion in cGMP in a book chapter runs the risk of becoming dated. What this section attempts to do is to outline some of the changes to pharmaceutical processing which have impacted upon GMP in recent years. These changes have generally centred on technological changes. Regulatory Authorities have been keen for pharmaceutical manufacturers to adopt the latest technologies for reasons of accuracy and increased assurance of time-to-result. This direction was captured by the FDA when they issued a document entitled "Pharmaceutical cGMPS for the 21st Century – A Risk-Based Approach"[103] (originally in 2003 and with subsequent updates). The objective was to enhance and modernise the regulation of pharmaceutical manufacturing and quality. The methodology was to use risk-based and science-based approaches for regulatory decision-making throughout the entire life-cycle of a product and to create a framework that will streamline the quality review of many products, allowing pharmaceutical organisations to use their valuable resources in a more efficient manner. This initiative set forth a plan to enhance and modernise FDA's regulations governing pharmaceutical manufacturing and product quality for human and veterinary drugs and human biological products.

With the guidance the FDA has sought to encourage:

- The early adoption of new technological advances by the pharmaceutical industry. This includes applications such as single-use disposable technologies for aseptic manufacturing, as part of a contamination control strategy[104].

- Facilitate industry application of modern quality management techniques, including implementation of quality systems approaches, to all aspects of pharmaceutical production and quality assurance.

- Encourage implementation of risk-based approaches that focus both industry and Agency attention on critical areas.

- Ensure that regulatory review and inspection policies are based on state-of-the-art pharmaceutical science.

- Enhance the consistency and coordination of FDA's drug quality regulatory programmes, in part, by integrating enhanced quality systems approaches into the Agency's business processes and regulatory policies concerning review and inspection activities.

Furthermore, the desire to improve analytical methods was sufficient to trigger a revision to Annex 16 of the EU GMP Guide and for the FDA to issue a statement in 2011 concerning the 'modernisation of regulatory science'. The FDA placed an emphasis on several key areas[105]:

- Modernising toxicology (the study of chemical, biological or physical agents that can be harmful), calling for an improvement of the ability of tests, models and measurements to predict product safety issues. This included the development of new methods that could reduce or replace animal testing.

- Supporting new and improved manufacturing methods by researching how new technologies affect product safety, effectiveness and quality.

- The evaluation of innovative and emerging technologies.

Process Analytical Technology

The conventional approach for pharmaceutical manufacturing centred on batch processing with laboratory testing conducted on collected samples to evaluate quality. This is mostly based upon agreeing how a process will be conducted in advance and then evaluating the success or failure of this through end product testing.

In recent years there has been a regulatory move, particularly by the FDA[106], towards encouraging manufactures to adopt Process Analytical Technology (PAT). PAT, drawn from the ICH definition, refers to a system for designing, analysing, and controlling manufacturing through timely measurements (that is, during processing) of critical quality and performance attributes of raw and in-process materials and processes with the goal of ensuring final product quality. The FDA summarised the goals of PAT as:

- To improve the scientific basis for establishing regulatory specifications

- To promote continuous improvement

- To improve manufacturing while maintaining or improving the current level of product quality

PAT is concerned with mechanisms to design, analyse, and control pharmaceutical manufacturing processes through the measurement of Critical Process Parameters for gaining process understanding[107]. Much of this is based on 'real time' monitoring (for example on-line total organic carbon measurements to assess the quality of a Water-for-Injections system). Real-time monitoring is discussed below.

Real time release testing

PAT can be applied to real-time testing in relation to manufacturing. The 2004 FDA guidance (and the 2008 general principles)[108], as well as outlining PAT, also addressed the concept of real-time release. This was defined as "the ability to evaluate and ensure the acceptable quality of in-process and/or final product based on process data". For this, both PAT and parametric release (used for products sterilised using terminal sterilisation methods) are presented as methods that can be used for real-time release for the examination of critical quality attributes (these are the physical, chemical, biological, or microbiological properties or characteristics that should be within an appropriate limit, range, or distribution to ensure the desired product quality). PAT also includes risk assessment approaches, centred on identifying critical control points (utilising risk assessment methods such as HACCP, which are discussed below)[109].

PAT tools typically enable non-destructive testing and provide the opportunity for enhanced monitoring. This is achieved through utilising technology such as on- or in-line analysers. Examples include near-infrared; particle-size analysis by laser diffraction and by ultrasonic extinction; and light-induced fluorescence instrumentation. Although most applications are for chemical testing, some PAT methods extend into microbiological testing, such as on-line endotoxin detection for pharmaceutical water systems. PAT allows for[110]:

1. Making timely measurements during processing,

2. Measuring critical quality parameters,

3. Measuring performance attributes,

4. Assuring acceptable end product quality,

5. Using designed sensors, probes and other devices or equipment to obtain real-time data

6. Conducting data collection and measurements in-line or on-line,

7. Monitoring or controlling individual operations or all operations in a process.

The other application of PAT, as indicated above, relates to parametric release. The purpose of parametric release is to provide a system for elimination or reduction of end product tested based upon real-time measurements, monitoring, and analysis of a controlled validated process. Furthermore, PAT techniques can also be used for concurrent validation that applies on-going or in-line measurements to provide a continuum of validation and eliminate or reduce periodic validation.

Following on from the FDA document, the International Conference on Harmonisation (ICH) adopted ICH Q8(R2) "Pharmaceutical Development", which used the term "real time release testing" (RTRT)[111]. The definition of this term moved the emphasis from the decision to release a batch to the measurements themselves and towards: "the ability to evaluate and ensure the quality of in-process and/or final product based on process data, which typically include a valid combination of measured material attributes and process controls." This philosophy represents a move away from traditional release testing conducted on small samples after batch manufacture is complete[112].

One of the reasons that regulatory authorities are promoting real-time release is because quality can also be improved through higher yields or lower rework or rejection rates. Studying the process in real time allows for greater product and process understanding.

Real-time-release is not applicable for all tests or all types of products (aseptically filled products still require an end products sterility test, for example. However, here there are advantages afforded by some rapid methods which speed up the time-to-release[113]). Furthermore, there is difference between regulatory agencies as to which aspects of real-time monitoring they will accept. However, for non-sterile pharmaceutical manufacturing, such as tablets, real-time analysis can assess a range of essential parameters including size, moisture content and blend uniformity.

Electronic records

The move away from traditional paper records to electronic records allows pharmaceutical manufacturers to more easily review data and provides a higher level of data security. Nonetheless, electronic data also present problems in terms of control, security and safety. A central part of cGMP concerns electronic data management.

cGMP points in relation to electronic records include:

- limiting system access to authorised individuals

- use of operational system checks

- use of authority checks

- use of device checks

- determination that persons who develop, maintain, or use electronic systems have the education, training, and experience to perform their assigned tasks

- establishment of and adherence to written policies that hold individuals accountable for actions initiated under their electronic signatures

- appropriate controls over systems documentation

Guidance on electronic records is provided by 21 CFR Part 11, 73 Electronic Records, Electronic Signatures, ISO/IEC 17799[114], Good Automated Manufacturing Practice[115] and the FDA document Part 11, Electronic Records, Electronic Signatures[116].

Risk management and the ICH Q8, Q9 and Q10 documents

Risk management has always been central to the manufacture of pharmaceuticals, even if it was not explicitly stated in regulatory documents[117]. In the 1990s, risk management began to feature in regulatory guidance, most notably in Annex 15 of the EU GMP guide. Annex 15 mandates the manufacturers to formally evaluate the risk during manufacturing related to activities of validation and change control. Annex 15 is specifically concerned with qualification and validation activities, and states that a "risk assessment approach should be used to determine the scope and extent of validation", and that the likely impact of changes "should be evaluated, including risk analysis". The use of more systematic tools represents the shift away from tick box approaches to more scientific assessments of risk, such as those encouraged in the US FDA document "Pharmaceutical cGMPs for the 21st Century: A Risk-Based Approach" and in the ICH Q9 Quality Risk Management guideline[118 e].

In November, 2005 the ICH published a guideline Quality Risk Management (numbered as ICH Q9) which was a significant milestone in the development of quality risk management activities within pharmaceutical industrial and regulatory activities[119]. The series of quality documents issued by the ICH have since been adopted as part of supranational GMP by the EU (Annex 20) and as national GMP by the FDA. The ICH quality documents consist of three parts[120]:

1   Q8 (second revision) Pharmaceutical Development (includes the Q8 parent guidance (Part I) and the annex (Part II), which provides further clarification of the Q8 parent guidance and describes the principles of quality by design)[121].

2   Q9 Quality Risk Management[122]

3   Q10 Pharmaceutical Quality Systems[123]

These ICH-Q documents form a tripartite approach to total quality management system for the pharmaceutical industry.

Quality risk management is a systematic process for the assessment, control, communication and review of risks to the quality of the product across the product's life cycle[124]. The concept of risk is built on fundamental concepts of severity (how bad could the risk be) and by chance, likelihood, or probability (with probability, the concept is typically applied by risk managers as a combination of data-based measures of probability and a subjective "degree of belief" meaning of probability)[125].

Quality Risk Management, when applied to pharmaceutical manufacturing, allows process and products, and in some cases personnel, to be better protected and it can be used to help to meet regulatory expectations and to meet cost demands or to seek process efficiencies.

Risk assessment involves identifying risk scenarios either prospectively or retrospectively. With the former, this involves determining what can go wrong in the system and all the associated consequences and likelihoods; with the latter this looks at what has gone wrong and using risk assessment to assess the process, product or environmental risk and to aid in formulating the appropriate actions to prevent the incident from re-occurring[126]. Risk analysis is also highly beneficial in that it can also be used to identify and justify process improvements. Furthermore, the use of risk assessments can allow manufacturers to explore weaknesses and to construct scientific and data based rationales. Two important considerations of QRM, as outlined in the PIC/S guidance on ICH Q9[127], are:

*   the evaluation of the risk to quality is based on scientific knowledge, experience with the process and ultimately links to the protection of the patient

*   the level of effort, formality and documentation of the quality risk management process is commensurate with the level of risk

There are two groups of approaches to the risk analysis process. These are **qualitative** and **quantitative** methods. Of these, the qualitative approach is the more simple and involves the identification of risks to which a process or a procedure is exposed. For the process the optimal approach is to gather the relevant experts together, come up with a list of risks and then qualify which risks are worth acting upon. This process is more intuitive and is enhanced by answering three basic questions[128]:

- What could happen?

- How likely is it to occur?

- What is the impact?

The downside with the qualitative approach is that it is difficult to rank risks relative to each other or to place the risks in context. It is also difficult to assess the effectiveness of an action or to measure if a risk previously assessed at a 'high level', for example, has decreased to a lower level.

Quantitative approaches normally involve the use of numbers so that one risk can be ranked or measured against another and compared with a pre-determined scale. Such an approach allows a decision to be taken about the cost-benefits of investing in a risk reduction strategy and to allow the costs of monitoring risks relative to the likelihood of the risk occurring to be taken.

The ICH documents outline the following quality risk management steps[129]:

- Clearly identify the process being assessed and what it is attempting to achieve, i.e., what the harm/risk is and what the impact could be on the patient

- Be based on systematic identifications of possible risk factors. This requires the use of risk assessment tools, such as: FMEA (Failure Mode Effect Analysis), FTA (Fault Tree Analysis), HAZOP (Hazard Operability Analysis) and HACCP (Hazard Analysis and Critical Control Points)[130]

- Take full account of current scientific knowledge

- Be conducted by people with experience in the risk assessment process and the process being risk assessed

- Use factual evidence supported by expert assessment to reach conclusions

- Do not include any unjustified assumptions

- Identify all reasonably expected risks simply and clearly, along with a factual assessment and mitigation where required

- Be documented to an appropriate level and controlled/approved

- Ultimately be linked to the protection of the patient

- Should contain an objective risk mitigation plan

With the methodologies there are four basic steps in a risk assessment:

- Step 1 – Risk Identification – Identify the hazards – (what might go wrong?)

- Step 2 – Risk Analysis – Evaluate the risks (how bad? how often?) and decide on the precautions (is there a need for further action?)

- Step 3 – Record the findings and any proposed risk control or risk reduction (action to mitigate against the risk).

- Step 4 – Review assessments and update as necessary.

cGMP requires that in undertaking risk assessments, it is important to attempt risk mitigation and to attempt to lower the risk until the risk can be lowered no further. This involves identifying actions to reduce the probability of the event and its severity. When this can be taken no further, the focus should move towards providing a more reliable detection method designed to initiate a reliable response to a risk event. A further important consideration is that risk-assuming actions should be periodically re-assessed.

Method validation

Method validation plays an important part of GMP, as has been discussed above. Whilst validation has long been an established part of pharmaceutical processing, it was not described in great detail in the early GMPs and it was not until the 1990s that validation processes were described in significant detail.

Validation is necessary in order to show that processes and equipment are functioning according to the required parameters (and, following this, re-validation concerns confirming that the equipment or the process continues to meet the validated parameters). Validation is also applicable to laboratory methods, with one of the key requirements being the assessment of whether methods can be shown to be reproducible and repeatable.

Like other aspects of GMP, new validation standards continue to be issued. In 2012, for example, the EMA issued more robust guidance on bioanalytical method validation, focusing on bioanalytical methods generating quantitative concentration data used for pharmacokinetic and toxicokinetic parameter determinations[131].

Process understanding

Understanding the pharmaceutical manufacturing process and being able to make controlled modifications in order to improve the quality of the product is a further dimension of cGMP. Such an understanding can also connect with real-time release for understanding sources of variability and their impact on downstream processes or processing, in-process materials, and drug product quality can provide an opportunity to shift controls upstream and minimise the need for end product testing.

Quality by Design

The principles of Quality by Design (QbD) require that when new products are being developed the product development should be evaluated. QbD is not a new concept in general manufacturing, but it has been promoted by GMP regulators more often since the mid-2000s. QbD is concerned with the development of a drug product formulation and the design of a manufacturing process using all current available knowledge to produce a pharmaceutical product with inherent qualities and properties for its intended use. QbD permits studies to be undertaken with the aim of increasing product knowledge for the production of a pharmaceutical with all the necessary attributes for its intended use[132]. A key element of QbD is the concept of "design space". Design space is a multidimensional space encompassing combinations of product design and processing variables that provides assurance of suitable product performance.

QbD can also be applied to existing products, as part of continuous improvement philosophies.

1.3.7   The advantages of GMP compliance

The text in this chapter has demonstrated many of the advantages of GMP compliance for the pharmaceutical manufacturer. As way of providing an overview of the chapter, these advantages are:

- Producing a product which is safe and efficacious for the patient. This means that the product meets the requirements in terms of identity, safety, purity, efficacy and potency

- Stability and consistency

- Gaining approval for a new drug

- Meeting the expectations of regulatory authorities and, in doing so, maintaining the licence to manufacture

- Lowering operating costs as rework and penalties due to non-compliance reduce and efficiencies increase

- Increasing process efficiency by having less waste, rejects, reworks, complaints and recalls

- Satisfying the expectations of patients, employees, stockholders, regulators and competitors

In essence, GMP is ultimately about primarily protecting the patient and, secondly, avoiding expensive errors during manufacturing. To continue to manufacture requires a regulatory licence or authorisation. For this, the maintenance of GMP and the regular demonstration that GMP standards are being upheld (through audits and inspections) is essential.

1.4   Summary

The GMPs have been part of the pharmaceutical industry since the phrase was first coined by the FDA in the 1940s. GMPs, in the form that is recognisable today, were developed in the 1970s, although they have since then undergone frequent revisions. The philosophy of GMP was to develop safe and efficacious medicines. From this foundation, GMPs have become wider, extending into all aspects of pharmaceutical manufacturing.

It is interesting to speculate how GMPs will develop. The FDA 21st century GMPs have set an important marker, especially with regard to process analytical technologies and the emphasis upon risk assessments.

It is likely that the future direction will see the extension of the process which led to the internationally harmonised and agreed "GMPs for Active Pharmaceutical Ingredients", issued by the International Conference for Harmonisation. Thus, the next phase of GMP development could be greater collaboration between regulatory agencies and more guidance designed to cut across international territories.

1.5   References

1   Sweetman, S.C. (Ed.). *Martindale: The Complete Drug Reference*, 34th Edition, Pharmaceutical Press: London, 2005.

2   Angelucci, L.M. Current Good Manufacturing Practice Design Trends in Active Pharmaceutical Ingredients Facilities, *Drug Information Journal* 1999; **33**: 739–746,

3   Woodcock, J. The concept of pharmaceutical quality. *American Pharmaceutical Review* 2004; **7**(6): 10-15.

4   ISO/FDIS 9001. Quality management systems – Requirements, 2000. International Standards Organisation, Geneva.

5   Whyte, W.; Eaton, T. Assessing microbial risk to patients from aseptically manufactured pharmaceuticals. *European Journal of Parenteral & Pharmaceutical Sciences* 2004; **9**(3): 71–77.

6   Tsim, Y.C.; Yeung, V. W. S.; Leung, E. T. C. "An adaptation to ISO 9001: 2000 for certified organisations", *Managerial Auditing Journal* 2002: **17**(5).

7   Brooker, C. Mosby's Dictionary of Medicine, Nursing and Health Professions. Elsevier, Edinburgh, 2010.

8   ISPE GAMP® 5: *A Risk-Based Approach to Compliant GxP Computerised Systems*, International Society for Pharmaceutical Engineering (ISPE), Fifth Edition, February 2008, www.ispe.org.

9   ISO 14155. Clinical investigation of medical devices for human subjects – Good clinical practice, 2011. Interrnational Standards Organisation, Geneva.

10   Angell, M. Ethical imperialism? Ethics in international collaborative clinical research. *The New England Journal of Medicine*, 1988; **319**(16): 1081–3.

11   Johnson, B. *Managing Operations*, Blackwell: Oxford, 1998.

12   Crosby, P. *Quality Is Free*, McGaw-Hill: New York, 1979.

13   Famulare, J. Quality Systems Guidance. PDA/ FDA Joint Regulatory Conference, Washington, DC, September 2004.

14   Anon. Guide to good manufacturing practice. 3rd edition, London: HM Stationery Office, 1983.

15   Bynum, W. F. *Science and the practice of Medicine in the Nineteenth Century*, Cambridge: Cambridge, 1994.

16   Hindle, T. *Guide to Management ideas and Gurus*, Profile Books: London, 2008

17   Deming W.E.: *Out of the Crisis*. Cambridge, MA: Massachusetts Institute of Technology Center for Advanced Engineering Study, 1986.

18   Vorley, G. and Tickle, F. *Quality Management: Tools and Techniques*, 1st edition. QMT Publications: Surrey, 2006.

19   Spenley, P. World Class Performance through Total Quality. Chapman and Hall: London, 1992.

20   Boddy, D. *Management: An Introduction*, 2nd edition, Prentice Hall: Harlow, 2002

21   Juran, J. Quality Control Handbook. 4th Edition. Mcgraw-Hill: London, 1988.

22   Voss, C.A. Alternative paradigms for manufacturing strategy, *International Journal of Operations & Production Management*, 1995; **15**(4): 5–16.

23   Kroemer, K.H.E. and Grandjean, E. Fitting The Task To The Human: A textbook of Occupational Ergonomics. Fifth Edition, CRC Press: London, 1997.

24   Oude Ophuis, P.A.M. & van Tripp, H.C.M. Perceived quality: a market driven and consumer oriented approach. *Food Quality and Preference*, 1995; **6**: 177–183.

25   "Pharmaceutical cGMPs for the 21st Century — A Risk-Based Approach — Final Report"; Department of Health and Human Services, U.S Food and Drug Administration, September 2004.

      Available at http://www.fda.gov/cder/gmp/gmp2004/GMP_ finalreport2004.htm.

26   Brathwaite, J. and Drahos, P. Global Business Regulation. Cambridge Univrsity Press: Cambridge, 2000.

27   Clothier Report. Report of the Committee Appointed to Inquire into the Circumstances, Including the Production, which Led to the Use of Contaminated Fluids in the Devonport Section of Plymouth General Hospital (C.M.Clothier, Chairman). London: Her Majesty's Stationery Office, 1972.

28   Matthews, B.R.. The Davonport Incident, the Clothier Report, and Related Matters—30 Years On. PDA *Journal of Pharmaceutical Science & Technology,* 2002; **56**(3): 137–149.

29   Shaw, J. Law of the European Union, 2nd Edition. Macmillan: Basingstoke, 1996.

30   Kibbe, A. H. (Ed.), *Handbook of Pharmaceutical Excipients.* (American Pharmaceutical Association,Washington, DC, 2000), p. 665.

31   Anon. Guide to good pharmaceutical manufacturing practice. 2nd edition. London: MS Stationery Office, 1977.

32   Kaufman, B. and Novack, G.D. Compliance Issues in Manufacturing of Drugs, *The Ocular Surface* 2003; **1** (2): 80-85.

33   Gilbert, P. and Allison, D. G. (1996): 'Redefining the 'Sterility' of Sterile Products', *European Journal of Parenteral Sciences*, 1996; **1**: 19–23.

34   Brown MRW and Gilbert P : Increasing the probability of sterility of medicinal products, *J. of Pharmacy and Pharmacology*, 1997; **27**: 484–491.

35   Massa, T. Life Cycle Management for Process and System Control: An Industry Proposal. Manufacturing Subcommittee, *Advisory Committee for Pharmaceutical Science.* July 20, 2004.

36   Todd JI. Performing your original search, good manufacturing practice. *Rev Sci Tech.* 2007; **26**(1):135-45.

37   FDA History: FDA Commissioners and Their Predecessors, US Food and Drug Administration, Rockville, MD, rev. 6 April 2000, www.fda.gov/opacom/morechoices/comm1.html.

38   Immel, B. K. A Brief History of the GMPs. Regulatory Compliance Newsletter. Winter 2005.

39   Center for Drug Evaluation and Research, Time Line: Chronology of Drug Regulation in the United States (FDA, Rockville, MD), www.fda.gov/cder/about/history/time1.htm.

40   "Milestones in US Food and Drug Law History," FDA Backgrounder: Current and Useful Information from the Food and Drug Administration (FDA, Rockville, MD, August 1995), www.fda.gov/opacom/backgrounders/miles.html.

41 Federal Register, "Drugs; Current Good Manufacturing Practice in Manufacture, Processing, Packing or Holding," Part 133, 28 FR 6385, June 20, 1963.

42 Jennings, R. W. Revised Good Manufacturing Practice Regulations, *Law Journal*, No. 107, 1970.

43 Nordenberg, T. "Protecting Against Unwanted Pregnancy: A Guide to Contraceptive Choices," *FDA Consumer* **31**(3) (April 1997), www.fda.gov/fdac/features/1997/397_baby.html.

44 Anon. The Medicines Act, London: HM Stationery Office, 1968.

45 Sharp, J. Good Manufacturing Practice: Philosophy and Applications. Interpharm Press: USA, 1991.

46 Slopecki, A.. Smith, K. and Moore, S. The value of Good Manufacturing Practice to a Blood Service in managing the delivery of quality, *Vox Sanguinis*, 2007 ; **92** (3): 187–196.

47 Code of Federal Regulations, Food and Drugs, "General: Good Laboratory Practice for Nonclinical Laboratory Studies," revised April 2000, Title 21 Part 58 (US Printing Office, Washington, DC).

48 Agalloco, J., *Validation of Aseptic Pharmaceutical Processes*. Marcel Dekker, 1986.

49 "Guide to Inspections of Cleaning Validation", U.S Food and Drug Administration, July 1993. Available at http://www.fda.gov/ora/inspect_ref/igs/valid.html.

50 International Conference on Harmonisation, Guide for Industry: E6, Good Clinical Practice — Consolidated Guidance (Geneva, Switzerland), April 1996. Published in the Federal Register, 62(90) 9 May 1997, pp. 25691-25709.

51 EMA. Enhancing GMP Inspection Cooperation between EMA and FDA: Moving from confidence-building to reliance upon. European Medicines Agency, London. Available at: http://www.fda.gov/downloads/InternationalPrograms/FDABeyondOurBordersForeignOffices/EuropeanUnion/EuropeanUnion/EuropeanCommission/UCM283088.pdf.

52 Abraham, J. and Reed, T. Trading Risks for Markets: the international harmonisation of pharmaceuticals regulations. *Health, Risks and Society*, 2001; 3(1): 113-118.

53 WHO. Quality assurance of pharmaceuticals A compendium of guidelines and related materials Volume 2, 2nd updated edition Good manufacturing practices and inspection. World Health Organization, Geneva, 2007.

54 Health Products and Food Branch Inspectorate. Good Manufacturing Practices (GMP) Guidelines – 2009 Edition, Version 2 GUI-0001. Health Canada, 2011.

55 State Foods and Drug Administration (SFDA). Guide for the Validation of Drug Production. Publishing House of Chinese Medicinal Science; 2003.

56 The IPEC Good Distribution Practice Guide for Pharmaceutical Excipients. The International Pharmaceutical Excipients Council, 2006.

57 Good Trade and Distribution Practices for Pharmaceutical Starting Materials. WHO Technical Report Series, No. 917, 2003. World Health Organization, Geneva.

58 Schmidt, P. Integrating accreditation, good laboratory practice and good manufacturing practice in an industrial analytical laboratory, *Journal for Quality, Comparability and Reliability in Chemical Measurement*, 1999; **4**(4): 129-132.

59 Grazal, J.D. and Earl, D.S. EU and FDA GMP Regulations: Overview and Comparison. *Quality Assurance Journal*, 1997; **2**: 55-60.

60 Taylor, J., Munro, G. Lee, G. and McKendrtrick, A. Good manufacturing practice and good distribution practice: An analysis of regulatory inspection findings for 2001-02. *Pharmaceutical Journal*. 2003; **270**: 127-129.

61 FDA, Good Manufacturing Practice for Finished Pharmaceuticals, 21 CFR 211.100 (a), Food and Drug Adminstration, Rockville: MD.

62 Lubert, A. and Simutid, R. Measurement and Control, in Ratledge, C. and Kristiansen, B. *Basic Biotechnology*, 2nd edition, Cambridge University Press: Cambridge, 2001.

63 Denyer S.P. Filtration Sterilisation, in Russell A.D. (Ed.) Principles and Practice of Disinfection, *Preservation and Sterilisation*, Blackwell Scientific Publications, Oxford (UK), 1982.

64 US FDA. Current Good Manufacturing Practice for Active Pharmaceutical Ingredients: Guidance for Industry—Manufacturing, Processing, or Holding Active Pharmaceutical Ingredients. Washington, DC:US Food and Drug Administration, 1998.

65 US FDA. Current Good Manufacturing Practice for Finished Pharmaceuticals. Washington, DC:US Food and Drug Administration, 1995.

66 O'Donnell, K. and Greene, A. A Risk Management Solution Designed To Facilitate Risk-Based Qualification, Validation, and Change Control Activities within GMP and Pharmaceutical Regulatory Compliance Environments in The EU. Part 1: Fundamental Principles, Design Criteria, Outline of Process. *Journal of GXP Compliance*. 2006; **10** (4): 12–35.

67 Dorresteijn, R.C., Wieten, G., van Santen, P.T.E., Philippi, M.C., de Gooijer, C.D., Tramper, J. and Beuvery, E.C. Current good manufacturing practice in plant automation of biological production processes, *Cytotechnology*, 2000; **23** (1-3): 19-28.

68 Velagaleti, R., Burns,P.K., Gill, M., and Prothro, J. Impact of current good manufacturing practices and emission regulations and guidances on the discharge of pharmaceutical chemicals into the environment from manufacturing, use, and disposal, *Environ Health Perspect*, 2002; **110**(3): 213–220.

69 Jenkins, K. M. and A. J. Vanderwielen, Cleaning Validation: An Overall Perspective, *Pharmaceutical Technology*, 1994; **18**(4): 60-73.

70   Supplementary guidelines on good manufacturing practices: validation. In: WHO Expert Committee on Specifications for Pharmaceutical Preparations. Fortieth report. Geneva, World Health Organization, 2006 (WHO Technical Report Series, No. 937, Annex 4).

71   Morris BG, Avis KE. Quality-control plan for intravenous ad mixture programs I: Visual inspection of solutions and environmental testing. *Am J Hosp Pharm* 1980; **37**: 189–95.

72   Nail, S. L.; Guazzo, D.; Rajagopalan, N.; Stickelmeyer, M.; Williams, T. Visual inspection of parenteral products in a development environment. *Am. Pharm. Rev.* 2006; **9**: 96–101.

73   Good manufacturing practices for pharmaceutical products: main principles. In: WHO Expert Committee on Specifications for Pharmaceutical Preparations. Thirty-seventh report. Geneva, World Health Organization, 2003 (WHO Technical Report Series, No. 908, Annex 4).

74   PIC/S recommendations on validation master plan, installation and operational qualification, non-sterile process validation, cleaning validation. TRS961.indd 213 TRS961.indd 213 02.05.11 22:30 02.05.11 22:30214 Geneva, Pharmaceutical Inspection Cooperation Scheme, 2007 (Document PI 006-3).

75   Tran, N.L., Hasselbalch, B., Morgan, K., Claycamp, G. Elicitation of Expert Knowledge About Risks Associated with Pharmaceutical Manufacturing Process. *Pharmaceutical Engineering*, July/August 2005: 24-38.

76   Underwood, E. (2008) Good Manufacturing Practice, in Russell, Hugo & Ayliffe's Principles and Practice of Disinfection, Preservation & Sterilisation, Fourth Edition (eds A. P. Fraise, P. A. Lambert and J.-Y. Maillard), Blackwell Publishing Ltd, Oxford, UK.

77   Russell, A.D. "Theoretical Aspects of Microbial Inactivation," in Morrissey, R.F. and Phillips, G.B. (Eds..) *Sterilisation Technology*, Van Nostrand Reinhold, New York, USA, 1993.

78   ISO Standard 14644 – Cleanrooms and associated controlled environments – Part 1: Classification of air cleanliness, International Standards Organization: Geneva, 1999.

79   Ljungqvist, B. and Reinmuller, B. Clean Room Design: Minimizing Contamination Through Proper Design. Interpharm Press: Buffalo Grove, IL, 1997.

80   Whyte, W. (2001) Cleanroom Design, John Wiley and Sons: Chichester.

81   Sandle, T. (2011): 'Environmental Monitoring' in Saghee, M.R., Sandle, T. and Tidswell, E.C. (Eds.) (2011): Microbiology and Sterility Assurance in Pharmaceuticals and Medical Devices, New Delhi: Business Horizons, pp293-326.

82   Chandler SW, Trissel LA, Wamsley LM, Lajeunesse JD, Anderson RW. Evaluation of air quality in a sterile-drug preparation area with an electronic particle-counter. *Am J Hosp Pharm* 1993; **50**: 2330–4.

83   Denyer, S.P. and Baird, R.M. Guide to microbiological control in pharmaceuticals and medical devices, CRC Press: Boca Raton, USA, 2007.

84   Krämer I. Good Manufacturing Practice in der aseptischen Herstellung applikationsfertiger Parenteralia in der *Apotheke. Pharm Ind* 1998; **60**: 787–94.

85   Cundell, A.M. Microbial Testing in Support of Aseptic Processing, Pharmaceutical Technology, June 2004, pp 56-66.

86   FDA. Food and Drug Administration Guidance for Industry, Sterile Drug Products Produced by Aseptic Processing—Current Good Manufacturing Practice. Rockville, MD, 2004.

87   Langer, U. And Kramer, I. GMP in Hospital Pharmacy: Environmental Monitoring of the CIVA Preparation Area, *Environmental Health Perspectives*, **7** (3): 97-107

88   Guazzo, D.M. Current Approaches in Leak Testing Pharmaceutical Products, *J. Pharm. Sci. Technol*, 1996; **50**(6): 378–385.

89   Friedman, R.; Mahoney, S. Risk factors in aseptic processing. *American Pharmaceutical Review* 2003; **6**(1): 44-46, 92.

90   Oji, R. T. Regulatory aspects concerning the quality controls of microbiological and non-viable particulate contamination in pharmaceutical manufacturing. *American Pharmaceutical Review* 2004; **7**(1): 18-22, 68.

91   Chestnut, W. and Waggener, J.W. Good Laboratory Documentation: An Introduction. *Journal of GXP Compliance*. 2002; **6**(3): 18-23.

92   Willig, S. Good manufacturing practices for pharmaceuticals: a plan for total quality control from manufacturer to consumer, CRC Press: Boca Raton, USA, 2000.

93   Ways, J.P., Preston, M.S., Baker,D., Huxsoll, J. and Bablak, J. Good manufacturing practice (GMP) compliance in the biologics sector: plasma fractionation, *Biotechnology and Applied Biochemistry*, 1999; **30**(3): 257–265.

94   Ljungqvist, B.; Reinmuller, B. Cleanroom dressed people as a contamination source; some calculations, *European Journal of Parenteral & Pharmaceutical Sciences* 2004; **9**(3): 83–8.

95   WHO guidelines for sampling of pharmaceutical products and related materials. In: WHO Expert Committee on Specifications for Pharmaceutical Preparations. Thirty-ninth report. Geneva, World Health Organization, 2005 (WHO Technical Report Series, No. 929, Annex 4).

96   Soncin, S., Lo Cicero, V., Astori, G., Soldati, G., Gola, M., Sürder, D. and Tiziano Moccetti. A practical approach for the validation of sterility, endotoxin and potency testing of bone marrow mononucleated cells used in cardiac regeneration in compliance with good manufacturing practice. *Journal of Translational Medicine*, 2009; **7**: 78.

97   Kella, D.B. and Sonnleitner, B. GMP — good modelling practice: an essential component of good manufacturing practice, *Trends in Biotechnology*, 1995; **13**(11): 481–492.

98 Chirino, A.J. and Mire-Sluis, A. Characterizing biological products and assessing comparability following manufacturing changes, *Nature Biotechnology*, 2004 ; **22**: 1383–1391.

99 Crawford SY, Narducci WA, Augustine SC. National survey of quality assurance activities for pharmacy prepared sterile products in hospitals. *Am J Hosp Pharm* 1991; **48**: 2398–413.

100 Keith, P. Effective Corrective and Preventative Action. *Global BioPharmaceutical Resources Newsletter*, September 2011: 1-5, at: http://www.gbprinc.com/pdf/GBPRSeptember2011EffectiveCAPA.pdf.

101 Good Manufacturing Practice Guide for Active Pharmaceutical Ingredients Q7 (previously coded Q7A), September 2001, International Conference on Harmonisation, at: www.ich.org/LOB/media/MEDIA433.pdf.

102 Directive 2004/27/EC amending Directive 2001/83/EC on medicinal products for human use, *Offical Journal*, 2004, L136, 34-57 at: http://ec.europa.eu/enterprise/pharmaceuticals/eudralex/vol1/dir_2004_27/dir_2004_27_en.pdf.

103 FDA. Pharmaceutical cGMPs for the 21st Century: A Risk-Based Approach, 2003. Food and Drug Adminstration, Rockville: MD.

104 Sandle, T. and Saghee, M. R. Some considerations for the implementation of disposable technology and single-use systems in biopharmaceuticals, *Journal of Commercial Biotechnology*, 2011; **17**(4): 319–329.

105 FDA. Advancing Regulatory Science at FDA, Food and Drug Administration, Rockville: MD, 2011.

106 FDA, Guidance for industry: PAT – A framework for innovative pharmaceutical development, manufacturing and quality assurance; September 2004. Food and Drug Adminstration, Rockville: MD.

107 Scott, B. and Wilcock, A. Process analytical technology in the pharmaceutical industry: A toolkit for continuous improvement. *PDA Journal of Pharmaceutical Science and Technology*, 2006; **60**(1): 17–53.

108 FDA, Draft Guidance for Industry, Process Validation: General Principles and Practices, Food and Drug Administration: Rockville, MD, 2008.

109 Jahnke, M. Use of the HACCP concept for the risk analysis of pharmaceutical manufacturing processes. *European J Parenteral Sci* 1997; **2**(4): 113–117.

110 FDA (CDER). Guidance for Industry. Draft Guidance. PAT, A Framework for Innovative Pharmaceutical Manufacturing and Quality Assurance. US Department of Health and Human Services, FDA, CDER/CVM/ORA. August 2003.

111 Pharmaceutical Development – Q8 (R2), International Conference on Harmonisation of Technical Requirements for Registration of Pharmaceuticals for Human Use (ICH), www.ich.org.

112   Drakulich, A. Real Time Release Testing, *Pharmaceutical Technology*, 2011; **35**(2): 42–49.

113   Miller, M. J. Rapid Microbiological Methods. In Prince, R. (ed.). Microbiology in Pharmaceutical Manufacturing; DHI Publishing: River Grove, IL and PDA: Bethesda, MD, 2008; **2**: 171–221.

114   ISO/IEC 17799. Information technology – Code of practice for information security management, 2000. Interrnational Standards Organisation, Geneva.

115   The Good Automated Manufacturing Practice (GAMP) Guide for Validation of Automated Systems, GAMP 4 (ISPE/GAMP Forum, 2001) (http://www.ispe.org/gamp/).

116   FDA. Guidance for Industry Part 11, Electronic Records; Electronic Signatures — Scope and Application. Food and Drug Administration, Rockville: MD, 2003.

117   Tidswell, E. C. Risk profiling pharmaceutical manufacturing processes. *European J Parenteral Pharm Sci* 2004; **9**(2): 49–55.

118   Phoenix., K. and Andrews, J. Adopting a Risk – Based Approach to 21 CFR Part 11 Assessments, *Pharmaceutical Engineering*, 2003; **23**(4).

119   Agalloco, J., Akers, J., Baseman, H., Boeh, R., Madsen, R. Ostrove, S. and Pavell, A. Risk Management, cGMP, and the Evolution of Aseptic Processing Technology. *PDA Journal of Pharmaceutical Science and Technology*, **63**(1): 8–10

120   FDA. Guidance for Industry Q8, Q9, and Q10 Questions and Answers (R4). Food and Drug Administration, Rockville: MD, 2011.

121   ICH Q8: Guideline on pharmaceutical development. Geneva, International Conference on Harmonisation of Technical Requirements for Registration of Pharmaceuticals for Human Use, 2005.

122   ICH Q9: Guideline on quality risk management. Geneva, International Conference on Harmonisation of Technical Requirements for Registration of Pharmaceuticals for Human Use, 2005.

123   ICH Q10: Guideline on pharmaceuical quality system. Geneva, International Conference on Harmonisation of Technical Requirements for Registration of Pharmaceuticals for Human Use, 2008.

124   Saghee, M.R. and Sandle, T. 'Embracing quality risk management: The new paradigm', Express Pharma, Pharma Technology Review, 16th – 30th September 2010, On-line paper at: http://www.expresspharmaonline.com/20100930/pharmatechnologyreview02.shtml (accessed 30th September 2010).

125   Claycamp, H.G. Probability Concepts in Quality Risk Management. *PDA Journal of Pharmaceutical Science and Technology*, 2012; **66**(1): 78–89.

126   Sandle, T. Risk Management in Microbiology In: Saghee, M.R., Sandle, T. and Tidswell, E. (eds.) *Microbiology and Sterility Assurance in Pharmaceuticals and Medical Devices*, 2010, Business Horizons, India

127 PIC/S. Quality Risk Management. Implementation of ICH Q9 in the pharmaceutical field an example of methodology from PIC/S, PIC/S PS/INF 1/2010.

128 Jackson S. Successfully implementing total quality management tools within healthcare: what are the key actions? *Int J Health Care Qual Assur* 2001;**14**(4):157–63.

129 Keith, P. Quality Risk Management: 16 Important Issues to Consider. Global BioPharmaceutical Resources Newsletter, January 2011: 1-2 (http://www.gbprinc.com/pdf/QualityRiskManagement.pdf).

130 World Health Organization (2003): Application of Hazard Analysis and Critical Control Point (HACCP) Methodology to Pharmaceuticals, WHO Technical Report Series No. 908, Annex 7, World Health Organisation, Geneva, 2003.

131 EMA. Guideline on bioanalytical method validation. European Medicines Agency, EMEA/CHMP/EWP/192217/2009, Committee for Medicinal Products for Human Use (CHMP).

132 Federal Register, Vol. 70, No. 134; 2005. US Department of Health and Human Services, Food and Drug Administration. Submission of Chemistry, Manufacturing, and Controls Information in a New Drug Application Under the New Pharmaceutical Quality Assessment System; Notice of Pilot Program.

a   Thalidomide was a sleeping pill widely used in pregnancy. Due to insufficient testing it was not known that the pill was toxic to fetuses and thus led to birth defects.

b   The Clothier report also covered events linked to Evans Medical (Speke, UK) when on 6th April 1971 deaths occurred due to a 5% Sterile Dextrose Solution (Lot D1192).

c   The forerunner to the FDA was Bureau of Chemistry, US Department of Agriculture.

d   The EMEA later became the European Medicines Agency (EMA).

e   Other similar approaches are: ISO/IEC Guide 73:2002 – Risk Management – Vocabulary – Guidelines for use in Standards; ISO/IEC Guide 51:1999 – Safety Aspects – Guideline for their inclusion in standards; AS/NZS 4360:2004 – Risk Management; WHO Technical Report Series No 908, 2003, Annex 7 Application of Hazard Analysis and Critical Control Point (HACCP) methodology to pharmaceuticals; EN ISO 14971: Application of risk management to medical devices; Pharmaceutical Development (ICH Q8) and Annex (ICH Q8(R1) and FDA Guidance for Industry PAT – A Framework for Innovative Pharmaceutical Development, Manufacturing and Quality Assurance; Pharmaceutical Quality Systems (ICH Q10) and FDA Guidance for Industry Quality Systems Approach to Pharmaceutical cGMP Regulation

CHAPTER 2

FDA drug regulation and enforcement

Seth A Mailhot
Michael Best and Friedrich LLP, USA

2.1 Introduction

This chapter provides an overview of US Food and Drug Administration (FDA) regulation and enforcement of drug products. The chapter begins with a basic overview of the regulatory requirements that apply prior to the first sale of a drug product in the US (pre-market regulations), and the regulatory requirements that apply after a drug is first marketed in the US (post-market regulations). The chapter next provides an overview of FDA's enforcement authority over drug products. As part of this overview, the chapter provides summaries of FDA's recent enforcement activities to illustrate where FDA currently focuses its drug enforcement efforts.

Objectives:

* Present an overview of the pre-market and post-market regulatory controls applicable to US drug products;

* Review FDA's regulatory authority and enforcement powers; and

* Summarise recent drug enforcement activities undertaken by FDA.

2.2 Overview of the US Food and Drug Administration

Drug products are subject to extensive pre- and post-market regulation in the United States by the US Food and Drug Administration (FDA). The FDA enforces the Federal Food, Drug and Cosmetic Act (FFDCA), which is the law that governs the testing, manufacturing, distribution, safety, efficacy, approval, labeling, storage, record keeping, reporting, advertising and promotion of drug products. In addition to the FFDCA, the FDA has implemented regulations that expand on and explain the requirements of the FFDCA. Failure to comply with applicable laws and regulations may result in, among other things, warning letters, clinical holds, civil or criminal penalties, recall or seizure of products, injunction, debarment, partial or total suspension of production or withdrawal of the product from the market.

The group within the FDA assigned to handle issues dealing with pre- and post-market drug regulation is the Center for Drug Evaluation and Research (CDER).

Within CDER, the Office of Compliance deals with enforcing drug regulations. Pre-market review of drug products at CDER is handled by the Office of New Drugs (for new drug products), and the Office of Generic Drugs (for generic drug products).

2.2.1 Regulatory Differences Between Foreign and Domestic Companies

There is a marked difference between how the FFDCA handles drug companies in the US, and those operating outside of the US The FDA has expanded power over foreign manufacturers, granted through Section 801(a) of the FFDCA[1]. The procedures for imports "are strikingly different from [those for] products of domestic origin[2]." Domestic products regulated by the FDA that are adulterated are subject to seizure and condemnation through the judicial process.

By contrast, "the *mere appearance* of adulteration is enough to compel refusal to admit" an FDA-regulated product into the United States[3]. The product does not actually have to be in violation of the FFDCA to be refused entry into the US, so long as there is information to suggest there may be a violation. Further, "[t]here is no provision for judicial review," and action by the FDA "is committed to agency discretion by law[4]." This makes challenging a determination of adulteration exceptionally difficult.

Section 810(a) authorises the FDA to refuse to admit possibly adulterated imported drug products based on information obtained from the testing and examination of samples "or otherwise." This language has been interpreted to permit FDA to refuse drug product shipments based on poor results from a foreign inspection.

The FDA's authority to inspect foreign drug manufacturing facilities is found in Section 801, and the registration and premarket approval requirements of the FFDCA[5]. Foreign drug companies have the option to refuse FDA's request for an inspection. The refusal to permit an FDA inspection, however, may be considered by FDA to be evidence of an appearance of adulteration.

Unlike Section 704 of the FFDCA, which applies to inspections of domestic manufacturers, Section 801 does not require FDA to collect stringent documentary evidence to establish violations at foreign sites[6]. The FDA does, however, direct its foreign investigators to collect sufficient records to substantiate their findings and to aid in further review[7].

Where the FDA observes significant deviations from current good manufacturing practice requirements (cGMP), unsanitary conditions or other instances of noncompliance during a foreign inspection, it may refuse to admit articles offered for import into the United States. When such deviations are observed, the FDA will often impose a "Detention Without Physical Examination" order, otherwise known as an automatic detention[8]. An automatic detention is a notice sent to FDA's field offices notifying them of shipments suspected of being adulterated. Shipments covered by an automatic detention order are not automatically refused by FDA. The FDA will first detain the shipment, giving importers a chance to provide information to refute the appearance of adulteration. If the information is insufficient to demonstrate the lack of an appearance of adulteration (or if no information is received by the FDA), then FDA will refuse the shipment.

An FDA automatic detention order may identify a specific manufacturing location, multiple manufacturing locations, specific products from one or more manufacturers, or specific products from a particular country. CDER's Office of Compliance approves the action based upon review of the inspection report and other evidence.

Generally, to support an automatic detention following a foreign drug inspection, the violations observed must be more significant than those that would typically warrant a warning letter. The violations do not, however, need to rise to the level required for a domestic seizure or injunction. In general, the more complex the drug, the more difficult it is to avoid the appearance of adulteration following a failing inspection.

When a foreign manufacturer's products are subject to automatic detention, the shipper or importer must prove to the FDA that the product meets its requirements before they can be released by the US customs agents. In instances when inspectors find significant deviations from drug CGMPs, favourable sample test results alone are unlikely to help gain the product's admission into the United States. The manufacturer typically must also change its operations and procedures, often requiring independent confirmation.

Foreign companies awaiting US marketing approval for a drug have additional concerns. Failure to adequately complete a premarket inspection may result in the FDA's refusal to approve the product until violations observed during a premarket inspection are corrected. Such refusals (or notices of program disapproval) may be based on deviations that would not normally rise to the level of a warning letter, such as the failure to complete required premarket validation studies. When businesses are competing in the US marketplace, even short delays in obtaining marketing approval can be costly. Such delays can translate into diminished market share, expired products and lost patent protection.

2.2.2   NDA Approval Process

The FDA must approve any new drug, including a new dosage form or new use of a previously approved drug, prior to marketing in the United States. All applications for FDA approval must contain, among other things, information relating to safety and efficacy, pharmaceutical formulation, stability, manufacturing, processing, packaging, labeling and quality control.

At the beginning of the process, a sponsor of a new drug must conduct preclinical laboratory and animal testing and formulation studies in compliance with FDA's Good Laboratory Practice (GLP) regulations. This information is used in an Investigational New Drug exemption (IND), which requests permission from FDA to conduct human clinical testing in the United States. Approval by independent institutional review boards (IRBs) is also required at each clinical trial site before the trial may be initiated.

Preclinical tests include laboratory evaluation of product chemistry, formulation and stability, as well as studies to evaluate toxicity in animals. The results of preclinical tests, together with manufacturing information, analytical data and a proposed clinical trial protocol and other information, are submitted as part of an IND application to the FDA. The IND automatically becomes effective 30 days after receipt by the FDA, unless the FDA, within the 30-day time period, places the trial on a clinical hold because of, among other things, concerns about the conduct of the clinical trial or about exposure of human research subjects to unreasonable health risks. In such a case, the IND sponsor and the FDA must resolve any outstanding concerns before the clinical trial can begin.

The IND must become effective before human clinical trials may begin in the United States. While an IND is not required for foreign clinical testing, such testing must be performed under recognised international ethical standards. The ultimate goal of these human clinical trials is to establish the safety and efficacy of the proposed drug product for each intended use. To be acceptable to FDA, the studies must include adequate and well-controlled clinical trials conducted in accordance with good clinical practices (GCP).

The FDA requires sponsors to amend existing INDs for each successive clinical trial conducted during product development. Further, the IRBs covering each medical center proposing to conduct the clinical trial must review and approve the plan for any clinical trial and informed consent information for subjects before the clinical trial commences at that center, and the IRBs monitor the clinical trial until completed. The FDA, the IRB or the sponsor may suspend a clinical trial at any time on various grounds, including a finding that the subjects or patients are being exposed to an unacceptable health risk. The FDA will typically inspect one or more clinical sites to ensure compliance with GCP before approving an NDA.

Clinical trials involve the administration of the investigational new drug to human subjects under the supervision of qualified investigators in accordance with GCP requirements. The GCP requirements include the requirement that all research subjects provide their informed consent for their participation in any clinical trial. For purposes of obtaining marketing approval, human clinical trials are conducted in three pre-market sequential phases, which often overlap:

Phase 1:   sponsors initially conduct clinical trials in a limited population to test the product candidate for safety, dose tolerance, absorption, metabolism, distribution and excretion in healthy humans or, on occasion, in patients, such as cancer patients.

Phase 2:   sponsors conduct clinical trials generally in a limited patient population to identify possible adverse effects and safety risks, to determine the efficacy of the product for specific targeted indications and to determine dose tolerance and optimal dosage. Sponsors may conduct multiple Phase 2 clinical trials to obtain information prior to beginning larger and more extensive Phase 3 clinical trials.

Phase 3:   these include expanded controlled and uncontrolled trials, including pivotal clinical trials. When Phase 2 evaluations suggest the effectiveness of a dose range of the product and acceptability of such product's safety profile, sponsors undertake Phase 3 clinical trials in larger patient populations to obtain additional information needed to evaluate the overall benefit and risk balance of the drug and to provide an adequate basis to develop labeling.

Once a sufficient amount of clinical data is collected, the sponsor may submit a New Drug Application (NDA) to the FDA. FDA reviews the data collected in the NDA to determine if the drug is safe and effective. As part of this review process, FDA conducts a pre-approval inspection of the product's manufacturing facility or facilities to assess compliance with the FDA's cGMP regulations, and to ensure that the facilities, methods and controls are adequate to preserve the drug's identity, quality and purity. The review process may also require review by an FDA advisory committee. An FDA advisory committee is convened when there are questions associated with the NDA that require review by outside experts.

As part of the NDA review process, the sponsor may elect to conduct, or be required by the FDA to conduct, one or more clinical trials after approval to further assess the drug's safety or effectiveness after NDA approval. Such postmarket trials are typically referred to as Phase 4 clinical trials.

Sponsors submit the results of product development, preclinical studies and clinical trials to the FDA as part of the NDA. An NDA must also contain extensive information relating to the product's pharmacology, chemistry, manufacture, controls and proposed labeling, among other things. For drugs that present a particular risk that must be mitigated in order to be approved, the FDA may require a risk evaluation and mitigation strategy (REMS), which could include medication guides, physician communication plans, or restrictions on distribution and use, such as limitations on who may prescribe the drug or where it may be dispensed or administered.

Upon receipt, the FDA has 60 days to determine whether the NDA is sufficiently complete to initiate a substantive review. If the FDA identifies deficiencies that would preclude substantive review, the FDA will refuse to accept the NDA and will inform the sponsor of the deficiencies that must be corrected prior to resubmission. If the FDA accepts the submission for substantive review, the FDA typically reviews the NDA in accordance with established timeframes.

Under the Prescription Drug User Fee Act (PDUFA), which modified the FFDCA, the FDA agreed to specific goals for NDA review times through a two-tiered classification system, Priority Review and Standard Review. Priority Review is given to drugs that offer major advances in treatment, or provide a treatment where no adequate therapy exists. For a Priority Review application, the FDA aims to complete the initial review cycle in six months. Standard Review applies to all other drug applications. The FDA generally aims to complete Standard Review NDAs within a ten-month timeframe.

The review process often extends significantly beyond anticipated completion dates due to FDA requests for additional information or clarification, difficulties scheduling an advisory committee meeting, negotiations regarding REMS, or FDA workload issues. The FDA may refer the application to an advisory committee for review, evaluation and recommendation as to the application's approval. The recommendations of an advisory committee do not bind the FDA, but the FDA generally follows such recommendations.

Under PDUFA, NDA applicants must pay significant NDA user fees to the FDA upon submission of the application. In addition, manufacturers of approved prescription drug products must pay annual establishment and product user fees.

2.2.3 Abbreviated New Drug Applications and Section 505(b)(2) New Drug Applications

As an alternate path to FDA approval, an applicant may submit an NDA under Section 505(b)(2) of the FFDCA, or an Abbreviated New Drug Application

(ANDA) under Section 505(j) of the FFDCA. Section 505(b)(2) applications are generally used when seeking marketing approval for modifications to drug products previously approved by the FDA. An ANDA is a type of premarket drug application that is used for the review and approval of a generic drug product.

A generic drug product is one that is the same as a previously approved innovator drug product, which means it has the same active ingredient, dosage form, and strength, route of administration, quality, performance characteristics, and intended use. An ANDA is generally not required to include preclinical and clinical data to establish safety and effectiveness. Instead, generic applicants must scientifically demonstrate that their product is bioequivalent to the previously approved drug, which means that it performs in the same manner.

Section 505(b)(2) applications are similar to ANDAs in that they permit an applicant to reference some of the information required for approval from clinical trials not conducted by or for the applicant, and for which the applicant has not obtained a right of reference. The FDA permits applicants to rely upon FDA's previous findings of safety and effectiveness for an approved product, although FDA may also require additional clinical trials to support changes from the previously approved product.

Section 505(b)(2) applications and ANDAs are subject to any non-patent exclusivity period applicable to the referenced product, which may delay approval of the 505(b)(2) application or ANDA even if FDA has completed its substantive review and determined the drug should be approved. In addition, 505(b)(2) applications and ANDAs must include patent certifications to any patents listed in the Orange Book as covering the referenced product. If the applicant seeks to obtain approval before the expiration of an applicable listed patent, the applicant must provide notice to the patent owner and NDA holder of the referenced product. If the patent owner or NDA holder bring a patent infringement lawsuit within 45 days of such notice, the 505(b)(2) application or ANDA cannot be approved for 30 months, or until the applicant prevails, whichever is sooner. If the applicant loses the patent infringement suit, FDA may not approve the 505(b)(2) application or ANDA until the patent and remaining exclusivity expire.

2.2.4   Other Marketing Pathways

Many over-the-counter (OTC) drug products are marketed through compliance with a monograph. A monograph is a regulation, or draft regulation, that specifies allowable active and inactive ingredients, formulations, and required labeling statements for a class of OTC drug products. Provided an OTC drug product complies with a monograph, it may be marketed without FDA approval. Similarly, homeopathic drugs may be marketed without FDA approval provided that they comply with a standard listed in the Homeopathic Pharmacopeia of the United States (HPUS). In order to market a drug product that deviates from an OTC monograph or HPUS standard, FDA approval would be required (obtained through an NDA, an ANDA, or a Section 505(b)(2) application).

2.2.5   Post-Approval Requirements

After approval, drug companies must comply with comprehensive requirements governing, among other things, drug listing, recordkeeping, manufacturing, marketing activities, product sampling and distribution, annual reporting and adverse event reporting. There are also extensive US Drug Enforcement Agency (DEA) regulations applicable to marketed controlled substances.

If new safety issues are identified following approval, the FDA can require the drug company to revise the approved labeling to reflect the new safety information; conduct post-market studies or clinical trials to assess the new safety information; and implement a REMS program to mitigate newly-identified risks. The FDA may also require post-approval testing, including Phase 4 studies, and surveillance programs to monitor the effect of approved products which have been commercialised. The FDA has the authority to prevent or limit further marketing of a product based on the results of post-marketing programs. Drugs may be marketed only for approved indications and in accordance with the provisions of the approved label.

Drug manufacturers and other entities involved in the manufacture and distribution of approved drugs are required to register their establishments with the FDA, and are subject to periodic inspections by the FDA for compliance with cGMP requirements. Changes to the manufacturing process are strictly regulated and often require prior FDA approval before being implemented. FDA regulations also require investigation and correction of any deviations from cGMP and impose reporting and documentation requirements upon drug companies.

If after approval the FDA determines that the product does not meet applicable regulatory requirements or poses unacceptable safety risks, the FDA may take other regulatory actions, including initiating suspension or withdrawal of the NDA approval. Later discovery of previously unknown problems with a product, including adverse events of unanticipated severity or frequency, or with manufacturing processes, or failure to comply with regulatory requirements, may result in, among other things:

- restrictions on the marketing or manufacturing of the product, complete withdrawal of the product from the market or product recalls;

- fines, warning letters or holds on post-approval clinical trials;

- refusal of FDA to approve pending applications or supplements to approved applications, or suspension or revocation of product license approvals;

- product seizure or detention, or refusal to permit the import or export of products; or

- injunctions or the imposition of civil or criminal penalties.

The FDA strictly regulates marketing, labeling, advertising and promotion of drug products that are placed on the market. These regulations include standards and restrictions for direct-to-consumer advertising, industry-sponsored scientific and educational activities, promotional activities involving the internet, and off-label promotion. While physicians may prescribe for off label uses, drug companies may only promote for the approved indications and in accordance with the provisions of the approved label. The FDA has very broad enforcement authority under the FFDCA, and failure to abide by these regulations can result in penalties, including the issuance of a warning letter directing entities to correct deviations from FDA standards, a requirement that future advertising and promotional materials be pre-cleared by the FDA, and state and federal civil and criminal investigations and prosecutions.

2.3   Overview of FDA Drug Enforcement 2010

FDA's inspectional authority over drug manufacturers differs based on whether the manufacturer is located within the US or outside of the US For any manufacturer located within the US, FDA has the authority "to inspect, at reasonable times and within reasonable limits and in a reasonable manner." 21 U.S.C. § 374(a)(1)(B). Manufacturers located outside of the US are not mandated to submit to an inspection, as are US manufacturers. Foreign manufacturers are generally compelled to submit to FDA inspection, however, subject to FDA's authority to refuse any import into the US for the appearance of an adulteration. The refusal to permit an inspection in this case may subject any exports to import detention and refusal.

Following any inspection, the FDA investigator may issue a list of observations on a Form FDA 483 ("Inspectional Observations"). The Form FDA 483 generally only represents the observations made by the investigator, and does not constitute a final determination that the regulated entity is in violation of FDA's laws and regulations. The Form FDA 483 is not generally made public unless released through a Freedom of Information Request, and will often be redacted.

Following the inspection, the investigator prepares an Establishment Inspection Report (EIR), which is used internally by FDA to document the inspection and observations in the FDA 483. EIRs are similarly not made publicly available unless released through a Freedom of Information Request, and will also be redacted. The inspected facility will not receive a copy of the EIR until after FDA has determined that the matters discovered during the inspection do not require further regulatory action.

FDA has a variety of regulatory tools at its disposal in handling firms that are not in compliance with FDA's laws and regulations. These tools generally follow the following order from least severe to most severe: advisory actions, recalls, administrative actions, and judicial actions.

Advisory actions include untitled letters, Warning Letters, and regulatory meetings. Untitled letters provide notice to regulated entities that certain conditions present violations of FDA's laws and regulations. Untitled letters are used when the violation is not particularly significant, or where the application of the regulation is so recent that the regulated entity may not have had adequate notice regarding the requirements. Untitled letters are generally not publicly posted (with the exception of untitled letters regarding drug advertising and promotion issues). Warning Letters are used when the violation or violations are significant, and that FDA may take further action if the issues are not resolved. Unlike untitled letters, Warning Letters are publicly posted on FDA's website. Regulatory meetings may be held in advance of the issuance of an untitled or Warning Letter, and, in some cases, may replace the need for FDA to issue an untitled or Warning Letter. The matters discussed during regulatory meetings are kept confidential.

Recalls are voluntary or FDA-mandated removals or corrections of distributed regulated products. The majority of recalls are voluntarily taken by regulated entities to correct a violation of the FFDCA. In certain situations, FDA may order a regulated entity to perform a recall to correct a violation, although such an action is rare[9].

Administrative actions include notices issued under section 305 of the FFDCA prior to FDA instituting a criminal proceeding, and civil money penalties. A section 305 notice is issued to provide notice to the person against whom a criminal proceeding is contemplated, and an opportunity to present his views, either orally or in writing, with regard to the contemplated criminal proceeding. Civil money penalties are fines levied by FDA subject to violations of the FFDCA. Some civil money penalties are specifically set out in the FFDCA, and some are agreed upon payments made by regulated entities to avoid further regulatory action.

Judicial actions involve civil and criminal charges of violations of the FFDCA, seizures, and injunctions. Civil and criminal penalties are possible under the FFDCA for violations, and, as the FFDCA is a strict-liability statute, an individual does not need to have intended to break the law in order to be subject to charges under the FFDCA. Further, officers of a company with authority over its operations may be liable for violations committed by employees under what is known as the **Park** doctrine. Seizures are judicial actions against goods that are in violation of the FFDCA. In a seizure action, individuals and companies may join in the action to disprove FDA's charge that the goods in question are in violation of the FFDCA, or may choose to allow FDA to assert its authority unchallenged. An injunction is a court order requiring an individual or company to comply with the FFDCA.

As an alternative to the judicial actions possible under the FFDCA, a regulated entity and the FDA may enter into what is known as a consent decree. A consent decree is an agreement, entered into between FDA and a regulated entity, and overseen by a court. Consent decrees may involve the payment of civil money penalties, as well as any other number of activities promised to correct violations identified by FDA.

This section and the next section provide a detailed analysis of FDA's enforcement of drug and dietary supplement companies. The following section generally focuses on the Warning Letters issued by the agency in 2010, while this section examines data made available through the Inspections Classification Database, and "FDA-TRACK" (providing publicly available information on the agency's performance measures and results).

Where possible, comparisons are made to previous years to provide some general trending. Also, comparisons are made between domestic and foreign firms and results across different FDA districts.

2.3.1   FDA's Inspectional Workload

FDA-TRACK, which started capturing data in fiscal year 2010, provides monthly and quarterly snapshots of key center priorities presented in "Dashboards." The two sets of Dashboards that are most relevant to drug manufacturing compliance are the Dashboards reporting results for the CDER Office of Compliance ("CDER/OC") and the Office of Regulatory Affairs ("ORA") Regional Offices.

CDER/OC has responsibility for enforcing the laws and regulations under CDER's authority. Investigations and enforcement actions can either be performed directly by CDER/OC staff, such as its enforcement actions against unapproved prescription pharmaceuticals, or through some other FDA organization. When working with other FDA offices and centers on investigations and enforcement, CDER/OC generally provides scientific review and guidance on enforcement actions, and coordinates CDER's inspectional programs with those offices and centers.

ORA Regional Offices perform the majority of FDA's inspections and enforcement work. ORA develops an annual workplan in conjunction with the centers, and carries out those inspections with its inspectional staff in each of its Regional Offices. Each Regional Office, which is further divided into District Offices and Resident Posts, generally handle regulated industry within their assigned geographic area.

Enforcement and compliance actions, for the most part, are initiated and handled by ORA compliance staff in each region. In these situations, compliance staff coordinates and seeks scientific guidance from the appropriate center[10]. Certain issues may only be initiated by center compliance staff. These issues generally deal with the failure to obtain the appropriate marketing authorization for a regulated product, or the failure to file a necessary report with the FDA (such as an adverse event report)[11].

As a general point of practice, it is important to understand the roles and authorities of each office and group within FDA when handling an enforcement issue. This will allow for better targeted responses, and may help when identifying the appropriate individuals at FDA who should be contacted to resolve the matter.

The FDA-TRACK dashboards provide useful information regarding the general compliance of industry and FDA enforcement. Despite these dashboards having only started in fiscal year 2010, limited trending can be conducted based on similarities to the information covered in FDA's prior enforcement publication, "The Enforcement Story," which last covered fiscal year 2008.

Recalls are frequently used by the agency to assess industry compliance with regulatory requirements, as well as to monitor the general quality of regulated products. Overall there has been a marked increase in the number of recalls involving CDER regulated products, with CDER reporting a total of 212 recalls in fiscal year 2010 impacting 764 products. By comparison, CDER reported a total of only 162 recalls effecting 379 products in fiscal year 2008.

DA-TRACK CDER Office of Compliance Dashboard 2010 (Fiscal and Calendar)

Enforcements activities to reduce risks associated with unsafe, ineffective, and poor quality drugs

Total number of selected compliance actions reviewed, either approved or not-approved, in the month, and routed appropriately within the agency for additional review, closure, or issuance

Measure	Target	Oct-09	Nov-09	Dec-09	Jan-10	Feb-10	Mar-10	Apr-10	May-10	Jun-10	Jul-10	Aug-10	Sep-10	Oct-10
Total number of Warning Letters recommendations reviewed in the month	N/A	8	53	13	10	12	17	14	12	10	8	14	9	12
Number of Seizure actions reviewed in the month	N/A	0	0	0	1	0	1	0	0	0	1	1	0	0
Number of Disqualifications reviewed in the month	N/A	1	0	1	0	0	0	1	0	1	0	1	0	0
Number of injunction packages reviewed in the month	N/A	0	0	0	0	1	3	2	0	1	0	1	1	0

Unapproved Dugs initiative actions taken in the quarter to remove drugs from the marketplace that have not gone through the FDA approval process and present a risk to public health

Measure	Target	Oct-09	Nov-09	Dec-09	Jan-10	Feb-10	Mar-10	Apr-10	May-10	Jun-10	Jul-10	Aug-10	Sep-10	Oct-10
Number of class actions taken in the quarter regarding marketed unapproved drugs	N/A		1			1			1			1		
Number of products involved	N/A		7			6			1			N/A		
Number of manufacturers involved	N/A		4			2			1			N/A		
Number of Warning Letters issued in the quarter for unapproved drugs that are not part of class actions	N/A		11			11			3			0		
Number of seizures or injunctions issued in the quarter for unapproved drugs	N/A		0			0			0			0		

Drug Quality Defects related to processing recall actions in the month

Measure	Target	Oct-09	Nov-09	Dec-09	Jan-10	Feb-10	Mar-10	Apr-10	May-10	Jun-10	Jul-10	Aug-10	Sep-10	Oct-10
Total number of recall events in the month for which the classification process was completed	N/A	20	23	5	9	8	37	24	16	32	20	26	38	15
Number of Class 1 recalls classified by Office of Regulatory Affairs after recommendation by CDER Compliance. [1]	N/A	0	2	1	0	1	3	3	3	4	2	2	8	3
Number products impacted by Class 1 recalls	N/A	0	2	1	0	1	7	8	12	66	2	30	29	4
Number of Class II recalls classified	N/A	12	9	2	2	3	19	17	8	17	11	19	19	9
Number products impacted by Class II recalls	N/A	21	88	2	2	3	25	30	21	42	17	48	90	18
Number of Class III recalls classified	N/A	8	12	2	7	4	15	4	5	11	7	5	11	3
Number products impacted by Class III recalls	N/A	8	14	2	187	6	27	5	14	20	10	7	21	3

[1] Authority to designate a drug recall as Class 1 currently rests with the Associate Commissioner for Regulatory Affairs (ACRA).

The overall increase in recalls may be a response by industry to the increase in FDA's inspectional resources, and the agency's more rigorous enforcement policies. Conducting a recall provides a way to correct violations of the Federal Food, Drug and Cosmetic Act while limiting the likelihood of further agency enforcement.

Class I recalls involve products that will cause serious adverse health consequences or death. CDER reported that it classified 22 drug recalls as Class I in fiscal year 2010 impacting 131 products. While this nearly matches the number of Class I drug recall events occurring in fiscal year 2008 (21 recalls), significantly fewer products were impacted by those recalls (36 products).

Class II recalls generally involve products that may cause temporary or medically reversible adverse health consequences. CDER classified 117 drug recalls as Class II in fiscal year 2010, involving 321 products. This compares to 72 recalls in 2008 impacting 176 products.

Another measure of FDA enforcement (as well as industry compliance) are injunctions and seizures. An injunction under the Federal Food, Drug and Cosmetic Act is a court order requiring an individual or company to refrain from violating the Act. Alternatively, a seizure is a court action against goods found to be in violation of the Act. Unlike other actions under the Federal Food, Drug and Cosmetic Act, a seizure is not specifically directed at an individual or company. Those companies and individuals with a possessory interest in the goods to be seized may voluntarily join the action to assert a claim over the goods (and establish that such goods are not in violation of the Act). Those with a possessory interest may also elect to stay out of the action, thereby potentially surrendering its interest in the goods to the agency.

The FDA infrequently elects to use seizure or injunction as an enforcement tool, and the decision to pursue such an action is largely based on fact-specific criteria. As a result, seizure and injunction actions are not as useful a benchmark of compliance and enforcement as other tools, such as Warning Letters. Overall, there has been an increase in the total number of seizures and injunctions. In fiscal year 2010, CDER was involved in 3 seizures (compared to 5 seizures in 2008) and 7 injunctions (compared to one injunction in 2008). It is not entirely clear whether the change in the number of injunctions (CDER actually conducted more seizures in fiscal year 2008 than it did in 2010) is a shift to more "personal" types of enforcement actions, or whether this is merely due to the particular facts of the cases arising in 2010.

DA-TRACK ORA Regional Offices Dashboard 2010 (Fiscal and Calendar)

ffice of Regional Operations Program Measures

Inspectional and Regulatory Action Measures: Foreign Inspections

Assess the number and outcome of inspections that the foreign cadres accomplish

Measure	Target	Oct-09	Nov-09	Dec-09	Jan-10	Feb-10	Mar-10	Apr-10	May-10	Jun-10	Jul-10	Aug-10	Sep-10	Oct-10
. Total number of drug foreign inspections n FY-10	TBD	N/A	N/A	N/A	N/A	N/A	N/A	N/A	N/A	N/A	N/A	N/A	N/A	N/A
. Number of drug foreign inspections ompleted at the end of the month	TBD	N/A	N/A	N/A	6	25	18	14	31	15	29	13	23	21

Foreign inspection findings

Measure	Target	Oct-09	Nov-09	Dec-09	Jan-10	Feb-10	Mar-10	Apr-10	May-10	Jun-10	Jul-10	Aug-10	Sep-10	Oct-10
. Number of foreign inspections completed n the month resulting in an initial lassification of No Action Indicated	N/A	N/A	N/A	N/A	7	16	12	11	23	13	38	11	5	15
. Percentage of foreign inspections in the nonth resulting in an initial classification f No Action Indicated														18%
. Number of foreign inspections completed n the month resulting in an initial lassification of Voluntary Action Indicated	N/A	N/A	N/A	N/A	6	21	15	19	27	30	30	29	18	41
. Percentage of foreign inspections in the nonth resulting in an initial classification f Voluntary Action Indicated														49%
. Number of foreign inspections completed n the month resulting in an initial lassification of Official Action Indicated	N/A	N/A	N/A	N/A	0	9	2	8	19	15	10	12	6	14
. Percent of foreign inspections in the nonth resulting in an initial classification f Official Action Indicated														17%
. Number of foreign inspections completed n the month resulting in a category of lassification called OTHER	N/A	N/A	N/A	N/A	0	0	0	0	0	0	0	0	24	14
. Percent of foreign inspections in the nonth resulting in an initial classification f Official Action Indicated														17%

onsolidated Regional Program Measures

Inspectional and Regulatory Action Measures

Track progress toward performance

Measure	Target	Oct-09	Nov-09	Dec-09	Jan-10	Feb-10	Mar-10	Apr-10	May-10	Jun-10	Jul-10	Aug-10	Sep-10	Oct-10
. Total number of the highest risk product omestic inspections in the FY10 erformance Goals	10,206	679	679	679	1,022	1,022	1,022	1,022	1,022	1,022	679	680	680	10,607
. Number of highest risk product domestic nspections completed in the month	N/A	683	989	916	888	873	1,175	927	966	1,368	850	947	896	N/A
. Cumulative annual highest risk product omestic inspections completed by the nd of the month	N/A	683	1581	2,497	3,385	4,258	5,433	6,360	7,326	8,694	9,544	10,491	11,387	N/A
. Cumulative annual highest risk product omestic inspections completed at the nd of the month	N/A	7%	15%	25%	33%	41%	53%	62%	72%	85%	94%	103%	112%	N/A

B. *Identify compliance of high-risk firms*

Measure	Target	Oct-09	Nov-09	Dec-09	Jan-10	Feb-10	Mar-10	Apr-10	May-10	Jun-10	Jul-10	Aug-10	Sep-10	Oct-1
1. Number of highest risk domestic inspections in the month resulting in a classification of Official Action Indicated	N/A	29	46	28	36	37	39	36	44	55	42	44	64	45
a. Percent of highest risk domestic inspections in the month resulting in a classification of Official Action Indicated	N/A	4%	5%	3%	4%	4%	3%	4%	5%	4%	5%	5%	7%	5%
2. Number of highest risk domestic inspections in the month resulting in a classification of Voluntary Action Indicated	N/A	310	376	368	334	350	450	372	345	482	351	338	353	339
a. Percent of highest risk domestic inspections in the month resulting in a classification of Voluntary Action Indicated	N/A	45%	42%	40%	38%	40%	38%	40%	36%	35%	41%	36%	39%	40%
3. Number of highest risk domestic inspections in the month resulting in a classification of No Action Indication	N/A	340	474	513	510	483	673	515	569	828	452	556	463	422
a. Percent of highest risk domestic inspections in the month resulting in a classification of No Action Indication	N/A	50%	53%	56%	57%	55%	57%	56%	59%	61%	53%	59%	52%	50%
4. Number of highest risk product domestic inspections in the month resulting in a category of classifications called OTHER	N/A	4	2	7	9	3	13	5	8	3	5	9	16	40
a. Percent of highest risk product domestic inspections in the month resulting in a category of classification called OTHER	N/A	1%	0%	1%	1%	0%	1%	1%	1%	0%	1%	1%	2%	5%

The Inspections Classification Database, which covers the 2009 and 2010 fiscal years, provides information on the results of inspections performed in those fiscal years. This provides a convenient way to view the overall results of inspections. In past years, FDA did not always make this information available, and when it was available, it was never issued in a consistent way. The CDER Inspections Classification Database information in the following tables is broken down by project area, inspection classification, and year.

The project areas represented in 2009 and 2010 are "Unapproved and Misbranded," "Bioresearch Monitoring," "Drug Quality Assurance," "Postmarket Surveillance and Epidemiology," and "Special Field Assignment." While the FDA does not provide any explanation on what each program area represents, the Drug Quality Assurance Program covers cGMP inspections, generally, and Bioresearch Monitoring covers inspections of investigators, sponsors and IRBs in their respective roles in the clinical research of drugs. The Unapproved and Misbranded program area covers inspections to determine whether drug products are marketed under an appropriate approval or monograph, and whether the labeling conforms to such approval or monograph. The Postmarket Surveillance and Epidemiology program area focuses on adverse event reporting, the conduct of postmarket studies, and reporting responsibilities associated with drug approvals. Finally, the Special Field Assignment program area serves as a catch-all to cover instances where CDER needs to issue an inspection assignment for reasons not otherwise planned or accounted for in the other program areas (such as a follow-up to a recall or reports of problems across multiple manufacturers of the same product). As multiple project areas may be covered in a single inspection, the sum of the project areas does not equal the total number of inspections. Separate district decisions on the results of an inspection are tracked for each project area covered.

Inspections are classified by districts in one of three ways: OAI, VAI and NAI. OAI, or "Official Action Indicated" are the most serious, and cover inspections where the district recommends taking some type of enforcement action as the result of the inspection. VAI, or "Voluntary Action Indicated," represents situations where the district noted problems during the inspection, but that the problems could be adequately addressed by the company without the need of an enforcement action. NAI, or "No Action Indicated," represents inspections where no problems or very minor problems were observed, requiring no action on the part of FDA or the company.

Examining the number of facility inspections in the Inspections Classification Database (not counting each separate project area covered) shows a marked decline in the total number of inspections conducted since 2004. Based on numbers provided in The Enforcement Story, the FDA conducted 21,805 inspections in 2004. This total had declined to 15,245 inspections by 2008. Following the numbers based on the Inspections Classification Database, this decline accelerated significantly in 2009, dropping to a mere 12,634 inspections. While the trend reversed in 2010, with the total number of inspections rising to 14,901, the number of inspections has yet to rise past the level it was at in 2008.

Although the current number of inspections is far behind the level it was in 2004, it is important to note the upward trend in the last year. While some may point to the ongoing budget debate as a sign that the number of inspections will begin to fall again, there are indications that FDA's inspectorate is being underutilised. Further, the tightening labor market and overall decline in the stock market (which has an impact on federal retirement savings) will likely keep more employees in the agency.

Some immediate trends can be examined, despite the limited data in the Inspections Classification Database. There was a healthy increase in the number of CDER inspections conducted in 2010 (an 8.4 % increase over 2009). Overall, CDER's representation in the total number of inspections fell slightly from 12% in 2009 to 11% in 2010. With respect to project areas, the FDA began to shift significant resources into the drug quality assurance project area, which increased nearly 35% from 2009. By comparison, inspections covering unapproved and misbranded drugs, bioresearch monitoring, and postmarket surveillance and epidemiology project areas fell just under 25% each, on average.

Drug Inspections (All)		Percentage
Total for 2009 & 2010		
Total	3329	100.0%
NAI	1697	51.0%
VAI	1470	44.2%
OAI	162	4.9%
Total for 2009		
Total	1605	48.2%
NAI	810	50.5%
VAI	700	43.6%
OAI	95	5.9%
Total for 2010		
Total	1724	51.8%
NAI	887	51.5%
VAI	770	44.7%
OAI	67	3.9%
Percent increases 2009 to 2010		
Total	7.4%	
NAI	9.5%	
VAI	10.0%	
OAI	-29.5%	

Drug Inspections (Unapproved and Misbranded)		Percentage
Total for 2009 & 2010		
Total	58	100.0%
NAI	16	27.6%
VAI	34	58.6%
OAI	8	13.8%
Total for 2009		
Total	33	56.9%
NAI	9	27.3%
VAI	18	54.5%
OAI	6	18.2%
Total for 2010		
Total	25	43.1%
NAI	7	28.0%
VAI	16	64.0%
OAI	2	8.0%
Percentage of Total		
2009 & 2010	1.7%	
2009	2.1%	
2010	1.5%	
Percent increases 2009 to 2010		
Total	-24.2%	

Drug Inspections (Bioresearch Monitoring)		Percentage
Total for 2009 & 2010		
Total	1050	100.0%
NAI	527	50.2%
VAI	475	45.2%
OAI	48	4.6%
Total for 2009		
Total	604	57.5%
NAI	304	50.3%
VAI	268	44.4%
OAI	32	5.3%
Total for 2010		
Total	446	42.5%
NAI	223	50.0%
VAI	207	46.4%
OAI	16	3.6%
Percentage of Total		
2009 & 2010	31.5%	
2009	37.6%	
2010	25.9%	
Percent increases 2009 to 2010		
Total	-26.2%	

Drug Inspections (Drug Quality Assurance)		Percentage
Total for 2009 & 2010		
Total	2068	100.0%
NAI	1059	51.2%
VAI	907	43.9%
OAI	102	4.9%
Total for 2009		
Total	882	42.6%
NAI	445	50.5%
VAI	381	43.2%
OAI	56	6.3%
Total for 2010		
Total	1186	57.4%
NAI	614	51.8%
VAI	526	44.4%
OAI	46	3.9%
Percentage of Total		
2009 & 2010	62.1%	
2009	55.0%	
2010	68.8%	
Percent increases 2009 to 2010		
Total	34.5%	

Drug Inspections (Postmarket Surv and Epidermology)		Percentage
Total for 2009 & 2010		
Total	152	100.0%
NAI	95	27.6%
VAI	53	58.6%
OAI	4	13.8%
Total for 2009		
Total	86	56.9%
NAI	52	27.3%
VAI	33	54.5%
OAI	1	18.2%
Total for 2010		
Total	66	43.1%
NAI	43	28.0%
VAI	20	64.0%
OAI	3	8.0%
Percentage of Total		
2009 & 2010	4.6%	
2009	5.4%	
2010	3.8%	
Percent increases 2009 to 2010		
Total	-23.3%	

Drug Inspections (Special Field Assignment)		Percentage
Total for 2009 & 2010		
Total	1	100.0%
NAI	0	0.0%
VAI	1	100.0%
OAI	0	0.0%
Total for 2009		
Total	0	0.0%
NAI	0	0.0%
VAI	0	0.0%
OAI	0	0.0%
Total for 2010		
Total	1	100.0%
NAI	0	0.0%
VAI	1	100.0%
OAI	0	0.0%
Percentage of Total		
2009 & 2010	0.0%	
2009	0.0%	
2010	0.1%	

Despite the overall increase to FDA's inspectional resources, there was a marked decline in the number of drug inspection project areas categorised as OAI. From 2009 to 2010, the percentage of inspection program areas designated as OAI dropped from 5.9% to 3.9%. It is difficult to know exactly what factors led to this decline. One possible explanation is an adjustment made by the drug industry to the recent inspectional and enforcement changes introduced by the agency over the last few years.

2.4   Drug Warning Letters

2.4.1   Warning Letter Overview

This review examines a very broad cross-section of Warning Letters (180 in total). Due to the breadth of issues covered in this section, the totals presented may not compare to totals from prior years or to totals for 2010 presented by other sources. This section reviews Warning Letters issued for a variety of issues, including:

- manufacturing of prescription and OTC finished pharmaceuticals;

- manufacturing of finished dietary supplements;

- manufacturing of active pharmaceutical ingredients ("APIs");

- marketing of prescription drugs; and

- distribution and sale of unapproved new drugs.

The following table provides a breakdown of the Warning Letters covered in this section by issue. Individual Warning Letters may deal with multiple issues and be represented in more than one total.

Number of Warning Letters in 2010 by Issue

Lack Approval Practice	Current Good Manufacturing	Labeling[16]	IND/ Subject Protection	Registration	Reporting Requirements (21 C.F.R. 314.80 and 314.81)
109	68	41	18	11	6

Warning Letters concerning dietary supplement current good manufacturing practice ("cGMP") requirements set out in 21 C.F.R. part 111 are included, as well as dietary supplements promoted as drugs or containing a known API. While in many cases these letters are issued by the Center for Food Safety and Applied Nutrition ("CFSAN"), they deal closely with issues related to the marketing of OTC and prescription drug products. Some dietary supplement products, for example, are positioned to compete for the same consumer dollar as OTC drug products, such as in the cough and cold category. Dietary supplements containing known prescription APIs present a risk of consumer injury and are a method of drug diversion. Further, the dietary supplement cGMP closely approximate current drug manufacturing making their inclusion relevant to examining how FDA reviews drug manufacturing processes. Given the relevance and importance of these issues to drug marketing generally, they have been analyzed in this report.

The following table gives a breakdown of the Warning Letters that were issued in 2010 for products that are not prescription drugs, but that are analyzed in this report due to their relationship to the regulation of drugs by the FDA[17]. The product type is based on the presumed marketing intent of the recipient of the Warning Letter, and not on the finding by the FDA[18]. As with drug issues, a single Warning Letter may deal with more than one product type and may be counted in multiple categories.

Number of Warning Letters in 2010 by Product Type

Drug[19]	Dietary Supplement[20]	OTC	API	Homeopathic Drug	Conventional Food	Medical Food
127	48	20	11	6	4	1

This report also examines issues related to labeling and advertising through the Internet. A total of sixty-seven (67) Warning Letters were issued in 2010 that make reference to labeling and/or advertising on the Internet. These Warning Letters reveal FDA's expanding understanding of the use of the Internet as a promotional and sales tool. For example, eleven (11) Warning Letters cite the use of metadata on websites to infer an intent to market products as new drugs. The FDA also cites statements made on a blog, and links provided to other websites as evidence of the intent to market a product for an unapproved new use. These Warning Letters also document FDA's continuing partnership with the US Federal Trade Commission ("FTC"), as three (3) Warning Letters were jointly issued with the FTC.

The use of metadata to support regulatory findings is a particularly important development. Metadata in general terms is defined as "data about data," although in purely technical terms, the word is used in a much broader context. Metadata, as it is used in the context of FDA's Warning Letters, generally refers to "guide metadata" which is data used to find specific items and is usually expressed as a set of keywords in a natural language. These keywords are generally not represented in the visible text of a website, but may be found by search engines such as Google. This development suggests that the agency may soon take advantage of even more powerful tools in its enforcement methods (if it has not already done so), such as data mining or stripping.

2.4.2   cGMP Manufacturing

2.4.2.1   Introduction

The regulatory schemes for manufacturing finished drug products, dietary supplements (covered by 21 C.F.R. part 111) and APIs (following "FDA's Guidance for Industry: Q7A Good Manufacturing Practice Guidance for Active Pharmaceutical Ingredients") share similarities in their structures. The earliest regulatory scheme governs the manufacturing of finished drug products, and is covered by 21 C.F.R. parts 210 and 211. While these regulations were implemented prior to the development and widespread adoption of the quality system approach, the FDA has found the regulatory scheme flexible enough to cover some modern quality system concepts. Despite this flexibility, the FDA has considered modernizing the finished drug product regulations for some time.

The regulatory schemes governing the manufacturing of dietary supplements and APIs are much more modern. While dietary supplements are governed by regulations found at 21 C.F.R. part 111, APIs have a less definite source. On the most basic level, API requirements are based on section 501(a)(2)(B) of the Federal Food, Drug and Cosmetic Act, which requires that the methods used in, or the facilities or controls used for, an API's manufacture, processing, packing, or holding must conform to and be operated and administered in conformity with current good manufacturing practice. The FDA defines current good manufacturing practice for APIs based on international guidelines, specifically International Conference on Harmonisation of Technical Requirements for Registration of Pharmaceuticals for Human Use (ICH) harmonised guideline Q7, "Good Manufacturing Practice Guide For Active Pharmaceutical Ingredients" ("ICH Q7A"). This guideline has been adopted as a formal guidance by the FDA ("Guidance for Industry Q7A Good Manufacturing Practice Guidance for Active Pharmaceutical Ingredients") under agreement with ICH. Of note, while API cGMP Warning Letters clearly reference ICH Q7A (in some places even quoting the language in ICH Q7A), the references are never directly attributed to their source. This is almost certainly due to the fact that FDA guidance is not permitted to carry the same weight as FDA laws or regulations, despite FDA's recognition of ICH Q7A as FDA guidance.

Despite their more recent development, there are several key differences between 21 C.F.R. part 111 and ICH Q7A. These differences make 21 C.F.R. part 111 a less rigorous scheme than what is proposed in ICH Q7A. One of the more significant differences is that 21 C.F.R. part 111 does not require validation, even of laboratory methods[21]. Validation is a requirement for several activities in 21 C.F.R. parts 210/211[22]. Further, 21 C.F.R. part 111 does not require periodic audits of the quality system[23]. The following table provides a comparison of the three regulatory schemes for finished pharmaceuticals, dietary supplements, and APIs.

Side-by-Side Comparison of 21 C.F.R. Parts 111 and 210/211 and ICH Q7A

Regulations / guidance	Part 111 – Current good manufacturing practice in manufacturing, packaging, labeling, or holding operations for dietary supplements	Part 210, part 211 – Current good manufacturing practice for finished pharmaceuticals	Guidance for industry Q7A good manufacturing practice guidance for active pharmaceutical ingredients guidance
General Provisions	**Subpart A – General Provisions** § 111.1 – Who is subject to this part? § 111.3 – What definitions apply to this part? § 111.5 – Do other statutory provisions and regulations apply?	**Subpart A – General Provisions** § 211.1 – Scope. § 211.3 – Definitions.	**I. INTRODUCTION (1)** A. Objective (1.1) B. Regulatory Applicability (1.2) C. Scope (1.3) GLOSSARY (20)
Personnel	**Subpart B – Personnel** § 111.8 – What are the requirements under this subpart B for written procedures? § 111.10 – What requirements apply for preventing microbial contamination from sick or infected personnel and for hygienic practices? § 111.12 – What personnel qualification requirements apply? § 111.13 – What supervisor requirements apply? § 111.14 – Under this subpart B, what records must you make and keep?	**Subpart B – Organization and Personnel** § 211.22 – Responsibilities of quality control unit. § 211.25 – Personnel qualifications. § 211.28 – Personnel responsibilities. § 211.34 – Consultants.	**III. PERSONNEL (3)** A. Personnel Qualifications (3.1) B. Personnel Hygiene (3.2) C. Consultants (3.3)
Building and Facility	**Subpart C – Physical Plant and Grounds** § 111.15 – What sanitation requirements apply to your physical plant and grounds? § 111.16 – What are the requirements under this subpart C for written procedures? § 111.20 – What design and construction requirements apply to your physical plant? § 111.23 – Under this subpart C, what records must you make and keep?	**Subpart C – Buildings and Facilities** § 211.42 – Design and construction features. § 211.44 – Lighting. § 211.46 – Ventilation, air filtration, air heating and cooling. § 211.48 – Plumbing. § 211.50 – Sewage and refuse. § 211.52 – Washing and toilet facilities. § 211.56 – Sanitation. § 211.58 – Maintenance.	**IV. BUILDINGS AND FACILITIES (4)** A. Design and Construction (4.1) B. Utilities (4.2) C. Water (4.3) D. Containment (4.4) E. Lighting (4.5) F. Sewage and Refuse (4.6) G. Sanitation and Maintenance (4.7)
Equipment	**Subpart D – Equipment and Utensils** § 111.25 – What are the requirements under this subpart D for written procedures? § 111.27 – What requirements apply to the equipment and utensils that you use? § 111.30 – What requirements apply to automated, mechanical, or electronic equipment? § 111.35 – Under this subpart D, what records must you make and keep?	**Subpart D – Equipment** § 211.63 – Equipment design, size, and location. § 211.65 – Equipment construction. § 211.67 – Equipment cleaning and maintenance. § 211.68 – Automatic, mechanical, and electronic equipment. § 211.72 – Filters.	**V. PROCESS EQUIPMENT (5)** A. Design and Construction (5.1) B. Equipment Maintenance and Cleaning (5.2) C. Calibration (5.3) D. Computerized Systems (5.4)
Production and Process Controls	**Subpart E – Requirement to Establish a Production and Process Control System** § 111.55 – What are the requirements to implement a production and process control system? § 111.60 – What are the design requirements for the production and process control system? § 111.65 – What are the requirements for quality control operations? § 111.70 – What specifications must you establish? § 111.73 – What is your responsibility for determining whether established specifications are met? § 111.75 – What must you do to determine whether specifications are met? § 111.77 – What must you do if established specifications are not met? § 111.80 – What representative samples must you collect? § 111.83 – What are the requirements for reserve samples? § 111.87 – Who conducts a material review and makes a disposition decision? § 111.90 – What requirements apply to treatments, in-process adjustments, and reprocessing when there is a deviation or unanticipated occurrence or when a specification established in accordance with 111.70 is not met? § 111.95 – Under this subpart E, what records must you make and keep?	**Subpart F – Production and Process Controls** § 211.100 – Written procedures; deviations. § 211.101 – Charge-in of components. § 211.103 – Calculation of yield. § 211.105 – Equipment identification. § 211.110 – Sampling and testing of in-process materials and drug products. § 211.111 – Time limitations on production. § 211.113 – Control of microbiological contamination. § 211.115 – Reprocessing.	**VIII. PRODUCTION AND IN-PROCESS CONTROLS (8)** A. Production Operations (8.1) B. Time Limits (8.2) C. In-process Sampling and Controls (8.3) D. Blending Batches of Intermediates or APIs (8.4) E. Contamination Control (8.5)

Regulations / guidance	Part 111 – Current good manufacturing practice in manufacturing, packaging, labeling, or holding operations for dietary supplements	Part 210, part 211 – Current good manufacturing practice for finished pharmaceuticals	Guidance for industry Q7A good manufacturing practice guidance for active pharmaceutical ingredients guidance
Quality Controls	**Subpart F – Production and Process Control System: Requirements for Quality Control** § 111.103 – What are the requirements under this subpart F for written procedures? § 111.105 – What must quality control personnel do? § 111.110 – What quality control operations are required for laboratory operations associated with the production and process control system? § 111.113 – What quality control operations are required for a material review and disposition decision? § 111.117 – What quality control operations are required for equipment, instruments, and controls? § 111.120 – What quality control operations are required for components, packaging, and labels before use in the manufacture of a dietary supplement? § 111.123 – What quality control operations are required for the master manufacturing record, the batch production record, and manufacturing operations? § 111.127 – What quality control operations are required for packaging and labeling operations? § 111.130 – What quality control operations are required for returned dietary supplements? § 111.135 – What quality control operations are required for product complaints? § 111.140 – Under this subpart F, what records must you make and keep?		**II. QUALITY MANAGEMENT (2)** A. Principles (2.1) B. Responsibilities of the Quality Unit(s) (2.2) C. Responsibility for Production Activities (2.3) D. Internal Audits (Self Inspection) (2.4) E. Product Quality Review (2.5) **XIII. CHANGE CONTROL (13)**
Component, Packaging and Labeling Controls Product Received for Packaging	**Subpart G – Production and Process Control System: Requirements for Components, Packaging, and Labels and for Product That You Receive for Packaging or Labeling as a Dietary Supplement** § 111.153 – What are the requirements under this subpart G for written procedures? § 111.155 – What requirements apply to components of dietary supplements? § 111.160 – What requirements apply to packaging and labels received? § 111.165 – What requirements apply to a product received for packaging or labeling as a dietary supplement (and for distribution rather than for return to the supplier)? § 111.170 – What requirements apply to rejected components, packaging, and labels, and to rejected products that are received for packaging or labeling as a dietary supplement? § 111.180 – Under this subpart G, what records must you make and keep?	**Subpart E – Control of Components and Drug Product Containers and Closures** § 211.80 – General requirements. § 211.82 – Receipt and storage of untested components, drug product containers, and closures. § 211.84 – Testing and approval or rejection of components, drug product containers, and closures. § 211.86 – Use of approved components, drug product containers, and closures. § 211.87 – Retesting of approved components, drug product containers, and closures. § 211.89 – Rejected components, drug product containers, and closures. § 211.94 – Drug product containers and closures.	**VII. MATERIALS MANAGEMENT (7)** A. General Controls (7.1) B. Receipt and Quarantine (7.2) C. Sampling and Testing of Incoming Production Materials (7.3) D. Storage (7.4) E. Re-evaluation (7.5)
Master Production and Control Record	**Subpart H – Production and Process Control System: Requirements for the Master Manufacturing Record** § 111.205 – What is the requirement to establish a master manufacturing record? § 111.210 – What must the master manufacturing record include?	**Subpart J – Records and Reports** § 211.186 – Master production and control records.	**VI. DOCUMENTATION AND RECORDS (6)** D. Master Production Instructions (Master Production and Control Records) (6.4)
Batch Production Record	**Subpart I – Production and Process Control System: Requirements for the Batch Production Record** § 111.255 – What is the requirement to establish a batch production record? § 111.260 – What must the batch record include?	**Subpart J – Records and Reports** § 211.188 – Batch production and control records.	**VI. DOCUMENTATION AND RECORDS (6)** E. Batch Production Records (Batch Production and Control Records) (6.5) G. Batch Production Record Review (6.7)

Regulations / guidance	Part 111 – Current good manufacturing practice in manufacturing, packaging, labeling, or holding operations for dietary supplements	Part 210, part 211 – Current good manufacturing practice for finished pharmaceuticals	Guidance for industry Q7A good manufacturing practice guidance for active pharmaceutical ingredients guidance
Laboratory Controls	**Subpart J – Production and Process Control System: Requirements for Laboratory Operations** § 111.303 – What are the requirements under this subpart J for written procedures? § 111.310 – What are the requirements for the laboratory facilities that you use? § 111.315 – What are the requirements for laboratory control processes? § 111.320 – What requirements apply to laboratory methods for testing and examination? § 111.325 – Under this subpart J, what records must you make and keep?	**Subpart I – Laboratory Controls** § 211.160 – General requirements. § 211.165 – Testing and release for distribution. § 211.166 – Stability testing. § 211.167 – Special testing requirements. § 211.170 – Reserve samples. § 211.173 – Laboratory animals. § 211.176 – Penicillin contamination.	**XI. LABORATORY CONTROLS (11)** A. General Controls (11.1) B. Testing of Intermediates and APIs (11.2) C. Validation of Analytical Procedures – See Section 12. (11.3) D. Certificates of Analysis (11.4) E. Stability Monitoring of APIs (11.5) F. Expiry and Retest Dating (11.6) G. Reserve/Retention Samples (11.7)
Validation		**Subpart D – Equipment** § 211.68 – Automatic, mechanical, and electronic equipment. **Subpart E – Control of Components and Drug Product Containers and Closures** § 211.84 – Testing and approval or rejection of components, drug product containers, and closures. **Subpart F – Production and Process Controls** § 211.113 – Control of microbiological contamination. **Subpart I – Laboratory Controls** § 211.165 – Testing and release for distribution.	**XII. VALIDATION (12)** A. Validation Policy (12.1) B. Validation Documentation (12.2) C. Qualification (12.3) D. Approaches to Process Validation (12.4) E. Process Validation Program (12.5) F. Periodic Review of Validated Systems (12.6) G. Cleaning Validation (12.7) H. Validation of Analytical Methods (12.8)
Manufacturing Operations	**Subpart K – Production and Process Control System: Requirements for Manufacturing Operations** § 111.353 – What are the requirements under this subpart K for written procedures? § 111.355 – What are the design requirements for manufacturing operations? § 111.360 – What are the requirements for sanitation? § 111.365 – What precautions must you take to prevent contamination? § 111.370 – What requirements apply to rejected dietary supplements? § 111.375 – Under this subpart K, what records must you make and keep?		
Packaging and Labeling Operations	**Subpart L – Production and Process Control System: Requirements for Packaging and Labeling Operations** § 111.403 – What are the requirements under this subpart L for written procedures? § 111.410 – What requirements apply to packaging and labels? § 111.415 – What requirements apply to filling, assembling, packaging, labeling, and related operations? § 111.420 – What requirements apply to repackaging and relabeling? § 111.425 – What requirements apply to a packaged and labeled dietary supplement that is rejected for distribution? § 111.430 – Under this subpart L, what records must you make and keep?	**Subpart G – Packaging and Labeling Control** § 211.122 – Materials examination and usage criteria. § 211.125 – Labeling issuance. § 211.130 – Packaging and labeling operations. § 211.132 – Tamper-evident packaging requirements for over-the-counter (OTC) human drug products. § 211.134 – Drug product inspection. § 211.137 – Expiration dating.	**IX. PACKAGING AND IDENTIFICATION LABELING OF APIs AND INTERMEDIATES (9)** A. General (9.1) B. Packaging Materials (9.2) C. Label Issuance and Control (9.3) D. Packaging and Labeling Operations (9.4)

Regulations / guidance	Part 111 – Current good manufacturing practice in manufacturing, packaging, labeling, or holding operations for dietary supplements	Part 210, part 211 – Current good manufacturing practice for finished pharmaceuticals	Guidance for industry Q7A good manufacturing practice guidance for active pharmaceutical ingredients guidance
Holding and Distribution	**Subpart M – Holding and Distributing** § 111.453 – What are the requirements under this subpart for M written procedures? § 111.455 – What requirements apply to holding components, dietary supplements, packaging, and labels? § 111.460 – What requirements apply to holding in-process material? § 111.465 – What requirements apply to holding reserve samples of dietary supplements? § 111.470 – What requirements apply to distributing dietary supplements? § 111.475 – Under this subpart M, what records must you make and keep?	**Subpart H – Holding and Distribution** § 211.142 – Warehousing procedures. § 211.150 – Distribution procedures.	**X. STORAGE AND DISTRIBUTION (10)** A. Warehousing Procedures (10.1) B. Distribution Procedures (10.2)
Returned and Salvaged Products	**Subpart N – Returned Dietary Supplements** § 111.503 – What are the requirements under this subpart N for written procedures? § 111.510 – What requirements apply when a returned dietary supplement is received? § 111.515 – When must a returned dietary supplement be destroyed, or otherwise suitably disposed of? § 111.520 – When may a returned dietary supplement be salvaged? § 111.525 – What requirements apply to a returned dietary supplement that quality control personnel approve for reprocessing? § 111.530 – When must an investigation be conducted of your manufacturing processes and other batches? § 111.535 – Under this subpart N, what records must you make and keep?	**Subpart K – Returned and Salvaged Drug Products** § 211.204 – Returned drug products. § 211.208 – Drug product salvaging.	**XIV. REJECTION AND RE-USE OF MATERIALS (14)** A. Rejection (14.1) B. Reprocessing (14.2) C. Reworking (14.3) D. Recovery of Materials and Solvents (14.4) E. Returns (14.5)
Product Complaints	**Subpart O – Product Complaints** § 111.553 – What are the requirements under this subpart O for written procedures? § 111.560 – What requirements apply to the review and investigation of a product complaint? § 111.570 – Under this subpart O, what records must you make and keep?	**Subpart J – Records and Reports** § 211.198 – Complaint files.	**XV. COMPLAINTS AND RECALLS (15)**
Records, Reports, Complaint Files	**Subpart P – Records and Recordkeeping** § 111.605 – What requirements apply to the records that you make and keep? § 111.610 – What records must be made available to FDA?	**Subpart J – Records and Reports** § 211.180 – General requirements. § 211.182 – Equipment cleaning and use log. § 211.184 – Component, drug product container, closure, and labeling records. § 211.192 – Production record review. § 211.194 – Laboratory records. § 211.196 – Distribution records.	**VI. DOCUMENTATION AND RECORDS (6)** A. Documentation System and Specifications (6.1) B. Equipment Cleaning and Use Record (6.2) C. Records of Raw Materials, Intermediates, API Labeling and Packaging Materials (6.3) F. Laboratory Control Records (6.6)

Regulations / guidance	Part 111 – Current good manufacturing practice in manufacturing, packaging, labeling, or holding operations for dietary supplements	Part 210, part 211 – Current good manufacturing practice for finished pharmaceuticals	Guidance for industry Q7A good manufacturing practice guidance for active pharmaceutical ingredients guidance
Controls Over Third Parties			**XVI. CONTRACT MANUFACTURERS (INCLUDING LABORATORIES) (16)** **XVII. AGENTS, BROKERS, TRADERS, DISTRIBUTORS, REPACKERS, AND RELABELLERS (17)** A. Applicability (17.1) B. Traceability of Distributed APIs and Intermediates (17.2) C. Quality Management (17.3) D. Repackaging, Relabeling, and Holding of APIs and Intermediates (17.4) E. Stability (17.5) F. Transfer of Information (17.6) G. Handling of Complaints and Recalls (17.7) H. Handling of Returns (17.8)
Production And Process Controls: Cell Culture / Fermentation			**XVIII. SPECIFIC GUIDANCE FOR APIs MANUFACTURED BY CELL CULTURE/FERMENTATION (18)** A. General (18.1) B. Cell Bank Maintenance and Record Keeping (18.2) C. Cell Culture/Fermentation (18.3) D. Harvesting, Isolation and Purification (18.4) E. Viral Removal/Inactivation steps (18.5)
Clinical Trials TRIALS (19)			**XIX. APIs FOR USE IN CLINICAL** A. General (19.1) B. Quality (19.2) C. Equipment and Facilities (19.3) D. Control of Raw Materials (19.4) E. Production (19.5) F. Validation (19.6) G. Changes (19.7) H. Laboratory Controls (19.8) I. Documentation (19.9)

The general similarities among these regulatory schemes allows for a consolidated review of cGMP cites by major category. The table below provides the total number of cites referencing each major category.

Warning Letter cGMP Cites by Category

cGMP Area	Number of Cites
General Provisions	2
Organization and Personnel[24]	36
Buildings and Facilities / Physical Plant and Grounds[25]	18
Equipment and Utensils[26]	26
Control of Components, Drug Product Containers and Closures, Packaging, and Labels[27]	20
Production and Process Controls[28]	79
Packaging and Labeling Control	6
Holding and Distribution	2
Laboratory Controls / Operations[29]	57
Records and Recordkeeping (not including master / batch production record and product complaint cites)[30]	47
Master Production and Control Records / Master Manufacturing Record[31]	13
Batch Production and Control Records / Batch Production Record	12
Product Complaints[32]	9
Records and Recordkeeping (combined)	81

The results in the preceding table suggest that the FDA tends to focus on areas that have a direct impact on product quality, such as Production and Process Controls and Laboratory Controls. This finding is consistent with FDA's risk-based enforcement strategy that generally targets manufacturers of products that are adulterated based on their failure to meet specifications, as opposed to more general cGMP failures. Recordkeeping is another area of focus, with particular attention paid to batch and master production records.

OTC Manufacturing

Of the twenty (20) Warning Letters related to OTC products, thirteen (13) involved violations of the cGMP requirements under 21 C.F.R. parts 210 and 211. A breakdown of the violations noted in these Warning Letters is provided in the following table.

OTC cGMP Cites[33]

Category	cGMP Cite	Subtotal	Count
		11	
Responsibilities of quality control unit.	211.22	8	–
	211.22(a)	–	2
	211.22(c)	–	3
	211.22(d)	–	3
Personnel qualifications.	211.25	3	–
	211.25(a)		3
		1	
Design and construction features.	211.42	1	–
	211.42(c)(1)	–	1
		11	
Equipment design, size, and location.	211.63	1	1
Equipment construction.	211.65	1	–
	211.65(a)	–	1
Equipment cleaning and maintenance.	211.67	4	–
	211.67(a)	–	1
	211.67(b)	–	3
Automatic, mechanical, and . electronic equipment	211.68	5	1
	211.68(a)	–	1
	211.68(b)	–	3
		8	
General requirements.	211.80	(+ 1)	–
Testing and approval or rejection of components, drug product containers, and closures.	211.84	7	–
	211.84(a)	–	1
	211.84(b) and 211.80(a)	–	1
	211.84(d)(2)	–	3
	211.84(e)	–	2
Retesting of approved components, drug product containers, and closures.	211.87	1	1
		13	
Written procedures; deviations.	211.100	9	1
	211.100(a)	–	7
	211.100(b)	–	1

Category	cGMP Cite	Subtotal	Count
Sampling and testing of in–process.	211.110	2	–
	211.110(a) & (c), 211.160(b)	–	1
	211.110(b)	–	1
Reprocessing.	211.115	2	–
	211.115(a)	–	1
	211.115(a) and (b)	–	1
Packaging and labeling operations.	211.130	1	–
	211.130(e)	–	1
Expiration dating.	211.137	2	1
	211.137 & 211.166	–	1
Warehousing procedures.	211.142	1	–
	211.142(b)	–	1
General requirements.	211.160	4 (+ 2)	–
	211.160(a)	–	1
	211.160(b)	–	3
Testing and release for distribution.	211.165	4	1
	211.165(a)	–	1
	211.165(e)	–	1
	211.165(f)	–	1
Stability testing.	211.166	6 (+ 1)	–
	211.166(a)	–	4
	211.166(a)(2)	–	1
	211.166(a)(3) and 211.160(b)	–	1
Reserve samples.	211.170	1	–
	211.170(b)	–	1
General requirements.	211.180	3	–
	211.180(a)	–	1
	211.180(e)	–	1
	211.180(e)(2)	–	1
Production record review.	211.192	11	11
Laboratory records.	211.194	2	–
	211.194(a)	–	2
Master production and control records.	211.186	1	–
	211.186(b)(4)	–	1

Category	cGMP Cite	Subtotal	Count
Batch production and control records.	211.188	3	1
	211.188(a)	–	1
	211.188(a) and (b)(4)	–	1
Complaint files.	211.198	4	2
	211.198(a)	–	1
	211.198(a) & (b)	–	1

The most frequently cited issues for OTC manufacturing related to recordkeeping, which is generally covered by subpart J (a total of 24 references). These references most regularly related to the review of production records, which is required by 21 C.F.R. 211.192. Eleven (11) out of thirteen (13) OTC manufacturing Warning Letters included this reference[34], and three (3) were a repeat violation from a previous inspection. Of the eleven (11) Warning Letters, nine (9) included discussion of a failure in the pre-release review of drug products, and four (4) involved a failure in a post-release review of a failure to meet specifications. Seven (7) of the eleven (11) references involved the failure to extend reviews of specification deviations to associated batches, and two (2) involved a failure to adequately investigate a failure during a validation run.

One of the more significant manufacturing problems to arise this past year involved the contamination of OTC drug products with 2, 4, 6 Tribromoanisole (TBA) at the Puerto Rico manufacturing facility of McNeil Healthcare LLC. The TBA contamination was suggested by the FDA to have occurred through the exposure of OTC drug products and bottles to 2, 4, 6 Tribromophenol (TBP), a pesticide and flame retardant used to treat wooden palates. TBA is a known degradant of TBP.

The key concern raised by the FDA in its Warning Letter involved the inability of the manufacturer to conduct a timely, comprehensive investigation. Despite having first encountered the issue in 2008 with complaints of a musty, mildew-type odor, the initial investigation was terminated by the manufacturer after microbiological contamination was ruled out as the source of the odor. The FDA noted that TBA contamination is detectible at parts per trillion using organoleptic means, clearly indicating that the manufacturer's initial investigation was not broad enough to address all possible sources of the odor. After multiple recalls were initiated and the source of the contamination had been traced to the palates, the manufacturer continued to limit its investigation of the problem by not extending its investigation to other drug products that were at risk for TBA contamination.

Along with the failure to adequately investigate the problems with the manufacturer's products, the FDA noted problems with the company's communication to the agency. For example, the FDA commented that, while the company was aware of the source of the contamination as early as September 2009, the company did not share its results with the FDA until after the inspection was initiated and several requests were made. The FDA also cited the company for the failure to timely issue an NDA-Field Alert Report (FAR) to the FDA. While complaints were received in 2008, the company did not submit a FAR until a year later in 2009.

The lack of communication with the FDA earned a strong rebuke of the company in the Warning Letter. The FDA cautioned that it was "concerned about the response of Johnson & Johnson" to the matter. By not taking appropriate actions to resolve the issues identified by the FDA, the FDA suggested that corporate management had failed to accept its responsibility to ensure the quality, safety, and integrity of its products by assuring the timely investigation and resolution of the issues.

The issue of management's role in the cGMP process was a common theme in OTC Warning Letters in 2010. The FDA even went so far as to raise the specter of a Corporate Warning Letter[35] in the Warning Letter issued to Shamrock Medical Solutions Group LLC. In the Shamrock Medical Solutions Group Warning Letter, the FDA stated that:

> It appears that you have not taken a global quality systems approach to corrective actions at your firm. Several of the CGMP deficiencies observed during the inspection of your Massachusetts facility were also documented during our inspection of your Ohio facility in 2007. . . . The Massachusetts and Ohio facilities share a single Corporate Quality Committee for decision making. It is your responsibility to review your operations at all facilities and apply appropriate corrective actions to address deficiencies present at multiple sites; i.e. the deficiencies cited by FDA in the inspection of your Ohio facility should have been noted and corrected at your Massachusetts facility.

It should be noted that the Shamrock Medical Solutions Group Warning Letter was issued by the New England District Office, which was also responsible for issuing the first Corporate Warning Letter to Boston Scientific Corporation in 2006. While previous Corporate Warning Letters have been issued to medical device manufacturers for Quality System regulation violations, the above captioned language raises the possibility that a Corporate Warning Letter may be issued to a drug manufacturer in the near future.

The role of management in the cGMP process was also raised in the April 29, 2010 Warning Letter issued to L. Perrigo Company. In the Warning Letter, the FDA cited the manufacturer for problems related to drug mix-ups, where tablets of one type found their way into finished product packaging for tablets of another type. The FDA noted that while the manufacturer had an ongoing program to address the issue that was started in 2005, the problem continued (as evidenced by product complaints and the FDA investigator's observations). After requesting additional details on how the manufacturer intended to implement, support, and sustain a comprehensive quality system, the FDA noted its expectation that corporate management would undertake a comprehensive evaluation of manufacturing operations to ensure compliance with cGMP.

Industry should anticipate a growing interest by the agency in the role that management, particularly executive management, plays in the cGMP process. The focus on the role of management in the quality process is consistent with the FDA's promise to increase its use of the *Park* Doctrine. The *Park* Doctrine stands for the proposition that senior management can be held responsible for Federal Food, Drug and Cosmetic Act violations committed by employees.

2.4.2.2   Active Pharmaceutical Ingredient (API) Manufacturing

The FDA issued eleven (11) Warning Letters to API manufacturers in 2010. As with letters issued in previous years, the FDA borrowed language from the International Conference on Harmonisation of Technical Requirements for Registration of Pharmaceuticals for Human Use (ICH) harmonised guideline Q7, "Good Manufacturing Practice Guide For Active Pharmaceutical Ingredients." The references to the ICH guideline in these Warning Letters are never attributed to their source. Even though the FDA formally adopted the ICH guideline as guidance, the ICH guideline / FDA guidance is not intended to carry the same weight as FDA's laws or regulations.

The most common violations listed in these Warning Letters relate to issues of quality management, documentation and records, and laboratory controls. Another common area cited in these Warning Letters is validation, which is generally split between validation of laboratory procedures and validation of production processes. A complete breakdown of the cites in these Warning Letters is provided in the following table.

Category	ICH Cite	Subtotal	Count
2. QUALITY MANAGEMENT			10
	ICH Q7A 2.1 Principles		4
	ICH Q7A 2.2 Responsibilities of the Quality Unit(s)	4	2
	ICH Q7A 2.2 Responsibilities of the Quality Unit(s) [compliance with cGMPs and specifications]		1
	ICH Q7A 2.2 Responsibilities of the Quality Unit(s) [materials tested and results reported]		1
	ICH Q7A 2.3 Responsibility for Production Activities		2
3. PERSONNEL		1	
	ICH Q7A 3.2 Personnel Hygiene		1
4. BUILDINGS AND FACILITIES		4	
	ICH Q7A 4.1 Design and Construction		3
	ICH Q7A 4.3 Water		1
5. PROCESS EQUIPMENT		2	
	ICH Q7A 5.3 Calibration and 12.3 Qualification		1
	ICH Q7A 5.3 Calibration, 12.3 Qualification and 12.8 Validation of Analytical Methods		1
6. DOCUMENTATION AND RECORDS		8	
	ICH Q7A 6.1 Documentation System and Specifications		2
	ICH Q7A 6.2 Equipment Cleaning and Use Record		2
	ICH Q7A 6.4 Master Production Instructions (Master Production and Control Records)		1
	ICH Q7A 6.6 Laboratory Control Records	3	1
	ICH Q7A 6.6 Laboratory Control Records [complete data]		1
	ICH Q7A 6.6 Laboratory Control Records [raw data]		1
11. LABORATORY CONTROLS		8	
	ICH Q7A 11.1 General Controls	4	2
	ICH Q7A 11.1 General Controls [ensure sampling plans and test procedures scientifically sound]		1
	ICH Q7A 11.1 General Controls [out-of-specification]		1
	ICH Q7A 11.2 Testing of Intermediates and APIs		1
	ICH Q7A 11.5 Stability Monitoring of APIs		3

Category	ICH Cite	Subtotal	Count
12. VALIDATION		5	
	ICH Q7A 12.1 Validation Policy		1
	ICH Q7A 12.3 Qualification	(+ 2)	1
	ICH Q7A 12.7 Cleaning Validation		1
	ICH Q7A 12.8 Validation of Analytical Methods	(+ 1)	2
14. REJECTION AND RE-USE OF MATERIALS		2	
	ICH Q7A 14.1 Rejection		1
	ICH Q7A 14.2 Reprocessing		1
15. COMPLAINTS AND RECALLS		1	1

Out of the twelve (12) Warning Letters issued to API manufacturers, only two (2) were located in the United States or its territories (California and Puerto Rico).

2.4.2.3 Dietary Supplement Manufacturing

There were nine (9) Warning Letters issued to dietary supplement manufacturers for cGMP violations in 2010. Of these, eight (8) cited the dietary supplement cGMP regulations of 21 C.F.R. part 111, and one (1) cited the finished drug cGMP regulations of 21 C.F.R. parts 210 and 211. In the Warning Letter citing 21 C.F.R. parts 210 and 211, the FDA detected the APIs sildenafil and tadalafil in the manufacturer's products. Although the products in question were ostensibly marketed as dietary supplements, the presence of the APIs rendered those products drugs under section 201(ff)(3)(B) of the Act, requiring compliance with 21 C.F.R. parts 210 and 211.

It should be noted that the FDA does not consistently apply the finished drug cGMP regulations to dietary supplement manufacturers found producing products containing APIs or analogues of APIs. In a Warning Letter to another manufacturer where products were found to contain sulfoaildenafil, an analogue of sildenafil, the FDA cited the manufacturer for marketing unapproved new drugs, but chose to apply cGMP violations under 21 C.F.R. part 111.

The general categories of dietary supplement cGMP violations that were cited in the Warning Letters issued in 2010 are provided in the following table.

Category (21 C.F.R. Part 111 Subpart)[36]	Total
Production and Process Control System (Subpart E)	13
Batch Production Record (Subpart I)	5
Master Manufacturing Record (Subpart H)	4
Equipment and Utensils (Subpart D)	3
Components, Packaging, and Labels (Subpart G)	2
Holding and Distributing (Subpart M)	1
Laboratory Operations (Subpart J)	1
Manufacturing Operations (Subpart K)	1
Personnel (Subpart B)	1
Product Complaints (Subpart O)	1
General Provisions (Subpart A)	1

The majority of the violations cited the regulations found in Subpart E of 21 C.F.R. Part 111, which concern the requirement to establish a production and process control system. Of note, no violations of Subpart F, which deal with the requirements for quality control, were cited in the Warning Letters issued in 2010.

The lack of any cite to 21 C.F.R. Part 111, Subpart F appears to be inconsistent with its treatment in FDA's enforcement policy regarding the dietary supplement cGMPs. The Compliance Program Guidance Manual ("CPGM") Program 7321.008, "Dietary Supplements – Import And Domestic," lists questions that investigators are directed to include in their Establishment Inspection Reports to provide verification that the underlying issues were adequately covered by the investigator during the inspection[37]. More questions are identified for subpart F in the CPGM than for any other subpart of part 111[38].

One example of this lack of focus on subpart F is a Warning Letter to a dietary supplement manufacturer citing violations under 21 C.F.R. part 111. In the Warning Letter, the FDA cited the manufacturer for violations of subparts D, H and I, but only commented on an issue related to subpart F. FDA's reasoning for only commenting on subpart F is not explained in the Warning Letter. Based on the discussion in the Warning Letter, the issue related to subpart F had a direct impact on product quality, resulting in the use of expired dietary supplement ingredients in finished products.

The basis for the lack of subpart F violations can only be speculated. One possible reason is a general lack of significant experience among FDA's inspections and enforcement staff on implementing a quality system based cGMP for dietary supplement manufacturers. In implementing a quality system inspection program for medical devices, the FDA developed the "Quality System Inspection Technique," which provided an inspectional process designed to identify quality system deficiencies. No comparable systems-based approach appears to have been developed for dietary supplements.

Another possible reason is FDA's tendency to cite violations that are the most directly linked to product adulteration and misbranding. Subpart F relates more to responsibility than actual functions that must be performed. Finally, it is possible that the FDA is selectively choosing to enforce process requirements over systems-based requirements as industry becomes more familiar with the dietary supplement cGMP regulations. The dietary supplement cGMP regulations only became fully enforced during 2010.

It is anticipated that Subpart F will be cited more frequently as FDA and industry become more accustomed to the dietary supplement cGMP regulations.

2.4.3   Lack of Drug Approval

Another common basis for a drug Warning Letter is the lack of approval for a marketed product. These Warning Letters fall into three general (3) categories:

- Products that are not marketed as drugs (such as cosmetics, dietary supplements, electronic cigarettes or food) that use drug-based claims in labeling or contain APIs;

- Products marketed as nonprescription drugs or homeopathic drugs that are found not to comply with a drug monograph; and

- Products marketed as prescription drugs based on Compliance Policy Guide Sec. 440.100, "Marketed New Drugs Without Approved NDAs or ANDAs."

The following table provides a count of the number of Warning Letters that involve the more common categories of products.

Type of Unapproved Drug	Warning Letters
Dietary Supplement	43
Unapproved Rx Drug	26
OTC	14
Contains Undeclared API	9
Lipodissolve Product	7
Homeopathic	6
Electronic Cigarette	5
Food	5

Warning Letters were also issued to a pharmacy performing drug compounding and an Internet pharmacy.

2.4.3.1 Nonprescription Drugs Lacking Approval

As noted above, there were fourteen (14) Warning Letters that cited OTC drug products as lacking approval. While many instances involved the failure to comply with an existing monograph, the failure to adhere to a monograph does not automatically result in a finding that the product is unapproved. For example, one Warning Letter cited a manufacturer for misbranding an OTC external analgesic product because the manufacturer's website made a claim that the product could be used to treat "herpes simplex infections." The FDA determined that the use of the word "herpes" is too broad and that it may mislead consumers into thinking the product could treat genital herpes.

The FDA also issued three (3) Warning Letters to companies marketing mouthwash that made antiplaque / antigingivitis claims, but only contained sodium fluoride as an active ingredient. The three companies receiving Warning Letters were a brand name company (Johnson & Johnson), and two private label companies (CVS and Walgreens) that were marketing a private label version of the brand product. While FDA has, in the past, sometimes limited its warnings to the brand-name company, the FDA may continue to send warnings to major private label companies, as well.

Another category of products targeted by the FDA were OTC hormones. While OTC hormone products are permitted under the external analgesic tentative final monograph ("TFM"), the products cited in FDA's Warning Letters exceeded the allowable claims for those products.

2.4.3.2   Dietary Supplements and Internet Labeling

The data in the preceding table demonstrates that products marketed as dietary supplements are the largest category of unapproved products. Most of these companies have an Internet presence that was identified by the FDA in its Warning Letters. Of the forty-three (43) Warning Letters issued to companies marketing a dietary supplement that was found to make drug claims or contain a prescription API, thirty-nine (39) made reference to the company's website. Some claims were found on other websites that were related with the company's website by means of a click-through link. In one instance, cited claims were found on a website's blog. Among the three (3) Warning Letters issued jointly by the Federal Trade Commission (FTC) and FDA, all three involved claims made on Internet websites.

The FDA is clearly becoming very Internet-savvy in its enforcement, as there were eleven (11) Warning Letters that made reference to the use of metadata as a basis for a drug-like marketing claim. In one Warning Letter, the FDA's basis for citing the company for unapproved drug claims was almost solely based on the metatags that remained on the website after the direct claim had been removed. Companies will need to be careful about the tools that are used to direct consumers to their website, such as metatags, as the FDA is increasingly viewing those tools as part of a company's labeling for a product.

2.5   Conclusion

The manufacture and sale of pharmaceutical products is pervasively regulated by FDA, and FDA's regulatory scheme includes both pre-market and post-market controls. FDA has a number of enforcement powers to ensure compliance with its regulations. FDA's enforcement options differ depending on whether products are imported into the US or produced domestically.

Reviewing recent enforcement actions, the majority of Warning Letters issued by FDA regard the lack of drug approval, including twenty-six (26) Warning Letters involving unapproved prescription drugs. The second most common issue involves current Good Manufacturing Practice violations, which are more often related to production and process control issues and inadequate documentation.

Regulated entities should conduct a thorough review of the approval status of marketed drug products, and make necessary preparations for a continued crackdown by FDA on unapproved drug products. Regulated entities that operate multiple manufacturing sites should also consider developing company-wide policies and procedures to ensure regulatory consistency between different facilities. Some Warning Letters issued by FDA suggest that FDA may begin issuing "corporate Warning Letters" to drug manufacturers that have regulatory problems at multiple manufacturing sites.

References

1   "If it appears from the examination of such samples or otherwise that (1) such article has been manufactured, processed, or packed under insanitary conditions or, in the case of a device, the methods used in, or the facilities or controls used for the manufacture, packing, storage, or installation of the device do not conform to the requirements of § 520(f) [21 U.S.C. § 360j(f)], or (2) such article is forbidden or restricted in sale in the country in which it was produced or from which it was exported, or (3) such article is adulterated, misbranded, or in violation of § 505 [21 U.S.C. § 355], then such article shall be refused admission." 21 U.S.C. § 381(a).

2   Sugarman v. Forbragd, 267 F. Supp. 817, 823 (N.D. Cal. 1967), affirmed, 405 F.2d 1189 (1968), cert. denied, 395 U.S. 960 (1969).

3   *Id.*

4   *Id.* at 824.

5   21 U.S.C. § 355; FDA, Guide to Inspections of Foreign Pharmaceutical Manufacturers, available at http://www.fda.gov/ICECI/Inspections/InspectionGuides/ucm075021.htm; FDA, Div. of Field Investigations, Guide to International Inspections and Travel, § 301 (November 2002), available at http://www.fda.gov/ICECI/Inspections/ForeignInspections/default.htm.

6   FDA, Guide to Inspections of Foreign Pharmaceutical Manufacturers, available at http://www.fda.gov/ICECI/Inspections/InspectionGuides/ucm075021.htm.

7   See id.

8   FDA, Regulatory Procedures Manual: Chapter 9 Import Operations/Actions, Automatic Detention (January 2008), available at http://www.fda.gov/downloads/ICECI/ComplianceManuals/RegulatoryProceduresManual/UCM074300.pdf.

9   Although FDA may exert considerable pressure on a company to conduct a recall, including threats of further regulatory sanctions, such recalls are still considered "voluntary." FDA mandated recalls require FDA to follow a much more formal process to initiate, and are generally reserved for situations where the company completely refuses to cooperate or has gone out of business.

10 The degree to which ORA seeks input from the appropriate center depends upon whether ORA has "direct reference authority" over the issue. Where ORA has direct reference authority, it may issue a Warning Letter without the review and clearance by the appropriate center. Currently, ORA does not have direct reference authority over any CDER-related issue.

11 When multiple compliance issues arise from the same inspection that would primarily be handed by both CDER/OC and ORA, ORA will coordinate the enforcement action.

12 The increase in recalls likely can also be traced to the manufacturing issues noted by McNeil Healthcare LLC.

13 A fourth type of finding, RTC, or "Referred to Center," which is not represented in FDA's Inspection Classification Database, involves situations that require CDER's input to determine whether a violation has occurred.

14 Note that these totals only count CDER inspections, and do not include inspections conducted by the Center for Veterinary Medicine or the Center for Biologics Evaluation & Research.

15 This total does not include Warning Letters and untitled letters issued by the Division of Drug Marketing, Advertising, and Communications (DDMAC), which are not part of this report.

16 Note that each Warning Letter regarding an unapproved drug product typically includes a companion cite regarding the failure of labeling to bear adequate directions for use. These cites were not included in the total for labeling violations to better represent the true number of labeling problems identified by the agency.

17 For example, these numbers include conventional foods and dietary supplements identified by FDA to have been marketed as new drugs, as well as compliance with cGMP regulations in the manufacturing of dietary supplements.

18 If the count was based on the agency's finding, and not the presumed intent of the manufacturer or distributor, many of these products would be categorised as new prescription drugs. For example, dietary supplements that bear marketing language that extends beyond permitted structure/function claims into disease claims are generally determined by the FDA to be new prescription drugs lacking approval. The agency rarely addresses such situations as labeling issues.

19 This total excludes Warning Letters issued regarding dietary supplements and other foods that are classified as drugs by the FDA based on their claims.

20 In some cases, the intent to market a product as a dietary supplement was inferred from the facts presented in the Warning Letter. For example, a reference to "100% natural."

21 The FDA noted in the preamble to the final rule regarding the dietary supplement cGMP requirements that "[w]e . . . decline to define 'validation' . . . because the final rule does not establish any requirements that use [this term]." FDA, Final Rule, Current Good Manufacturing Practice in Manufacturing, Packaging, Labeling, or Holding Operations for Dietary Supplements 72 Fed. Reg. 34,752, 34,805 (Jun. 25, 2007).

22 Regulations for finished drug products require validation of supplier's test results [21 C.F.R. 211.84(d)(2) and (3)], all aseptic and sterilization processes [21 C.F.R. 211.113(b)], computerised systems where certain data (such as calculations performed in connection with laboratory analysis) are eliminated during the computerization process [21 C.F.R. 211.68(b)], and test methods that determine satisfactory conformance to final specifications [21 C.F.R. 211.165(e)].

23 Quality control personnel are only required to periodically review all records for calibration of instruments and controls, as well as records of calibrations, inspections, and checks of automated, mechanical, and electronic equipment [21 C.F.R. 111.30(c), 111.117(b) and (c)], and periodically conduct qualification of suppliers when relying on a certificate of analysis [21 C.F.R. 111.75(a)(2)(ii)(D)].

24 Includes cites based on sections 2, "Quality Management," and 3, "Personnel" of ICH Q7A.

25 Includes cites based on section 4, "Buildings and Facilities" of ICH Q7A.

26 Includes cites based on section 5, "Process Equipment" of ICH Q7A.

27 Includes cites based on section 14, "Rejection and Re-Use of Materials" of ICH Q7A.

28 Includes cites to 21 C.F.R. 111 subpart K, "Production and Process Control System: Requirements for Manufacturing Operations," as well as cites based on section 12, "Validation" of ICH Q7A, other than section 12.8, "Validation of Analytical Methods".

29 Includes cites based on section 11, "Laboratory Controls" and section 12.8, "Validation of Analytical Methods" of ICH Q7A.

30 Includes cites based on section 6, "Documentation and Records" of ICH Q7A, other than section 6.4, "Master Production Instructions (Master Production and Control Records)."

31 Includes cites based on section 6.4, "Master Production Instructions (Master Production and Control Records)" of ICH Q7A.

32 Includes cites based on section 15, "Complaints and Recalls" of ICH Q7A.

33 Tables following this format list the number of times a specific regulation was referenced as a violation. These lists exclude references to regulations in the discussion of the violation, such as references providing support for examples or identifying related regulations. Subtotals are provided in the first column and actual counts of violations are provided in the second column. Where a subtotal includes a number in parentheses starting with a plus sign (+ 1), the number represents instances where the regulation was listed as a secondary cite for a violation.

34 This differs from some other regulations that are referenced more than once in the same Warning Letter for violations of different subsections.

35 A Corporate Warning Letter is a type of Warning Letter issued to the parent corporation of multiple facilities with similar violations.

36 Where violations were cited to 21 C.F.R. Part 211, the comparable subpart in 21 C.F.R. Part 111 (using the table "Side-by-Side Comparison of 21 C.F.R. Parts 111 and 210/211 and ICH Q7A" provided above) was counted in generating the totals.

37 FDA, Dietary Supplements - Import and Domestic, CPGM 7321.008 (Jul. 8, 2010).

38 Subpart F lists twenty-one (21) required questions, while the average number of questions for other subparts is three (3). FDA, Dietary Supplements - Import and Domestic, CPGM 7321.008 (Jul. 8, 2010).

CHAPTER 3

System-based approach to inspections

TIM SANDLE, BIO PRODUCTS LABORATORY, UK
DAVID BARR, DIRECTOR QA COMPLIANCE, ALKERMES, USA

3.1   Introduction

The manufacture of drugs and medical devices is overseen by regulatory agencies appropriate to the country in which the pharmaceutical product is manufactured or to the territory in which the product is distributed (which requires registration by the manufacturing organisation with the regulator). Pharmaceutical manufacturers are required to follow Good Manufacturing Practice (GMP) which embraces each manufacturing step, from purchasing raw materials to the finished product[1]. In order to assess if manufacturing facilities are conforming to GMP requirements, regulatory authorities undertake GMP inspections of the manufacturers. In recent years there has been a change in approach by regulators from product specific inspections to system-based inspections.

System-based inspections reflects the current approach to the inspection of pharmaceutical facilities adopted by the major regulatory agencies. The method was first established by the US Food and Drug Administration (FDA) and later adopted, through the activities of the International Conference on Harmonisation (ICH), by the European Medicines Agency. Given the origination of the approach by the FDA and the wider availability of material under the US Freedom of Information Act, the examples within this chapter generally focus on the FDA inspection approach although most of the concepts are applicable to other inspectorate agencies.

The concept of a systematic approach was first set out in a guide for FDA inspectors issued in 2002: "Compliance Program Guidance Manual for FDA Staff: Drug Manufacturing Inspections", Program 7356.002"[2], which outlined the procedures to be followed by FDA inspectors. The concept was then expanded upon in the FDA documents "Pharmaceutical cGMPs for the 21st Century – A Risk-Based Approach"[3] and "Quality Systems Approach to Pharmaceutical CGMP Regulations"[4]. What these documents did was to change the way in which inspections were conducted, where an inspector would wander through the plant noting any non-conformances or would focus on a product stream. The change was towards the examination of *systems*. The idea was that an examination of representative components within a system would be indicative of the manufacturer's overall conformance to all aspects of that system. This approach merged the established science-based risk management process with an integrated quality systems approach.

Using a risk management approach, the FDA could match the level of effort required for an inspection against the magnitude of risk. However, due to resource limitations, the FDA was unable to undertake uniformly intensive coverage of all pharmaceutical products and production. For example, the risks associated with the manufacture of a sterile injectable product are greater than the risks associated with the manufacture of a solid dosage form.

In terms of a science-based approach, this reflected significant advances in the pharmaceutical sciences and in manufacturing technologies which have occurred since the late 1990s. This has led to inspectors needing to understand and evaluate the nature of the science in order to ensure that product quality is maintained and also to update regulatory guidance to reflect the scientific changes. The adoption of scientific innovations can also contribute significantly to assessment of process risk. This approach involves identifying risks and then assessing each risk for its severity and likelihood of impact. Once evaluated, attempts are made to mitigate the risk and when a risk cannot be adequately resolved, mechanisms should be put in place for the detection of the risk. Risk management also provides a mechanism for risks to be prioritised as part of the establishment of a risk management plan[5].

A further driver for change was reflected in a number of global changes on the regulatory and pharmaceutical manufacturing landscapes common to all pharmaceutical manufacturers. These included:

- Decreased frequency of FDA manufacturing inspections as a result of fewer resources available.

- FDA's accumulation of experience with, and lessons learned from various approaches to the regulation of product quality.

- Advances in pharmaceutical sciences and manufacturing technologies.

- Application of biotechnology in drug discovery and manufacturing.

- Advances in the science and management of quality.

- Globalization of the pharmaceutical industry

To undertake its regulatory requirements, the FDA developed two basic strategies:

- Evaluating through facility inspections, including the collection and analysis of associated samples, the conditions and practices under which drugs and drug products are manufactured, packed, tested and held.

- Monitoring the quality of drugs and drug products through surveillance activities such as sampling and analysing products in distribution.

What a systems based inspection means is simply that the inspection focuses on six key systems (**Figure 1**). These systems are[6]:

- Quality

- Facilities and Equipment

- Materials

- Production

- Packaging and Labelling

- Laboratory

The FDA introduced its system-based initiative to ensure more efficient use of resources and enable more focused inspections.

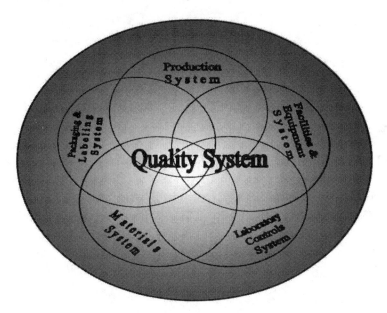

Figure 1: Six systems as outlined in the Food and Drug Administration's (FDA's) Drug Manufacturing Inspections compliance program[2]

The examination of systems also enables the FDA to undertake a full inspection whereby each system is thoroughly inspected, or an abbreviated inspection where two or more systems are inspected. This distinction, based on risk, is discussed below. The approach also means that when a firm responds to an inspection finding it is required to examine all aspects of the system which have been found to be deficient rather than just focusing on the finding which has been reported.

As captured by the list of systems above, the objective of system-based inspections is to focus on key operating systems. In this way, a regulatory agency can conduct a comprehensive review of all aspects of the production and distribution of drugs and drug products to assure that such products meet the requirements of the national or international legislation (with the FDA this is the Food, Drugs and Cosmetics Act).

This chapter examines the key elements of system-based inspections and focuses on the primary systems. Before undertaking a detailed analysis, the chapter presents an overview of FDA inspections and an overview of the systems approach in general. In undertaking the review of each system, which forms the core part of the chapter, the most important elements of each system are illustrated through a review of warning letters. In doing so, example observations have been selected and segregated according to the applicable GMP requirements, making them easier to review in a systematic fashion.

Whilst the warning letters selected were contemporaneous to the time when the chapter was written they are, in a sense, timeless since they outline a number of common failings and illustrate how these failings can build a case whereby the inspector extends agency concerns beyond the specific area being examined and across the entire system.

3.2   FDA inspections

The Food and Drug Administration (FDA) is an agency of the United States Department of Health and Human Services, one of the United States federal executive departments[7]. The FDA is responsible for protecting and promoting public health through the regulation and supervision of food safety, tobacco products, dietary supplements, prescription and over-the-counter pharmaceutical drugs (medications), vaccines, biopharmaceuticals, blood transfusions, medical devices, electromagnetic radiation emitting devices, veterinary products, and cosmetics[8]. The focus of this chapter, and indeed much of this book, is upon the FDA's regulatory role with respect to the regulation of pharmaceutical drugs.

Prior to a detailed overview of the system-based inspections, it is important to consider what FDA GMP inspections are for. With GMP inspections, inspectors assess whether the manufacturer is in compliance with GMP, ensuring that all manufacturing operations are performed in accordance with the relevant marketing authorisation and published guidelines.

For inspections, the FDA's Investigations Operations Manual (IOM) is the primary source of guidance regarding Agency policy and procedures for field investigators and inspectors[9]. This extends to all individuals who perform field investigational activities in support of the Agency's public mission. Accordingly, it directs the conduct of all fundamental field investigational activities.

The purpose of an inspection may vary somewhat in the details; however, all inspections are designed to:

- Determine if violations of law within FDA's jurisdiction are occurring, and if so,

- Obtain voluntary correction by the inspected entity, or

- Develop the necessary evidence to support FDA enforcement action if voluntary correction is not promptly forthcoming or is ineffective.

The laws enforced by the FDA in regulating the industry also establish the scope, certain limitations, and procedures FDA must follow.

At the end of each inspection of a manufacturer, the inspector conducts a closing meeting at which the deficiencies or failures to comply with GMP are presented formally to the representatives of the company (normally the technical management, including the key personnel and preferably some or all of the senior management) and may be discussed. The final meeting is a significant part of the inspection. At the end of the process the regulatory agency, in the case of the FDA, may take action. Such actions may include the issuing of a warning letter.

After inspecting a food, drug, medical device, or biologic establishment, the FDA prepares a written report of its inspection findings, following a debriefing. This report is intended primarily for internal FDA use and is not provided to the inspected institution at the conclusion of the on-site visit. To provide the facility with its own written list of discrepancies noted during the inspection, FDA developed form FDA 483, "Notice of Inspectional Observations," issued by the field investigator. Form 483 should contain only those observations that can be directly linked to a violation of regulations not suggestions, guidance, or other comments. Although the 483 does not contain references to the regulations, each observation should be directly traceable to a section of the applicable regulations. After receiving a form 483, the recipient should respond to the FDA notifying them about each item and agreement and should also mention a timeline for correction. Regardless of the number of observations, response of the recipient has to be submitted within fifteen calendar days[10].

A common statement from the FDA within a warning letter runs:

"Within fifteen working days of receipt of this letter, please notify this office in writing of the specific steps that you have taken to correct violations. Include an explanation of each step being taken to prevent the recurrence of violations and copies of supporting documentation. If you cannot complete corrective action within fifteen working days, state the reason for the delay and the date by which you will have completed the correction. Additionally, your response should state if you no longer manufacture or distribute the drug product(s) manufactured at this facility, and provide the date(s) and reason(s) you ceased production. Please identify your response with appropriate FEI # for each location."

Importantly, in terms of the review of 'systems', the FDA Form 483 is a report which does not include observations of questionable or unknown significance at the time of the inspection. There may be other objectionable conditions that exist at the firm that are not cited on the FDA Form 483. FDA investigators are instructed to note only what they saw during the course of the inspection. Companies are responsible to take corrective action to address the cited objectionable conditions and any related non-cited objectionable conditions that might exist.

When the FDA sends a "Warning Letter", which describes the manufacturer's violations of FDA regulations, the FDA requires a reply within fifteen days of receipt of the letter. The Warning Letter generally states that the firm has made products that are adulterated, violating the Food, Drug, and Cosmetic Act and that the firm has a very limited amount of time to address the problem(s) before the FDA takes further regulatory action against the firm, the adulterated product, and responsible individuals.

The general components of a warning letter are:

• The person to whom it is addressed

• The issue causing the compliance problem

• The recommended remedial action and time period

Once the company has responded, with the help of the appropriate centre, the district offices determine the adequacy of response of the warned firm. In the case of an adequate response, the district or appropriate centre notify other appropriate agency units and then re-inspection of the firm takes place.

Thus an FDA inspection is a fact finding exercise. The FDA conducts a biennial inspection of manufacturing sites (which additionally includes such facilities as repackaging companies and contract test laboratories). This inspection programme is usually performed under the Compliance Program 7356.002 regulation. This regulation constitutes the so-termed "Systems-Based Inspections".

For such inspections a facility is visited by one or more FDA inspectors. Drug inspections are conducted by FDA Investigators who generally have academic training and advanced degrees in the sciences (e.g. engineering, biology, chemistry, pharmacy, etc.). They are also trained by the FDA (and contracted schools of pharmacy and law) in investigative and inspectional techniques, US law, and in basics of pharmaceutical manufacturing. Depending upon the type of inspection, the inspector will work to the relevant FDA Compliance Program Guidance Manual. This includes:

- Drug Manufacturing Inspections 7356.002. This guide covers Pharmaceutical Systems Inspection

- Sterile Drug Process Inspections 7356.002A. This guide has specific guidance inspecting sterile processes (aseptic and terminally sterilised)

- Pre-Approval Inspections 7346.832. This guide covers the Pre-Approval inspections which are required before FDA will approve most pharmaceuticals (NDA and ANDA)

- Post-Approval Inspections 7346.843. This provides continuing coverage of approved drugs (NDA and ANDA), auditing for changes in the production and control practices that occur after approval and assure that approved applications have been appropriately supplemented

3.3   System-based inspections

Systems reflect the ways of manufacturing and controlling pharmaceutical products. Inspections of drug manufacturers are generally made and reported by focusing on systems. System-based inspections apply to different profile classes of drugs (where drugs are categorised into groups). The use of profile classes means that the inspection focuses on types of products rather than considering individual products. Therefore, system-based inspections result in a determination of acceptability/non-acceptability for all profile classes (e.g. sterile dugs, non-sterile drugs, inhalants, patches, etc.). Inspection coverage is considered to be representative of all the profile classes manufactured by the organisation.

Drugs and drug products are manufactured using many physical operations to bring together components and containers and closures into a product released for distribution. Activities found in drug organisations can be organised into systems that are sets of operations and related activities. Control of all systems helps to ensure the organisation will produce drugs that are safe, have the identity and strength, and meet the quality and purity characteristics as intended.

The FDA inspector is required to report on the system in sufficient detail, with specific examples selected, so that the system inspection outcome reflects the state of control in that system for every profile class. If a particular system is adequate, it should, in theory, be adequate for all profile classes manufactured by the organisation. For example, the way an organisation handles "materials" (that is receipt, sampling, testing, acceptance, etc.) should be common across all profile classes. For example, the investigator should not have to inspect the Material System for each profile class. Likewise in the Production System, there are general requirements like SOP use, charge-in of components, equipment identification, in-process sampling and testing which can be evaluated through selection of example products in various profile classes. Under each system there may be something unique for a particular profile class (for example, under the Materials System, the production of Water for Injection USP for use in manufacturing). Selecting unique functions within a system is at the discretion of the lead investigator.

System-based inspections relate to the Code of Federal Regulations (CFR) and the pharmaceutical manufacturer should understand the interaction as part of the strategy for preparing for and responding to inspections. For example, in relation Quality Systems the following aspects of the system inspection relate to the following CFRs:

- Failure Investigations: 211.22 ("Responsibilities of quality control unit")
- Training/Qualification: 211.25 ("Personnel qualifications")
- Validation/Computer: 211.68 ("Automatic, mechanical, and electronic equipment")
- Rejects: 211.89/110 ("Rejected Components, drug product containers, and closures"/"Sampling and testing of in-process materials and drug products")
- Validation/Manufacturing: 211.110 ("Sampling and testing of in-process materials and drug products")
- Change Control: 211.100/160 ("Written procedures; deviations"/"Lab controls – General requirements")
- Reprocessing/Rework: 211.115 ("Reprocessing")
- Quarantine Products: 211.142 ("Warehousing procedures")

- Stability Failures: 211.166 ("Stability testing")

- Product Reviewers: 211.180 ("Records and Reports – General Requirements")

- Validation/Lab. Method: 211.194 (Laboratory records")

- Complaint Files: 211.198 ("Complaint Files")

- Returns/Salvages: 211.204/208 ("Returned drug products"/"Drug product salvaging")

In this sense, the system based inspection involves looking for patterns and from these patterns the inspector draws inferences. For example, a review of the Quality System may indicate the following concerns which can be clustered into different patterns:

1) Failure to review/approve procedures.

2) Failure to document execution of operations as required.

3) Failure to review documentation.

4) Failure to conduct investigations and resolve discrepancies/ failures/ deviations/complaints.

5) Failure to assess other systems to assure compliance with GMP and SOPs.

With the above list, for any of the patterns, an inspector would have a concern about the robustness of the quality system.

3.3.1 Types of system-based inspections

Any given inspection need not cover every system. A drug manufacturing inspection is a facilities based inspection evaluating two or more systems (although this will normally always include the Quality System). This is performed in order to determine if manufacturing activities are occurring in a state of control. The System Based Drug Inspection is applied through one of the following options[11]:

- The Full Inspection Option

  The Full Inspection Option is a surveillance or compliance inspection intended to provide a broad and deep evaluation of the organisation's 'current' GMP (cGMP). This is done when little or no information is known about an organisation's cGMP compliance (for example, as applying to new organisations); or for organisations where there are concerns about the cGMP compliance within the organisation (for example, where the previous inspection has resulted in the issuing of a warning letter); or follow up to previous regulatory actions. During the course of a Full Inspection, verification of Quality System activities may require limited coverage in other systems, although at least four of the systems are examined.

• The Abbreviated Inspection Option

The Abbreviated Inspection Option is a surveillance or compliance inspection which is meant to provide an efficient update evaluation of an organisation's cGMP. The abbreviated inspection will provide documentation for continuing an organisation in a satisfactory cGMP compliance status. Generally this is done when an organisation has a record of satisfactory cGMP compliance, with no significant recall, or product defect or alert incidents, or with little shift in the manufacturing profiles of the organisation within the previous two years.

The Abbreviated Inspection Option will normally include an inspection audit of at least two of the systems, one of which must be the Quality System. Some organisations participate in a limited part of the production of a drug or drug product, e.g., a contract laboratory. Such organisations may employ only two of the systems defined. In these cases the inspection of the existing systems will comprise inspection of the entire organisation.

• Compliance Inspections

Compliance Inspections are done to evaluate or verify compliance corrective actions after a regulatory action has been taken. The coverage given in compliance inspections must be related to the areas found deficient and subjected to corrective actions. In addition, coverage is extended to quality systems because a determination must be made on the overall compliance status of the organisation after the corrective actions are taken. The organisation is expected to address all of its operations in its corrective action plan and not just the deficiencies noted in the FDA-483 (that is, the organisation must review the entire system within which a deficiency has been found). The Full Inspection Option is generally used for compliance inspections.

Compliance inspections also include "For Cause Inspections". "For Cause Inspections" are compliance inspections performed to investigate a specific problem that has come to the attention of the FDA. The problems may be indicated in Field Alert Reports, industry complaints, recalls, indicators of defective products, etc.

3.4   Review of the systems and review of FDA warning letters

This section of the chapter examines the systems which comprise the system based inspection and highlights some of the deficiencies listed in FDA warning letters from the period 2010-2012[a].

For the review of warning letters, a number of them were examined. These were selected as examples because they each covered one or more of the systems reviewed. To enable the analysis to be easier to follow, the warning letters have been coded alphabetically.

The warning letters are:

- Allure Labs, Inc. 5/24/11[12]

- Aurobindo Pharma Limited 5/20/11[13]

- NuSil Technology LLC 3/22/12[14]

- Wintac Limited 2/23/12[15]

- Merck KGaA 12/15/11[16]

- Ben Venue Laboratories, Inc. 16-Nov-07[17]

- Professional Disposables International 7/5/11[18]

- Miramar Cosmetics, Inc. 8/18/11[19]

- SmithKline Beecham Limited 10/7/11[20]

- Jenahexal Pharm GmbH 10/19/11[21]

- Novartis International AG 11/18/11[22]

- Gulf Pharmaceutical Industries 2/23/12[23]

- Bristol Myers Squibb Holdings Pharma., Ltd. 8/30/10[24]

- Cadila Healthcare 6/21/11[25]

- Gilead Sciences Inc. 9/21/10[26]

- Triad Alcohol Prep Pads, Alcohol Swabs, and Alcohol Swabsticks: Recall[27]

- Letter Luitpold Pharmaceuticals, Inc. 8/31/11[28]

- Sichuan Pharmaceutical Co., Ltd. 9/9/11[29]

Quality System

As indicated earlier, the inspection of the Quality System is the most important area of the inspection as it is the point from which all other systems within the organisation originate. The FDA employs the Quality System Inspection Technique (QSIT) to examine Quality Systems. For this, the investigator utilises a "top down" approach to inspections. The inspector commences the inspection by determining if procedures that address the requirements of the Quality System regulation have been defined and documented. The investigator then samples records to determine if the procedures are being followed.

The Quality System assures overall compliance with internal procedures and specifications. The system includes the quality control unit and all of its review and approval duties (for example, change control, reprocessing, batch release, annual record review, validation protocols, reports, etc.)[30]. It includes all product defect evaluations and evaluation of returned and salvaged drug products. This is set out in the FDA cGMP regulation: 21 CFR 211 Subparts B, E, F, G, I, J, and K.

Assessment of the Quality System is in two parts. The first phase evaluates whether the Quality Unit has fulfilled the responsibility to review and approve all procedures related to production, quality control and quality assurance, and assure the procedures are adequate for their intended use. This also includes the associated record keeping systems. The second phase is to assess the data collected to identify quality problems and may link to other major systems for inspectional coverage.

For each of the following, the organisation should have written and approved procedures and documentation. The organisation's adherence to written procedures should be verified through observation whenever possible. These areas are not limited to finished products, and may also incorporate components and in-process materials. These areas may indicate deficiencies not only in this system but also in other major systems that would warrant expansion of the scope of the inspection. All areas under this system are to be covered; however the depth of coverage may vary depending upon inspectional findings:

• Product reviews should be undertaken at least annually. Such reviews should include information from areas listed below as appropriate; batches reviewed, for each product, are representative of all batches manufactured; trends are identified; refer to 21 CFR 211.180(e)

• Complaint reviews (quality and medical): documented; evaluated; investigated in a timely manner; includes corrective action where appropriate

• Discrepancy and failure investigations related to manufacturing and testing: documented; evaluated; investigated in a timely manner; includes corrective action where appropriate

- Change Control: documented; evaluated; approved; need for revalidation assessed

- Product Improvement Projects: for marketed products

- Reprocess/Rework: evaluation, review and approval; impact on validation and stability

- Returns/Salvages: assessment; investigation expanded where warranted; disposition

- Rejects: investigation expanded where warranted; corrective action where appropriate

- Stability Failures: investigation expanded where warranted; need for field alerts evaluated; disposition

- Quarantine products

- Validation: status of required validation/revalidation (e.g. computer, manufacturing process, laboratory methods)

- Training/qualification of employees in quality control unit functions

Examples of inspection findings

An example of a quality system failure is found with **warning letter A**. One important part of the quality system is training, for without correctly trained staff, auditors cannot be satisfied that the medicine has been manufactured correctly and consistently. In relation to this, one FDA warning letter mentioned:

> *"Your firm has failed to ensure that each person engaged in the manufacture, processing, packing, or holding of a drug product has the education, training, and experience, or any combination thereof, to enable that person to perform their assigned functions [21 CFR § 211.25(a)]."*

Specifically:

> *"For example, the employees of your firm, including a member of your quality control staff, admitted to our investigators that they were unaware of and were not trained to follow your SOP for handling deviations. There were at least two instances in which an OOS investigation was not conducted."*

Such a citation undermines any confidence that the inspector will have in all aspects of the company's operations.

A further example of a training system issue is contained within **warning letter N**. It is important, within pharmaceutical manufacturing, that training has been carried out and that personnel are aware of the tasks required of them and that they can execute these tasks competently. It is also important that that training records have been completed and are up-to-date as only written evidence will be accepted by the inspector.

The warning letter related to this point remarked:

> *"The technician performing the air sampling held the probe close to the HEPA filter face rather than (b) (4) as specified in section 4.5 of your written procedure SOP/QC/049.*
>
> *During the inspection, the investigators were provided with retraining records for technicians performing active air sampling."*

Here it was clear that something was amiss and that the original error related to a training issue. If the inspector has insufficient confidence in the training system or with the way in which training has been performed, this can undermine confidence across the quality system. In this example, FDA guidance states the location at which air velocity should be measured and the range for the air velocity reading from a HEPA filter. The air velocity is an important indicator that the air speed is sufficient to remove any contamination from the air stream and to direct air away from the critical zone (an essential prerequisite for aseptic filling)[31].

The quality system extends beyond the manufacture and release of medicines. It also includes transport, distribution and the acknowledgement and processing of customer complaints.

One company (**warning letter K**) was criticised for its failure to respond adequately to customer complaints, which suggested a fundamental weakness in the quality system, as the FDA warning letter noted:

> *"Your Quality Unit has failed in the responsibility and authority to monitor Quality systems designed to assure the quality of drug products manufactured and packaged at your firm.*
>
> *This failure is evidenced in the Observations described below: (Failure to adequately investigate consumer complaints, Failure to assure your processes remain in a current validated state, Failure to conduct complete Annual Product Reviews, Failure to train employees within your operations and quality systems, Failure to extend investigations of known problems*

to all lots potentially affected, Failure to file adequate NDA Field Alerts in a timely manner and Failure to have an adequate number of trained personnel in your Quality Unit)."

This comment is relatively self-explanatory and suggests a failure to join together many of the important elements of quality assurance.

It is common for warning letters to indicate that the company needs to look beyond the specific issue(s) cited in the warning letter and to review its entire quality system. For example:

"The violations cited in this letter are not intended to be an all-inclusive statement of violations that exist at your facility. You are responsible for investigating and determining the causes of the violations identified above and for preventing their recurrence and the occurrence of other violations. If you wish to continue to ship your products to the United States, it is the responsibility of your firm to ensure compliance with all US standards for CGMP and all applicable US laws and regulations."

Ensuring that the drug has been formulated correctly and that the stability of the formulation of the drug has been proven over its lifetime is an essential aspect of quality control. Failure to conform to these aspects suggests a breakdown of the quality system in relation to the design, production and release of the drug. In relation to this, a warning letter (**letter J**) noted:

"Your firm has not established a written assessment of stability of homeopathic drug products based at least on testing or examination of the drug product for compatibility of the ingredients and marketing experience with the drug product to indicate that there is no degradation of the product for the normal or expected period of use [21 CFR § 211.166(c)(1)].

In your response you indicated that homeopathic drug products are exempt from the requirements under § 211.137. While expiration dating is not required for homeopathic products, evidence of the product stability over the expected shelf life is required. Your response failed to provide an assessment of the stability of the homeopathic injectable products that you manufacture."

In this example, there was a failure to understand what the status and requirement of the drug was a matter which should have been picked up at the design stage, and a failure to understand cGMP requirements. Such a failure would raise concerns about the robustness of the company's quality system.

Facilities and Equipment System

This system includes the measures and activities which provide an appropriate physical environment and resources used in the production of the drugs or drug products. It includes:

a) Buildings and facilities along with maintenance (which includes critical utilities like water systems and controlled environments, such as cleanrooms);

b) Equipment qualifications (installation and operation); equipment calibration and preventative maintenance; and cleaning and validation of cleaning processes as appropriate. Process performance qualification is evaluated as part of the inspection of the overall process validation which is done within the system where the process is employed; and,

c) Utilities that are not intended to be incorporated into the product such as HVAC, compressed gases, steam and water systems. This is set out in the FDA cGMP regulation: 21 CFR 211 Subparts B, C, D, and J[32].

For each of the areas listed below, the organisation should have written and approved procedures and documentation. The organisation's adherence to written procedures is often verified by the inspector, through observation. These areas may indicate deficiencies not only in this system but also in other systems that would warrant expansion of coverage.

Facilities

• Cleaning and maintenance

• Facility layout and air handling systems for prevention of cross-contamination (e.g. penicillin, beta-lactams, steroids, hormones, cytotoxics, etc.)

• Specifically designed areas for the manufacturing operations performed by the organisation to prevent contamination or mix-ups

• General air handling systems

• Control system for implementing changes in the building

• Lighting, potable water, washing and toilet facilities, sewage and refuse disposal

• Sanitation of the building, use of rodenticides, fungicides, insecticides, cleaning and sanitizing agents

Equipment

- Equipment installation and operational qualification where appropriate

- Adequacy of equipment design, size, and location

- Equipment surfaces should not be reactive, additive, or absorptive

- Appropriate use of equipment operations substances, (lubricants, coolants, refrigerants, etc.) contacting products/containers/etc.

- Cleaning procedures and cleaning validation

- Controls to prevent contamination, particularly with any pesticides or any other toxic materials, or other drug or non-drug chemicals

- Qualification, calibration and maintenance of storage equipment, such as refrigerators and freezers for ensuring that standards, raw materials, reagents, etc. are stored at the proper temperatures

- Equipment qualification, calibration and maintenance, including computer qualification/validation and security

- Control system for implementing changes in the equipment

- Equipment identification practices (where appropriate)

- Documented investigation into any unexpected discrepancy

Examples of inspection findings

One aspect of facilities within a pharmaceutical manufacturing organisation is the pharmaceutical grade water system, given that water is a key ingredient of most pharmaceutical preparations.

From one inspection (**warning letter C**), an FDA warning letter recorded the following finding:

"Your firm failed to adequately validate and monitor the treatment process for your purified water system.

For example, your firm uses purified water as a key ingredient in the manufacture of Simethicone Emulsion USP (30%), MED-341. Between September 2009 and January 2011, you have conducted microbiological tests and found recurrent Burkholderia cepacia contamination.

Your firm failed to assure that your water system is suitably designed and operated to produce appropriate water quality. Regarding the latter, your firm has not established and validated appropriate cleaning and sanitizing schedules for your purified water system.

You have hired a water process subject matter expert and taken other steps to strengthen monitoring of the purified water system. Your response is not acceptable because you have not demonstrated that your purified water system is capable of operating in a continuing state of control. Please provide a validation plan for your purified water system and describe any interim actions that your firm will take to ensure that purified water used in manufacture of your drug products meets its action limits."

The development and execution of a cleaning and sanitization programme as part of a total validation plan, which ultimately requires intensive sampling and testing over a prolonged period of time of the Purified Water System, is essential to maintaining the water. In not doing so, a fundamental failing was uncovered.

A second important aspect of processing is in relation to equipment intended to run sterilisation cycles. The failure of a sterilisation cycle can lead directly to patient harm.

In direct relation to this, one warning letter (**letter L**) stated:

"The revalidation of the (b) (4) sterilization cycles for machine parts is deficient in that the exact placements of the biological indicators (BIs) and (b)(4) are not documented and consequently cannot be confirmed as the worst case locations.

Your response indicates that BI locations are now identified in SOP QAS 296 and that retraining has occurred. However, SOP QAS 296, provided in your response, indicates that BIs are not attached with the (b)(4) in several locations including, but not limited to, (b)(4) of the (b)(4), within the (b)(4) attached to the (b)(4), and (b)(4) through the (b)(4) with (b)(4) and (b)(4). Your response is inadequate because it fails to include your rationale for not routinely placing BIs next to the (b)(4) in the areas you have designated as most difficult to sterilise."

The warning letter indicates that the firm did not fully understand how sterilisation is verified through the correct use of biological indicators. Biological validation is normally done by ensuring that the process inactivates several biological challenge test pieces, each carrying $\geq 10^6$ spores of *Geobacillus stearothermophilus* (D_{121}-values meeting the pharmacopeial criteria of equal to or greater than 1.5 min) placed at critical points within the device to be sterilised.

Biological Indicators (BI) should only be located in their respective positions after the autoclave has been temperature mapped with thermocouples. Thermocouples used as part of developing a load pattern are essential for determining the "cold" spots within the autoclave. Placing the BI within the "cold" spots" after locating them through the use of thermocouples will provide for the Sterility Assurance Level of 10^{-6} that is required to confirm sterility[33]. The Sterility Assurance Level applies to terminally sterilised products and is the minimum requirement that there must be no more than one chance in a million that viable contaminants survive in any one unit.

The warning letter also criticised the interpretation and understanding of the operators who were tasked with executing the biological indicator validation. The letter went on further to state:

> "Operators fail to follow the diagram of the validated (b)(4) load configuration provided in SOP QAS-296 entitled Validation/Re-validation of Sterilization Cycles in the (b)(4) and SOP PGSI-087: Injectables Machine Parts Preparation Before Sterilization and Batch Processing. SOP QAS-296 entitled Validation/Re-validation of Sterilization Cycles in the (b)(4) and SOP PGSI-087: Injectables Machine Parts Preparation Before Sterilization and Batch Processing provide a diagram of the load configurations; however, our inspection found that the operator performing this function for the parts to be used in the production of (b)(4) Injection lot (b)(4), cannot follow this load pattern due to insufficient space on the (b)(4) trolley. The current (b)(4) load configuration used by production operators to (b)(4) sterilise filling machine parts has not been validated.

> Your response states, "The subject employee inadvertently did not follow the validated load pattern for (b) (4) sterilization of filling machine parts." It should be noted that your firm's personnel explained that the failure to follow the validated load pattern was not inadvertent, but was the prevailing practice. Your response indicates an incident report was raised, all operators were retrained, and a sign-off sheet for load configuration was added to the batch record. However, your response is inadequate because it fails to address the disposition of (b)(4) Injection lot (b)(4) as well as any other batches that may have been impacted by this failure to follow validated sterilization load configurations."

This issue also raises concerns with the training system and thus the focus of the inspection is likely to extend back to the quality system.

A third example (also taken from the analysis of **warning letter L**) relating to facilities is with cleanrooms. A cleanroom is a controlled environment, certified to a certain level of airborne particulates. The way in which a cleanroom achieves a level of cleanliness is through a functioning Heating Ventilation and Air Conditioning (HVAC) system. One of the key components of the HVAC system is the High Efficiency Particulate Air (HEPA filter). Unfiltered and uncontrolled air contains a high number of particles. Many such particles will be what is described as non-viable (such as dust or debris). However, a number will be micro-organisms (bacteria and fungi) and a higher number will be an association of the two, such as bacteria being carried on rafts of skin or dust[34]. Within cleanrooms, because of the association with micro-organisms and the assumption that some particles will be micro-organisms, pharmaceutical manufacturing areas are designed to minimise the number of particles through physical controls and facilities are monitored for particulate levels using optical particle counters.

In relation to HVAC operation, there are several FDA findings which relate to inadequate testing, certification or verification. For example a warning letter addressed the following area:

"Your firm has not established or followed appropriate written procedures designed to prevent microbiological contamination of drug products purporting to be sterile. Such procedures include validation of the sterilization process [21 CFR § 211.113(b)]."

In relation to HVAC systems specifically:

"HEPA air velocity is not evaluated proximal to the working level.

SOP ECPI-021: Calibration Procedure for unidirectional Airflow Unit and Bench is deficient in that it only requires HEPA air velocity checks to be performed (b)(4) inches below the filter face, but does not require that the air velocity be evaluated proximal to the working level.

Your response is inadequate because your corrective action for your failure to evaluate air velocity proximal to the working level consisted of providing a revised procedure and training, but you have not yet evaluated the current air velocity at the working level."

Understanding that air supply velocity and air direction are important aspects of contamination control underlines the engineering support and operator understanding. Such an observation could lead to concerns being expressed about the adequacy of the facilities and equipment systems.

Cleanroom control and certification is an important FDA concern and which reflects the criticality given to the environment in which a product is manufactured. To draw upon an example from another facility (**warning letter O**):

"Your firm has failed to maintain adequate separate or defined areas or such other control systems for the firm's operation as are necessary to prevent contamination or mixups during the course of aseptic processing [21 CFR § 211.42(c)(10)]."

With the specific issue being:

"Your current practice for assessing the unidirectional airflow in the critical product path is inadequate to prevent contamination of the product. Specifically, procedure SDSOP-0325, Testing and Certification of HEPA Filters (Revision 30), only requires testing for airflow velocity uniformity testing of HEPA filters housed inside the [b4]. CGMP requires, at a minimum, that all unidirectional airflow systems must be tested for uniformity of air velocity. However, your SOP does not include instructions for uniformity of air velocity testing of other HEPA filters within the aseptic processing room, particularly those in the critical path of the product through the room.

Your response is inadequate. You stated in your response that you will update SDSOP-0325 to include the evaluation of all the filters within the ISO [b4] dynamic area by March 21, 2010. However, you provided no timeframe for conducting the velocity uniformity test of the untested filters. You also failed to define your term, "ISO [b4] dynamic area."

Such an issue would raise questions over the level of environmental control across the facility. The ISO references relate to the cleanroom classification standard ISO 14644[35] and, given the global reach of this standard, all pharmaceutical manufacturers should be conversant with its contents.

The maintenance and calibration of process equipment is important in order to have confidence that equipment is operating within validated design parameters and to ensure batch-to-batch consistency. Many warning letters frequently site inadequate calibration systems or failures to calibrate equipment in a timely manner or to plan.

For example (from **warning letter Q**):

"Your firm has failed to routinely calibrate, inspect, or check automatic equipment in accordance to an adequately written program designed to assure proper performance [21 CFR § 211.68(a)].

For example, the SOP, SOP 601.23, Validation of the (b)(4) for the annual qualification of the (b)(4) does not include a requirement to evaluate contamination particle sizes discovered in your products from past investigations or recovered from an external forensic laboratory. The (b)(4) are qualified to identify particle size ranges (b)(4) and (b)(4). However, past investigations have identified particles less than (b)(4) and greater than (b)(4). In some cases, the (b)(4)have only been qualified to evaluate (b)(4) particle size. For example, (b)(4) has been qualified only with (b)(4) sized particles in the most recent validation dated 10/21/10.

You response fails to provide assurance that the annual qualification of the (b)(4) demonstrate they can effectively remove vials with particulate matter, and do so across sufficiently representative vial sizes and types. We also note that your response, which states that an (b)(4) rejection rate of test vials on the (b)(4) is acceptable performance, contradicts the qualification acceptance criteria in your SOP 601.23, "(b)(4)," that requires (b)(4) to reject (b)(4) or more test vials with particulates."

Such a citing would again trigger a review of the organisation's entire calibration programme and would have wider implications in relation to equipment controls.

Materials system

This system includes measures and activities to control finished products, components, including water or gases, which are incorporated into the product, as well as containers and closures. It includes validation of computerised inventory control processes, drug storage, distribution controls, and records. This is set out in the FDA cGMP regulation: 21 CFR 211 Subparts B, E, H, and J. This is particularly important for sterile products. Products labelled as sterile are expected to be free from viable microbial contamination throughout the product's entire shelf life or dating period. The sterility of a sterile dosage form can only be guaranteed while it is protected from the surrounding non-sterile environment within a container made from materials impermeable to microbial penetration.

For each of the areas below, the organisation should have written and approved procedures and documentation relating to finished products, components and in-process materials. Important areas to examine include:

- Training/qualification of personnel
- Identification of components, containers, closures
- Inventory of components, containers, closures
- Storage conditions
- Storage under quarantine until tested or examined and released

- Representative samples collected, tested or examined using appropriate means

- At least one specific identity test is conducted on each lot of each component

- A visual identification is conducted on each lot of containers and closures

- Testing or validation of supplier's test results for components, containers and closures

- Rejection of any component, container, closure not meeting acceptance requirements. Investigate fully the organisation's procedures for verification of the source of components.

- Appropriate retesting/re-examination of components, containers, closures

- First in-first out use of components, containers, closures

- Quarantine of rejected materials

- Water and process gas supply, design, maintenance, validation and operation

- Containers and closures should not be additive, reactive, or absorptive to the drug product

- Control system for implementing changes in the materials handling operations

- Qualification/validation and security of computerised or automated processes

- Finished product distribution records by lot

- Documented investigation into any unexpected discrepancy

Examples of inspection findings

In relation to sampling and control of materials, an FDA warning letter (**letter A**) raised the following:

> *"Your firm has not established scientifically sound and appropriate specifications, standards, sampling plans, and test procedures designed to assure that components, in-process materials, and drug products conform to appropriate standards of identity, strength, quality, and purity [21 CFR § 211.160(b)]."*

For example:

> *"Your firm has failed to provide a scientific justification for how the samples of bulk drugs are representative of the lot when they are collected only from the top of the kettle."*

This citation highlights the need for a detailed and controlled sampling plan, and one based on sound scientific principles which can be justified to regulators.

Maintaining controls over materials extends to critical reagents such as disinfectants. Disinfectant agents are essential for contamination control. Although cleanrooms are designed to a high standard and are subject to environmental monitoring, the activities and people within the cleanrooms will pose an ever present contamination risk. Cleaning and disinfection practices are, therefore, an essential part of contamination control in the pharmaceutical industry. These practices can be divided into: the cleaning of equipment (which requires cleaning validation to verify the effectiveness of the cleaning) and the cleaning and disinfection of cleanrooms.

Therefore, cleaning and disinfection chemicals should be controlled and evaluatedxxxvi. In relation to evaluation, one warning letter (**letter l**) stated:

> *"Your firm has not established appropriate written procedures designed to prevent microbiological contamination of drug products purporting to be sterile [21 CFR § 211.113(b)]. For example:*
>
> *The qualification of your disinfectant (b)(4) failed to demonstrate that it is suitable and effective to remove micro-organisms from different surfaces. Specifically, this disinfectant failed to meet qualification criteria when challenged with multiple organisms."*

In this example the company had failed to demonstrate if the disinfectant was capable of achieving a four-log reduction of bacteria, as required by the United States Pharmacopeia, against a variety of surfaces (it is typical to evaluate such materials as vinyl, stainless steel, glass, and wall laminate).

The use of unqualified disinfectants is of a major concern to inspectors. This indicates that high levels of microbial contamination or objectionable micro-organisms may be present in the process areas. Depending on the type of product and its intended route of administration, the presence of certain micro-organisms will present a risk in terms of spoilage or patient harm[37]. The observation also indicates that the control of materials, from purchasing to laboratory release, is at fault through either failing to understand technical requirements or by allowing unauthorised release.

A sound and scientifically based sampling system is essential in order to ensure that the material presented for analysis is representative. If this is not done, then the material released may not have met its specification.

A related warning letter (**letter M**) to this stated:

> *"Your firm has not established scientifically sound and appropriate specifications, standards, sampling plans, and test procedures designed to assure that drug products conform to appropriate standards of identity, strength, quality, and purity [21 CFR § 211.160(b)].*
>
> *For example, your Firm established acceptance criteria for major defects in the control Form entitled, "Working Inspection Qualification, WIQ," using (b)(4). The acceptance criteria is inadequate because your QCD approved the specification without adequate justification or a scientifically sound statistical analysis."*

With this example, the issue related not only to the specific area seen by the inspector for adopting the system based approach; the FDA extended their concerns to other aspects of the materials system. The letter went on to state:

> *"The violations cited in this letter are not intended to be an all-inclusive statement of the violations that exist at your facility. You are responsible for investigating and determining the causes of the violations identified above and for preventing their recurrence, as well as occurrence of other violations. It is your responsibility to ensure that your firm complies with all requirements of federal law and FDA regulations."*

Production system.

This system includes measures and activities to control the manufacture of drugs and drug products including batch compounding, dosage form production, in-process sampling and testing, and process validation. It also includes establishing, following and documenting performance of approved manufacturing procedures. This is set out in the FDA cGMP regulation: 21 CFR 211 Subparts B, F, and J.

The key components of the production system are:

- Training/qualification of personnel
- Control system for implementing changes in processes
- Adequate procedure and practice for charge-in of components

- Formulation/manufacturing at not less than 100%

- Identification of equipment with contents, and where appropriate phase of manufacturing and/or status

- Validation and verification of cleaning/sterilization/ depyrogenation of containers and closures

- Calculation and documentation of actual yields and percentage of theoretical yields

- Contemporaneous and complete batch production documentation

- Established time limits for completion of phases of production

- Implementation and documentation of in-process controls, tests and examinations (e.g. pH, adequacy of mix, weight variation, clarity)

- Justification and consistency of in-process specifications and drug product final specifications

- Prevention of objectionable micro-organisms in non-sterile drug products

- Adherence to pre-processing procedures (e.g. set-up, line clearance, etc.)

- Equipment cleaning and use logs

- Master production and control records

- Batch production and control records

- Process validation, including validation and security of computerised or automated processes

- Change control; the need for revalidation evaluated

- Documented investigation into any unexpected discrepancy

Examples of inspection findings

In relation to production, one company (from **warning letter A**) was cited by the FDA for:

> *"Your firm does not have adequate written procedures for production and process controls designed to assure that the drug products you manufacture have the identity, strength, quality, and/or purity they purport or are represented to possess [21 CFR § 211.100(a)]."*

The primary reason for this was:

"Your firm's mixing operations for bulk drugs have not been validated to ensure homogeneity".

Ensuring that samples are representative underpins processing and batch release, otherwise the quality control of the materials cannot be guaranteed. In a similar way:

"Your firm has failed to evaluate the holding time and handling (e.g., transfer from mixing kettles to intermediate storage containers) of bulk drugs during and after intermediate storage to ensure the bulk drugs continue to meet established specifications prior to filling."

Each stage of the process must be validated. This is particularly important if the product is held for a substantial period of time during the process.

A key aspect of control over production systems relates to environmental monitoring. Failure to monitor means that information relating to environmental control is not available, as indicated by the levels of viable micro-organisms and airborne particulatesxxxviii. One company (**warning letter B**) was cited for inadequate environmental monitoring:

"Your firm has not established or followed appropriate written procedures designed to prevent microbiological contamination of drug products purporting to be sterile [21 CFR § 211.113(b)]."

This was in relation to:

"For example, during the December 2010 inspection, the investigators found that your SOPs related to your environmental programs failed to adequately identify (e.g., diagrams) the locations where active and passive environmental monitoring samples are to be collected from. The inspection also found that your procedure for environmental sampling does not require that employees be sampled (b)(4) time they exit the Class (b)(4) clean rooms.

This deficiency increases our concern regarding the reliability of the data generated and your ability to identify the source of your microbial contamination. We expect that SOPs related to Environmental Monitoring include sufficient instructions to ensure that the plates intended to detect microbial growth are appropriately located. These procedures should also include specific instructions for the collection of microbiological samples.

Your firm needs to establish a robust environmental monitoring program capable of generating meaningful data, and that would serve as an early warning system to detect possible environmental contaminants that may impact the sterility of the sterile APIs and finished drug products manufactured at your facility. There is no assurance that your current environmental monitoring program is capable of detecting microbiological contaminants."

Environmental monitoring programmes must be based on sound scientific principles with the frequencies of monitoring, locations for monitoring and times of monitoring outlined in a rationale.

An important part of processing is the maintenance of premises through cleaning regimes, especially of equipment in order to prevent cross-contamination.

One warning letter (**letter C**) noted:

"Your firm failed to establish procedures to adequately clean and store equipment and utensils to prevent contamination or carry-over of material that would alter the API beyond the established specifications."

Specifically this was in relation to:

"For example, our inspection established that you do not routinely perform equipment cleaning after each manufacturing run. Your cleaning procedure requires you to clean the equipment when it has been more than (b)(4) since you manufactured the last batch in the same equipment, yet your cleaning validation does not contain data to support a (b)(4) "dirty hold time." Your firm determined that failure to clean a transfer hose before use was the probable cause of the microbial contamination of the Simethicone Emulsion USP (30%).

In your response, you stated that you performed a study to evaluate the cleaning step for the transfer hose and that you will sanitise the cleaned hoses with a dry heat treatment prior to use for future batches. Your response is inadequate because it does not provide data to support that your cleaning procedure for the transfer hose cleans and sanitises the hose adequately. Please provide a validation report that includes microbial test data that evaluates whether your cleaning and sanitization procedures for your transfer hoses and other equipment is adequate."

Making conclusions without support data is a common occurrence observed within warning letters., especially in relation to validation studies.

A second example in relation to process controls and preventing cross contamination is captured in the following extract from a warning letter (**letter R**):

"Failure to have appropriate procedures in place to prevent cross-contamination.

From September 2008 to July 2009 your firm manufactured (b)(4) API in workshop (b)(4), which is adjacent to workshops (b)(4) and (b)(4) where you manufactured (b)(4) API and (b)(4) injection, respectively. However, you failed to have adequate controls and monitoring program to prevent cross-contamination between these adjacent workshops.

In addition, your firm manufactures a (b)(4) API ((b)(4) (API) in a facility that was previously used to manufacture (b)(4) without conducting adequate decontamination, renovation, and activation of the facility. Your firm has failed to conduct adequate assessment of the cross contamination risks.

Please note that analytical testing of a product for possible contamination with (b)(4) is not sufficient to ensure adequate conditions for (b)(4) manufacture. In your response to this letter include your plans for decontamination, renovation, and reactivation (if appropriate) of your facility including the decontamination agent, decontamination plans, analytical methodology for environmental and product testing, and the data obtained to support the effectiveness of the decontamination plan."

The implications of this warning letter are very serious and infer that there is a risk of cross contamination across lines and that cleaning practices and batch segregation issues are uncontrolled.

Regulatory agencies place considerable importance upon the production of sterile medicines, particularly those produced aseptically. Aseptic processing is one of most commonly cited deficiencies in FDA warning letters.

For example, one warning letter (**letter D**) stated:

"Your firm has not established or followed appropriate written procedures designed to prevent microbiological contamination of drug products purporting to be sterile. Such procedures shall include validation of all aseptic and sterilization processes [21 CFR § 211.113(b)]."

For example:

"Your in situ air pattern analysis (smoke studies) does not demonstrate unidirectional airflow and sweeping action over and away from the critical processing areas under dynamic conditions.

Your response indicates that dynamic smoke studies for the (b)(4) filling area would be completed by the end of November, 2011. However, your response is inadequate because you failed to include a detailed interim plan describing additional steps taken in an effort to ensure that these areas were suitable for aseptic manufacturing of sterile drug products. Please provide a summary of the complete smoke studies (i.e., copy of video) conducted to evaluate whether the (b)(4) filling area is suitable for aseptic manufacturing of sterile drug products, and your firm's risk assessment of the product released to the market prior to your completion of the smoke studies."

Airflow studies are an important part of understanding contamination risks and for implementing contamination control measures. Included within the FDA's comments has been the question of how a firm determines its risk assessment of all products released to the marketplace prior to a satisfactory evaluation of the risk and the development of an interim plan to assure all of the areas (ISO Class 5 or EU GMP Grade A) are acceptable for use. Including such responses in the initial response to the Form FDA 483 can assist in eliminating the follow-up warning letter.

Another important aspect of aseptic manufacturing is staff training, for if personnel do not behave in a fashion which safeguards the product then there is a risk of microbial contamination. If the FDA witness practices which indicate that the product is at risk, this undermines the production system.

For example, one FDA warning letter noted (**letter D**):

"Your firm demonstrates poor aseptic practices during the filling of sterile (b)(4) products including, but not limited to:

- *An operator performing critical aseptic operations with exposed skin at the forehead, posing an unreasonable risk of the product becoming contaminated.*

- *Operators moving very quickly in the aseptic area, which may create unacceptable turbulence in the area, and disrupt the unidirectional airflow.*

- *Operators leaning halfway in and out of the class 100 area while performing interventions over opened bottles.*

Your firm's response indicates that the operators are trained in aseptic practices. However, it fails to specifically address the observed deficiencies in the aseptic area during the inspection and the potential impact of these practices on drug products already distributed."

Training of operators in aseptic practices is only part of what management needs to assure. The other area that needs to be assured is that the operators practice what they have learned. A fundamental question regulators like the FDA will ask when a poor practice is observed "what is the potential impact of these practices on drug products already distributed?" In responses to FDA 483s, this must always be considered as part of any response.

Validation is an important part of the product system. Validation covers a range of different applications depending upon the particular process. In relation to aseptic filling, one of the most important parts of validation is the media fill (or process simulation). It is important that this part of the product system is undertaken adequately.

Media fills not only assess the process, they assess the personnel and their ability to intervene into the critical zone. This involves undertaking intervention exercises.

In light of this, an FDA warning letter noted (**letter E**):

"Your firm lacks records to establish that aseptic processing personnel performed interventions that are representative of the operation.

For example, during the inspection, the investigator noted that your firm's media fill, conducted February 24, 2011 on Line, (b)(4) failed to require performance of interventions, document these interventions, or document when the individuals performing aseptic processing operations entered and exited the room."

Interventions to include their documentation should be noted as being performed. While each individual is not required to perform each intervention, all required interventions should nevertheless be performed to assure that all are completed. In addition, documenting when personnel performing aseptic processing operations both enter and leave the ISO Class 5 environment is essential to assure that all required monitoring is achieved, to assure that the consequences (total number of personnel within the clean room) is known and to demonstrate that the gowning performs as intended.

Record keeping, along with other types of documentation, is an important means of showing process controls. However, the FDA frequently cites companies for failing to keep detailed or accurate records.

For example (from warning **letter F**):

> *"The media fill manufacturing batch records do not accurately account for the number of vials that are filled and there is no reconciliation to assure that all media filled vials are accounted for. The CAPA TRK #107148, dated 10/15/10 provides Regulatory Deviation Reports, for example TRK #105923, dated 7/6/10 describes, "During the (b)(4) and (b)(4) day reads it was discovered the documented amount of units received for incubation varied from the amount actually incubated. There is currently no specification for accountability of media filled units." The TRK chronology of event document a similar concern occurred for (b)(4) media fills dated from 5/27/10 to 8/20/10. In addition, the aforementioned accountability for media filled vials applies to the following media fill lots;*
>
> *i. 12/31/10-#1012-60-2017196, < (b)(4) media filled vials of (b)(4)*
>
> *ii. 03/08/11-#1038-71-2036541, <(b)(4) media filled vials of (b)(4)."*

In relation to this specific observation, the FDA's Guidance for Industry Sterile Drug Products (September 2004), section IX, which relates to the incubation and examination of media-filled units states "appropriate criteria should be established for yield and accountability (reconciliation of filled units)[39]. Media fill record reconciliation documentation should include a full accounting and description of units rejected from a batch".

Packaging and labelling system

This system includes measures and activities that control the packaging and labelling of drugs and drug products. It includes written procedures, label examination and usage, label storage and issuance, packaging and labelling operations controls, and validation of these operations. This is set out in the FDA cGMP regulation: 21 CFR 211 Subparts B, G, and J.

Important aspects of the packaging and labelling system include:

- Training/qualification of personnel
- Acceptance operations for packaging and labelling materials
- Control system for implementing changes in packaging and labelling operations

- Adequate storage for labels and labelling, both approved and returned after issued

- Control of labels which are similar in size, shape and colour for different products

- Finished product cut labels for immediate containers which are similar in appearance without some type of 100% electronic or visual verification system or the use of dedicated lines

- Gang printing of labels is not done, unless they are differentiated by size, shape or colour

- Control of filled unlabelled containers that are later labelled under multiple private labels

- Adequate packaging records that will include specimens of all labels used

- Control of issuance of labelling, examination of issued labels and reconciliation of used labels

- Examination of the labelled finished product

- Adequate inspection (proofing) of incoming labelling

- Use of lot numbers, destruction of excess labelling bearing lot/control numbers

- Physical/spatial separation between different labelling and packaging lines

- Monitoring of printing devices associated with manufacturing lines

- Line clearance, inspection and documentation

- Adequate expiration dates on the label

- Conformance to tamper-evident (TEP) packaging requirements (see 21CFR 211.132 and Compliance Policy Guide, 7132a.17)

- Validation of packaging and labelling operations including validation and security of computerised processes

- Documented investigation into any unexpected discrepancy

Examples of inspection findings

The incorrect packaging or mislabelling of drugs can have serious implications for the administration of the medicine and hence for the patient. In relation to this, one FDA inspectorate finding, drawn from warning **letter A**, was:

"Your firm manufactures and distributes numerous drug products for over-the-counter (OTC) use. During the inspection noted above, the investigators collected labeling for some of these drug products. As presently formulated,

labeled, and promoted, these products are unapproved and/or misbranded drugs in violation of Sections 505(a) and 502 of the Federal Food, Drug, and Cosmetic Act (the Act) [21 U.S.C. § 355 and 352]."

This finding is of importance because the company were making untested claims about the use and properties of the drug. Manufacturing process must always conform to marketing authorisations.

Materials used for products must be suitable for use. When the FDA note concerns this, it could suggest major design issues, which could range from general unsuitability to risk of microbial contamination.

In relation to contamination, one warning letter noted focused on CFR 211.065(3)(d) and 21 CFR 211.113(a), clauses which discuss the need to test components prior to their use as well as prevent objectionable micro-organisms from products not designed to be sterile. In the cause of this company contamination was detected in alcohol wipes which was shown to be an endospore forming bacterium Bacillus cereus, and therefore a risk to the user, especially when applied to wounds.

A related recall letter (**reference P**) noted:

"Specifically, MedWatch Report #456119 was filed on a patient with hemophilia who developed Bacillus spp. bacteremia after receiving Factor 8 infusion. The hospital reported to find Bacillus spp in PDI alcohol prep pads, Lot#11100055. Your company confirmed such finding and also found the same species, Bacillus cereus, in additional PDA Alcohol Prep Pad finished product, Lot# 11100515, 11003269 11003222 and 11002166. Your root cause analysis for Alcohol Prep Pad Product family dated 05/24/11 demonstrated that Bacillus cereus was recovered from the applicator used to manufacture the alcohol prep pads. However, your correction action and prevention action from your non-conformance report #NYH-11-0131 addressing the recovery of Bacillus cereus in the non-sterile PDI Alcohol Prep Pads does not propose any necessary changes to include microbiological testing on each lot of applicator use in the production of both non-sterile and sterile alcohol prep pads as your Work Instruction QAS-012 approved in July 2008 "QA Raw Material Inspection testing and Disposition" 3.1.13 states (b)(4). As the result none of the applicator lots used in the production PDI Alcohol Prep Pad finished Product, Lot# 11100055, 11100515, 11003269 11003222 and 11002166 had a microbiological testing preformed prior to making into the finished product."

The reason for the contamination was a failure to carry out adequate microbiological testing as part of batch release.

It is important that medicines are packaged in a way which will not break the protection of the medicine. It leads to a serious event if an inspector detects any errors relating to the primary packaging components. For example (from warning **letter O**):

"Your firm has not thoroughly investigated the failure of a batch or any of its components to meet its specifications whether or not the batch has already been distributed [21 CFR § 211.192].

For example, you conducted an inadequate investigation into the acceptable quality limits (AQL) failures of [b4] tablets process re-validation lots [b4], and [b4]. These lots repeatedly failed visual inspection for AQL of[b4] and [b4], and [b4] tablets. Instead of conducting an adequate investigation into this event, you culled out defective tablets and packaged the remainder of the lots.

Your investigation report of May 2009 concluded that this event impacted the process validation. However, in your report of September 2009 you concluded that the process validation was successful based on tablet [b4], and [b4] testing results.

You stated in your response that these "cosmetic defects" are not related to the changes in the process (change in the [b4] supplier and use of a new tablet press). Your response is inadequate because you did not provide the scientific rationale you used to make this determination including your rationale for naming these types of defects "cosmetic defects." In addition, your response did not include any assessment of the lot ([b4]) sent to your site in [b4] although this lot also exhibited these defects."

In this example, the inspection, packaging and labelling system was not up to standard. This would lead to wider questions about the robustness of the entire batch finalisation and release system.

After packaging and formal release, the distribution of medicines must be controlled and the company has ownership of the material until it is received by the purchaser under the requirements of Good Distribution Practice (GDP). Sometimes the controls established by companies are insufficient.

Laboratory control system

This system includes measures and activities related to laboratory procedures, testing, analytical methods development, microbiological tests and validation or verification, and the stability program. This is set out in the FDA cGMP regulation: 21 CFR 211 Subparts B, I, J, and K.

The main aspects of the laboratory control system are:

- Training/qualification of personnel

- Adequacy of staffing for laboratory operations

- Adequacy of equipment and facility for intended use

- Calibration and maintenance programs for analytical instruments and equipment

- Validation and security of computerised or automated processes

- Reference standards; source, purity and assay, and tests to establish equivalency to current official reference standards as appropriate

- System suitability checks on chromatographic systems (e.g., GC or HPLC)

- Specifications, standards, and representative sampling plans

- Adherence to the written methods of analysis

- Validation/verification of analytical methods

- Control system for implementing changes in laboratory operations

- Required testing is performed on the correct samples

- Documented investigation into any unexpected discrepancy

- Complete analytical records from all tests and summaries of results

- Quality and retention of raw data (e.g. chromatograms and spectra)

- Correlation of result summaries to raw data; presence of unused data

- Adherence to an adequate Out of Specification (OOS) procedure which includes timely completion of the investigation

- Adequate reserve samples; documentation of reserve sample examination

- Stability testing program, including demonstration of stability indicating capability of the test methods.

Examples of inspection findings

In relation to laboratory testing, an FDA warning letter (**letter A**) cited a company for:

> *"Your firm has failed to establish an adequate quality control unit with the responsibility and authority to approve or reject all components, drug product containers, closures, in-process materials, packaging material, labeling, and drug products [21 CFR § 211.22(a)]."*

This was for the following reasons:

a) Poor microbiological test controls:

> *"Failed to reject a lot of Honeysuckle component (lot (b)(4)) after it failed specifications for yeast, mold, and Aerobic Plate Count."*

This finding would suggest that the company did not have appropriate control over specifications and an inadequate system for following up on corrective and preventative actions.

The FDA went on to note that the company's microbial limit specifications for finished product permitted a bioburden in the drug product that could allow the presence of objectionable and potentially pathogenic organisms in the drug products. It was also noted that the company did not specifically test for objectionable micro-organisms like *Staphylococcus aureus* and *Pseudomonas aeruginosa*. This demonstrated a lack of understanding of the compendial methods. Microbial limit standards for some, but not all, non-sterile dosage forms are given in USP monographs. In setting appropriate limits for particular non-sterile products, the USP takes account of the significance of micro-organisms to different types of product, to the way in which the product is used, and to the potential hazard to the patient. For instance, the majority of the monographs defining microbial limits listed in the USP for oral dosage forms restrict micro-organisms (*E. coli* and/or *Salmonella*) that are pathogenic by gastrointestinal ingestion.

b) Failure to ensure good laboratory practices in terms of results reporting:

> *"Your QCU failed to detect an employee miscalculating the microbial results for three lots of (b)(4) Moisturizing Face Screen (lots (b)(4)) finished products by not applying a dilution factor of 10-2. The correct calculation of the results would have shown that these products were Out-of-Specification (OOS). The three lots of (b)(4) Moisturizing Face Screen were released and distributed by the QCU based on the erroneous results."*

And similarly:

> *"Your QCU failed to detect multiple discrepancies in sample weights and dilution factors between the analyst's notebook and the Calculation Sheet. As a result, incorrect data was recorded for multiple products and finished products not meeting specifications were released."*

Failures relating to calculation errors undermine the confidence in the entire laboratory and raise questions about all results reported from the laboratory. This probably arose because the laboratory did not require a second, independent person to review the raw data, calculations and records before releasing these lots for distribution.

A similar finding in relation to laboratory test controls and to the reporting of data, from a different company (warning **letter B**), stated:

> *"Your firm's laboratory records fail to include complete data derived from all tests necessary to assure compliance with established specifications and standards [21 CFR § 211.194]."*

This related to:

> *"On December 13, 2010, the FDA investigator observed a microbiological plate that contained one (1) large colony forming unit (CFU) of mold. However, your firm's laboratory documentation reported 0 CFU for the same microbiological plate."*

This error represented a very serious failing as it either represented an attempt to alter the data or a failure to comprehend the reporting of microbiological data. With this incident, the inspection found that the laboratory manager had documented "NIL," (i.e. no growth for this plate), while the same laboratory manager confirmed microbial growth in the presence of the investigators. Later during the same inspection, the FDA investigator asked to see the original plate and was told that it had been destroyed.

It is also important to look wider and consider any other batches which might be implicated from the cited error. Failure to do so can lead to problems in a follow-up inspection. For example, one company received the following warning (taken from **letter K**):

> *"Fourteen out of 79 (18%) laboratory investigations lacked documentation of either an investigation into other associated batches or products, or a*

corrective action" and "you failed to extend your investigation into other associated batches or establish a root cause."

In this specific case the company had released the pharmaceutical product into the marketplace without having investigated if other products were implicated. This led to the warning letter stating:

"It is important that your firm's investigation procedures ensure that a full investigation, including all associated lots and root cause identification, is performed prior to distribution. We will verify the implementation of your revised investigation procedures during a future inspection."

Another common laboratory failing cited by the FDA is failure to conduct adequate out-of-limits investigation. In relation to this, one FDA warning letter (**letter C**), aimed at a non-steriles manufacturer, noted:

"Your firm failed to have an adequate out-of-specification (OOS) procedure to conduct thorough and scientifically sound investigations including corrective actions."

Specifically, the letter stated:

"For example, the OOS report number (b)(4) for lot (b)(4) of Simethicone Emulsion USP (30%), MED-341, dated September 17, 2010, states that your firm performed two retests as part of the investigation of a microbial test result of> 11,000 CFU/g. You released the lot based on the results of retesting, which were within the specification of (b)(4). You did not justify the invalidation of the original results, nor identify the root cause of the failure. You relied solely on the acceptable retest results to release the lot. Your OOS Report procedure (SOP#(b)(4)), is deficient because it does not require an analysis of the data, assessment as to whether a significant problem exists, identification of root causes, or allocation of the tasks for corrective actions.

In your response, you stated that you have modified your SOP# (b)(4) to include testing of lots before and after a failing result in your procedure. However, your response is inadequate because you failed to provide evidence that your firm will scientifically justify and document retesting of a lot that fails to meet a microbiological specification, and conduct an appropriate investigation."

Not justifying the invalidation of the original test results or the identification of the root cause of a failure prior to the release of an API (Active Pharmaceutical

Ingredient) is a major quality issue. All non-conformances require investigation and identification prior to closing a batch record.

A similar issue relating to out-of-specification investigations was cited at another company. In this case, the warning letter (**letter I**) indicated:

"*Your firm fails to follow laboratory control procedures [21 CFR § 211.160(a)]*".

This related to:

"*The inspection revealed that the laboratory investigations FQ991_6058, FQ991_6239, FQ991_7049, FQ991_7050, and FQ991_7484 were conducted without having Form B completed and approved by your Quality Unit, as required by your procedure. Your firm's procedure QAA4 "Investigation Out of Specification (OOS) Test and Atypical Results Procedure" establishes that the Form B is intended to document any retest, root cause investigation, and whether any remedial corrective and preventative actions are required. Your firm's response indicates that although the Form B was not used, the quality of the investigations is equivalent to those investigations in which the Form B was completed. However, you provided no support for this conclusion. In addition, your response failed to justify the re-test that was conducted without authorization by your Quality Unit.*"

In this case, due to an apparent failure to conduct an adequate investigation, the company was required to "review all the OOS investigations for product within expiration date to determine if the investigation procedures were properly followed". This was because the inspector did not have confidence in the ability of the firm to conduct an adequate out-of-specification investigation.

Of the different types of laboratory analysis, the Sterility Test is arguably one of the most important since a failure to conduct it properly could lead to a contaminated product being released. It is also important that the laboratory system can deal adequately with the investigation of any sterility test failures which might occur. The pharmacopeial standard applying to sterile products is that they must be capable of passing a Test for Sterility[40]. A Test for Sterility is described in *US Pharmacopeia* (USP) under Section 71, the *European Pharmacopoeia* (Ph.Eur.) under Section 2.6.1, and in the *Japanese Pharmacopoeia* (4.06).

Such issues can reveal a great deal to the FDA inspector about good laboratory practices. For example (from **letter F**):

"The sterility failure found on 01/20/10 associated with (b)(4) lot# 2378-44-1157184 {TRK #93890) identified Paenibacillus woosongensisas as the contaminant. This lot was manufactured on 01/25/08 in fill room (b)(4) the North facility. The environmental monitoring program, from dates 12/10/07 through 07/17/08 identified the recovery of Paenibacillus from Class 100 personnel monitoring and from Class 10,000 area and it was not recovered from the sterility tests isolator. The Senior Microbiologist explained that they believe that the microbial contaminant was due to concerns with the sterility test isolator. The aforementioned locations were not considered as a source for the microbial contamination."

This is clearly an example of an attempt to release a contaminated product by ignoring a link to the process environment and instead drawing an unsupported link to the sterility testing isolator when no cause to draw that link was apparent.

Some warning letters relate to general inadequacies of laboratory testing in terms of failing to draw up plans to test sufficient quantities of material or to conduct the correct tests for batch release. In relation to this, one warning letter (**letter H**) ran:

"Your firm has failed to perform appropriate laboratory testing, as necessary, of each batch of drug product required to be free of objectionable micro-organisms [21 C.F.R § 211.165(b)].

For example, your firm failed to conduct microbiological testing for 32 batches of topical drug products (Sport Cream, Hongotrim Spray and Hongotrim Soap) prior to release.

This is a repeat observation from the 2007 inspection."

Microbial testing should be an integral part of any specifications for non-sterile products that may permit the growth of micro-organisms. To not perform microbiological testing of 32 lots prior to release suggests that the product may be contaminated/adulterated and not suitable for release. For this to be a repeat violation is considered to be a major FDA issue.

3.5   Conclusion

This chapter has outlined the reasons for FDA inspections and the ways in which FDA inspections are conducted. The chapter has also outlined the primary inspection approach: system-based inspections, and has illustrated some examples of non-conformances for each of the six systems through an analysis of pertinent FDA warning letters.

In summarising this, the objective of FDA inspections is to ensure compliance with current Good Manufacturing Practice. The key things that an FDA inspector will seek to ascertain are:

- Is the organisation following cGMP?

- Are the personnel involved knowledgeable and familiar with regulations and cGMP?

- Is documentation available demonstrating training, monitoring, and compliance programs?

Of the different approaches to inspections, systems based inspections represent the most common form of regulatory inspection, particularly for those GMP inspections which fall under the remit of the FDA. The six systems inspected as part of systems based inspections represent comprehensive and rational categories in relation to cGMP. The systems provide an indication to an inspector as to the overall performance of the organisation and across all areas of manufacturing.

This chapter has examined the key aspects of each system and has drawn upon some main issues raised in FDA inspections through an analysis of warning letters. The analysis has highlighted some of the typical non-conformances, with the aim of providing some guidance to those tasked with preparing for FDA inspections.

Appendix A: FDA Compliance Program Guidance Manual 7356.002 (Drug Manufacturing Inspections)

FIELD REPORTING REQUIREMENTS

Forward a copy of each *Establishment Inspection Report* (EIR) for inspections classified as OAI due to CGMP deficiencies as part of any regulatory action recommendation submitted to HFD-300. For all inspections that result in the issuance of a Warning Letter, forward an electronic copy of each letter to the Division of Manufacturing and Product Quality, Case Management and Guidance Branch (HFD-325). An e-mail account has been established to receive copies of Warning Letters. The account e-mail address is CDERCGMPWL.

This program provides guidance in evaluating compliance with CGMP requirements. As soon as the District becomes aware of any significant inspectional, analytical, or other information developed under this program that may affect the agency's new drug approval decisions with respect to a firm, the District should report the information immediately according to current FACTS procedures. This includes filing OAI notifications and removing the notifications.

However, if information is encountered pertaining to inadequate postapproval reporting (Annual Reports, Supplements, Field Alert Reports, Adverse Drug Experience Reports, etc.) the information should be described in accordance with directions provided in those applicable compliance programs and under separate captions in the EIR. Data system information about these inspectional activities should be reported using applicable separate PAC(s). Expansion of coverage under these programs into a CGMP inspection must be reported under this compliance program.

The Districts are requested to use this revised compliance program for **all** GMP inspections.

PART I – BACKGROUND

A primary mission of the Food and Drug Administration is to conduct comprehensive regulatory coverage of all aspects of production and distribution of drugs and drug products to assure that such products meet the 501(a)(2)(B) requirements of the Act. FDA has developed two basic strategies:

1. evaluating through factory inspections, including the collection and analysis of associated samples, the conditions and practices under which drugs and drug products are manufactured, packed, tested and held, and

2. monitoring the quality of drugs and drug products through surveillance activities such as sampling and analyzing products in distribution.

This compliance program is designed to provide guidance for implementing the first strategy. Products from production and distribution facilities covered under this program are consistently of acceptable quality if the firm is operating in a state of control. The Drug Product Surveillance Program (CP 7356.008) provides guidance for the latter strategy.

The inspectional guidance in this program is structured to provide for efficient use of resources devoted to routine surveillance coverage, recognizing that in-depth coverage of all systems and all processes is not feasible for all firms on a biennial basis. It also provides for follow-up compliance coverage as needed.

PART II – IMPLEMENTATION

OBJECTIVES

The goal of this program's activities is to minimise consumers' exposure to adulterated drug products.

Under this program, inspections and investigations, sample collections and analyses, and regulatory or administrative follow-up are made:

1.   to determine whether inspected firms are operating in compliance with applicable CGMP requirements, and if not, to provide the evidence for actions to prevent adulterated products from entering the market and as appropriate to remove adulterated products from the market, and to take action against persons responsible as appropriate;

2.   to provide CGMP assessment which may be used in efficient determination of acceptability of the firm in the pre-approval review of a facility for new drug applications;

3.   to provide input to firms during inspections to improve their compliance with regulations; and,

4.   to continue FDA's unique expertise in drug manufacturing in determining the adequacy of CGMP requirements, Agency CGMP regulatory policy, and guidance documents.

STRATEGY

A.   Biennial Inspection of Manufacturing Sites (includes repackaging, contract labs, etc.)

Drugs and drug products are manufactured using many physical operations to bring together components and containers and closures into a product that is released for distribution. Activities found in drug firms can be organised into systems that are sets of operations and related activities. Control of all systems helps to ensure the firm will produce drugs that are safe, have the identity and strength, and meet the quality and purity characteristics as intended.

Biennial inspections (every two years) conducted under this program:

1.   reduce the risk that adulterated products are reaching the marketplace;

2.   increase communication between the industry and the Agency;

3.   provide for timely evaluation of new manufacturing operations in the firm; and,

4.   provide for regular feedback from the Agency to individual firms on the continuing status of the firm's GMP compliance.

This program applies to all drug manufacturing operations.

Currently there are not enough FDA resources to audit every aspect of CGMP in every manufacturing facility during every inspection visit. Profile classes generalise inspection coverage from a small number of specific products to all the products in that class. This program establishes a systems approach to further generalise inspection coverage from a small number of profile classes to an overall evaluation of the firm. Reporting coverage for every profile class as defined in FACTS, in each biennial inspection, provides the most broadly resource-efficient approach. Biennial updating of all profile classes will allow for CGMP acceptability determinations to be made without delays resulting from revisiting the firm. This will speed the review process, in response to compressed time frames for application decisions and in response to provisions of the Food and Drug Administration Modernization Act of 1997 (FDAMA). This will allow for Pre-approval Inspections/Investigations Program inspections and Post-approval Audit Inspections Program inspections to focus on the specific issues related to a given application or the firmís ability to keep applications current.

The inspection is defined as audit coverage of 2 or more systems, with mandatory coverage of the Quality System (see system definitions below). Inspection options include different numbers of systems to be covered depending on the purpose of the inspection. Inspecting the minimum number of systems, or more systems as deemed necessary by the District, will provide the basis for an overall CGMP decision.

B.   Inspection of Systems

Inspections of drug manufacturers should be made and reported using the system definitions and organization in this compliance program. Focusing on systems, rather than profile classes, will increase efficiency in conducting inspections because the systems are often applicable to multiple profile classes.

One biennial inspection visit will result in a determination of acceptability/non-acceptability for all profile classes. Inspection coverage should be representative of all the profile classes manufactured by the firm.

The efficiency will be realised because multiple visits to a firm will not be needed to cover all profile classes; delays in approval decisions will be avoided because up-to-date profile class information will be available at all times.

Coverage of a system should be sufficiently detailed, with specific examples selected, so that the system inspection outcome reflects the state of control in that system for every profile class. If a particular system is adequate, it should be adequate for all profile classes manufactured by the firm. For example, the way a firm handles "materials" (i.e., receipt, sampling, testing, acceptance, etc.) should be the same for all profile classes. The investigator should not have to inspect the Material System for each profile class.

Likewise in the Production System, there are general requirements like SOP use, charge-in of components, equipment identification, in-process sampling and testing which can be evaluated through selection of example products in various profile classes. Under each system there may be something unique for a particular profile class: e.g., under the Materials System, the production of Water for Injection USP for use in manufacturing. Selecting unique functions within a system will be at the discretion of the lead investigator. Any given inspection need not cover every system. See Part III.

Complete inspection of one system may necessitate further follow up of some items within the activities of another/other system(s) to fully document the findings. However, this coverage does not constitute nor require complete coverage of these other systems.

C.   A Scheme of Systems for the Manufacture of Drugs/Drug Products

A general scheme of systems for auditing the manufacture of drugs and drug products consists of the following:

1.   **Quality System**. This system assures overall compliance with cGMPs and internal procedures and specifications. The system includes the quality control unit and all of its review and approval duties (e.g., change control, reprocessing, batch release, annual record review, validation protocols, and reports, etc.). It includes all product defect evaluations and evaluation of returned and salvaged drug products. See the CGMP regulation, 21 CFR 211 Subparts B, E, F, G, I, J, and K.

2.   **Facilities and Equipment System**. This system includes the measures and activities which provide an appropriate physical environment and resources used in the production of the drugs or drug products. It includes:

     a)   Buildings and facilities along with maintenance;

b) Equipment qualifications (installation and operation); equipment calibration and preventative maintenance; and cleaning and validation of cleaning processes as appropriate. Process performance qualification will be evaluated as part of the inspection of the overall process validation which is done within the system where the process is employed; and,

c) Utilities that are not intended to be incorporated into the product such as HVAC, compressed gases, steam and water systems.

See the CGMP regulation, 21 CFR 211 Subparts B, C, D, and J.

2. **Materials System**. This system includes measures and activities to control finished products, components, including water or gases, that are incorporated into the product, containers and closures. It includes validation of computerised inventory control processes, drug storage, distribution controls, and records. See the CGMP regulation, 21 CFR 211 Subparts B, E, H, and J.

4. **Production System**. This system includes measures and activities to control the manufacture of drugs and drug products including batch compounding, dosage form production, in-process sampling and testing, and process validation. It also includes establishing, following, and documenting performance of approved manufacturing procedures. See the CGMP regulation, 21 CFR 211 Subparts B, F, and J.

5. **Packaging and Labeling System**. This system includes measures and activities that control the packaging and labeling of drugs and drug products. It includes written procedures, label examination and usage, label storage and issuance, packaging and labeling operations controls, and validation of these operations. See the CGMP regulation, 21 CFR 211 Subparts B, G, and J.

6. **Laboratory Control System**. This system includes measures and activities related to laboratory procedures, testing, analytical methods development and validation or verification, and the stability program. See the CGMP regulation, 21 CFR 211 Subparts B, I, J, and K.

The overall theme in devising this scheme of systems was the subchapter structure of the CGMP regulation. Every effort was made to group whole subchapters together in a rational set of six systems which incorporates the general scheme of pharmaceutical manufacturing operations.

The organization and personnel, including appropriate qualifications and training, employed in any given system, will be evaluated as part of that system's operation. Production, control, or distribution records required to be maintained by the CGMP regulation and selected for review should be included for inspection audit within the context of each of the above systems. Inspections of contract companies should be within the system for which the product or service is contracted as well as their Quality System.

As this program approach is implemented, the experience gained will be reviewed to make modifications to the system definitions and organization as needed.

PROGRAM MANAGEMENT INSTRUCTIONS

A. Definitions

1. Surveillance Inspections

The Full Inspection Option

The Full Inspection Option is a surveillance or compliance inspection which is meant to provide a broad and deep evaluation of the firm's CGMP. This will be done when little or no information is known about a firm's CGMP compliance (e.g., for new firms); or for firms where there is doubt about the CGMP compliance in the firm (e.g., a firm whose history has documented short-lived compliance and recidivism); or follow up to previous regulatory actions. Based on findings of objectionable conditions as listed in Part V in one or more systems (a minimum of two systems must be completed), a Full Inspection may revert to the Abbreviated Inspection Option, with District concurrence. See Part III, Section B.1.

During the course of a Full Inspection, verification of Quality System activities may require limited coverage in other systems. The Full Inspection Option will normally include an inspection audit of at least four of the systems, one of which must be the Quality System (the system which includes the responsibility for the annual product reviews).

The Abbreviated Inspection Option

The Abbreviated Inspection Option is a surveillance or compliance inspection which is meant to provide an efficient update evaluation of a firm's CGMP. The abbreviated inspection will provide documentation for continuing a firm in a satisfactory CGMP compliance status. Generally this will be done when a firm has a record of satisfactory CGMP compliance, with no significant recall, or product defect or alert incidents, or with little shift in the manufacturing profiles of the firm within the previous two years. See Part III, Section B.2. A Full Inspection may revert to an Abbreviated Inspection based on findings of objectionable conditions as listed in Part V in one or more systems. The Abbreviated Inspection Option normally will include an inspection audit of at least two of the systems, one of which must be the Quality System (the system which includes the responsibility for the annual product reviews). The District drug program managers should ensure that the optional systems are rotated in successive Abbreviated

Inspections. During the course of an abbreviated inspection, verification of quality system activities may require limited coverage in other systems. Some firms participate in a limited part of the production of a drug or drug product, e.g., a contract laboratory. Such firms may employ only two of the systems defined.

In these cases the inspection of the two systems will comprise inspection of the entire firm and will be considered the Full Inspection Option.

Selecting Systems for Coverage

The selection of the system(s) for coverage will be made by the District Office based on such factors as a given firm's specific operation, history of previous coverage, history of compliance, or other priorities determined by the District Office.

2. **Compliance Inspections**

Compliance Inspections are inspections done to evaluate or verify compliance corrective actions after a regulatory action has been taken. First, the coverage given in compliance inspections must be related to the areas found deficient and subjected to corrective actions.

In addition, coverage must be given to systems because a determination must be made on the overall compliance status of the firm after the corrective actions are taken. The firm is expected to address all of its operations in its corrective action plan after a previously violative inspection, not just the deficiencies noted in the FDA-483. The Full Inspection Option should be used for a compliance inspection, especially if the Abbreviated Inspection Option was used during the violative inspection.

Compliance Inspections include For Cause Inspections. For Cause Inspections are compliance inspections which are done to investigate a specific problem that has come to the attention of some level of the agency. The problems may be indicated in Field Alert Reports (FARs), industry complaints, recalls, indicators of defective products, etc. Coverage of these areas may be assigned under other compliance programs, however, expansion of the coverage to a GMP inspection is to be reported under this program.

For Cause Inspections may be assigned under this program as the need arises.

3. State of Control

A drug firm is considered to be operating in a state of control when it employs conditions and practices that assure compliance with the intent of Sections 501(a)(2)(B) of the Act and portions of the CGMP regulations that pertain to their systems. A firm in a state of control produces finished drug products for which there is an adequate level of assurance of quality, strength, identity and purity.

A firm is out of control if any one system is out of control. A system is out of control if the quality, identity, strength and purity of the products resulting from that/those system(s) cannot be adequately assured. Documented CGMP deficiencies provide the evidence for concluding that a system is not operating in a state of control. See Part V. Regulatory/Administrative Strategy for a discussion of compliance actions based on inspection findings demonstrating out of control systems/firm.

4. Drug Process

A drug process is a related series of operations which result in the preparation of a drug or drug product.

Major operations or steps in a drug process may include mixing, granulation, encapsulation, tableting, chemical synthesis, fermentation, aseptic filling, sterilization, packing, labeling, testing, etc.

5. Drug Manufacturing Inspection

A drug manufacturing inspection is a factory inspection in which evaluation of two or more systems, including the Quality System, is done to determine if manufacturing is occurring in a state of control.

B. Inspection Planning

The Field will conduct drug manufacturing inspections and maintain profiles or other monitoring systems which will ensure that each drug firm receives biennial inspectional coverage, as provided for in the strategy.

The District Office is responsible for determining the depth of coverage given to each drug firm. CGMP inspectional coverage shall be sufficient to assess the state of compliance for each firm.

The frequency and depth of inspection should be determined by the statutory obligation, the firm's compliance history, the technology employed, and the characteristics of the products. When a system is inspected, the inspection of that system may be considered applicable to all products which use it.

Investigators should select an adequate number and type of products to accomplish coverage of the system. Selection of products should be made so that coverage is representative of the firm's overall abilities in manufacturing within CGMP requirements.

Review of NDA/ANDA files may assist in selecting significant drug processes for coverage in the various systems. Significant drug processes are those which utilise all the systems in the firm very broadly and/or which contain steps with unique or difficult manipulation in the performance of a step. Products posing special manufacturing features, e.g., low dose products, narrow therapeutic range drugs, combination drugs, modified release products, etc., and new products made under an approved drug application, should be considered first in selecting products for coverage.

The health significance of certain CGMP deviations may be lower when the drug product involved has no major systemic effect or no dosage limitations such as in products like calamine lotion or OTC medicated shampoos. Such products should be given inspection coverage with appropriate priority.

Inspections for this compliance program may be performed during visits to a firm when operations are being performed for other compliance programs or other investigations.

C. Profiles

The inspection findings will be used as the basis for updating all profile classes in the profile screen of the FACTS EIR coversheet that is used to record profile/class determinations. Normally, an inspection under this systems approach will result in all profile classes being updated.

PART III – INSPECTIONAL

INVESTIGATIONAL OPERATIONS

A. General

Review and use the CGMPs for Finished Pharmaceuticals (21 CFR 210 and 211) to evaluate manufacturing processes. Use Guides to Inspection published by the Office of Regional Operations for information on technical applications in various manufacturing systems.

The investigator should conduct inspections according to the STRATEGY section in Part II of this compliance program. Recognizing that drug firms vary greatly in size and scope, and manufacturing systems are more or less sophisticated, the approach to inspecting each firm should be carefully planned.

For example, it may be more appropriate to review the Quality System thoroughly before entering production areas in some firms; in others, the Quality System review should take place concurrently with inspection of another system or systems selected for coverage. The complexity and variability necessitate a flexible inspection approach; one which allows the investigator to choose the inspection focus and depth appropriate for a specific firm, but also one which directs the performance and reporting on the inspection within a framework which will provide for a uniform level of CGMP assessment. Furthermore, this inspection approach will provide for fast communication and evaluation of findings.

Inspectional Observations noting CGMP deficiencies should be related to a requirement. Requirements for manufacture of drug products (dosage forms) are in the CGMP regulation and are amplified by policy in the Compliance Policy Guides, case precedents, etc. CGMP requirements apply to the manufacture of distributed prescription drug products, OTC drug products, approved products and products not requiring approval, as well as drug products used in clinical trials. The CGMP regulations are not direct requirements for manufacture of APIís; the regulations should not be referenced as the basis for a GMP deficiency in the manufacture of Active Pharmaceutical Ingredients (APIs), but they are guidance for CGMP in API manufacture.

Guidance documents do not establish requirements. They state examples of ways to meet requirements.

Guidance documents are not to be referred to as the justification for an inspectional observation. The justification comes from the CGMPs. Current Guides to Inspection and Guidance to Industry documents provide interpretations of requirements, which may assist in the evaluation of the adequacy of CGMP systems.

Current inspectional observation policy as stated in the IOM says that the FDA-483, when issued, should be specific and contain only significant items. For this program, inspection observations should be organised under separate captions by the systems defined in this program. List observations in order of importance within each system. Where repeated or similar observations are made, they should be consolidated under a unified observation. For those Districts utilizing Turbo EIR, a limited number of observations can be common to more than one system (e.g. organization and personnel including appropriate qualifications and training). In these instances, put the observation in the first system reported on the FDA-483 and in the text of the EIR, reference the applicability to other systems where appropriate.

This is being done to accommodate the structure of Turbo EIR which allows individual citation once per FDA-483. Refrain from using unsubstantiated conclusions. Do not use the term "inadequate" without explaining why and how. Refer to policy in the IOM, Chapter 5, Section 512 and Field Management Directive 120 for further guidance on the content of Inspectional Observations.

Specific specialised inspectional guidance may be provided as attachments to this program, or in requests for inspection, assignments, etc.

B.   Inspection Approaches

This program provides two surveillance inspectional options, Abbreviated Inspection Option and Full Inspection Option. See the definitions of the inspection options in Part II of this program.

1.   **Selecting the Full Inspection Option**. The Full Inspection Option will include inspection of at least four of the systems as listed in Part II Strategy, one of which must be the Quality System.

  a.   Select the Full Inspection Option for an initial FDA inspection of a facility. A Full Inspection may revert to the Abbreviated Inspection Option, **with District concurrence**, based on finding of objectionable conditions as listed in Part V in one or more systems (a minimum of two systems must be completed).

  b.   Select the Full Inspection Option when the firm has a history of fluctuating into and out of compliance. To determine if the firm meets this criterion, the District should utilise all information at its disposal, such as, inspection results, results of sample analyses, complaints, DQRS reports, recalls, etc. and the compliance actions resulting from them or from past inspections. A Full Inspection may revert to the Abbreviated Inspection Option, **with District concurrence**, based on findings of objectionable conditions as listed in Part V in one or more systems (a minimum of two systems must be completed).

  c.   Evaluate if important changes have occurred by comparing current operations against the EIR for the previous Full Inspection. The following types of changes are typical of those that warrant the Full Inspection Option:

    1.   New potential for cross-contamination arising through change in process or product line.

    2.   Use of new technology requiring new expertise, significant new equipment, or new facilities.

  d.   A Full Inspection may also be conducted on a surveillance basis at the District's discretion.

  e.   The Full Inspection Option will satisfy the biennial inspection requirement.

  f.   Follow up to a Warning Letter or other significant regulatory actions should require a Full Inspection option.

2.   **Selecting the Abbreviated Inspection Option**. The Abbreviated Inspection Option normally will include inspection audit of at least two systems, one of which must be the Quality System. During the course of an Abbreviated Inspection, verification of quality system activities may require limited coverage in other systems.

a.   This option involves an inspection of the manufacturer to maintain surveillance over the firm's activities and to provide input to the firm on maintaining and improving the GMP level of assurance of quality of its products.

b.   A Full Inspection may revert to the Abbreviated Inspection option, **with District concurrence**, based on findings of objectionable conditions as listed in Part V in one or more systems (a minimum of two systems must be completed).

c.   An abbreviated inspection is adequate for routine coverage and will satisfy the biennial inspectional requirement.

Comprehensive Inspection Coverage

It is not anticipated that Full Inspections will be conducted every two years. They may be conducted at less frequent intervals, perhaps at every third or fourth inspection cycle. Districts should consider selecting different optional systems for inspection coverage as a cycle of Abbreviated Inspections is carried out to build comprehensive information on the firmís total manufacturing activities.

C.   System Inspection Coverage

QUALITY SYSTEM

Assessment of the Quality System is two phased. The first phase is to evaluate whether the Quality Control Unit has fulfilled the responsibility to review and approve all procedures related to production, quality control, and quality assurance and assure the procedures are adequate for their intended use. This also includes the associated recordkeeping systems. The second phase is to assess the data collected to identify quality problems and may link to other major systems for inspectional coverage.

For each of the following, the firm should have written and approved procedures and documentation resulting therefrom. The firmís adherence to written procedures should be verified through observation whenever possible. These areas are not limited to finished products, but may also incorporate components and in-process materials. These areas may indicate deficiencies not only in this system but also in other major systems that would warrant expansion of coverage. All areas under this system should be covered; however the depth of coverage may vary depending upon inspectional findings.

–   Product reviews: at least annually; should include information from areas listed below as appropriate; batches reviewed, for each product, are representative of all batches manufactured; trends are identified; refer to 21 CFR 211.180(e).

–   Complaint reviews (quality and medical): documented; evaluated; investigated in a timely manner; includes corrective action where appropriate.

–   Discrepancy and failure investigations related to manufacturing and testing: documented; evaluated; investigated in a timely manner; includes corrective action where appropriate.

–   Change Control: documented; evaluated; approved; need for revalidation assessed.

–   Product Improvement Projects: for marketed products

–   Reprocess/Rework: evaluation, review and approval; impact on validation and stability.

–   Returns/Salvages: assessment; investigation expanded where warranted; disposition.

–   Rejects: investigation expanded where warranted; corrective action where appropriate.

–   Stability Failures: investigation expanded where warranted; need for field alerts evaluated; disposition.

–   Quarantine products.

– Validation: status of required validation/revalidation (e.g., computer, manufacturing process, laboratory methods).

– Training/qualification of employees in quality control unit functions.

FACILITIES AND EQUIPMENT SYSTEM

For each of the following, the firm should have written and approved procedures and documentation resulting therefrom. The firmís adherence to written procedures should be verified through observation whenever possible. These areas may indicate deficiencies not only in this system but also in other systems that would warrant expansion of coverage. When this system is selected for coverage in addition to the Quality System, all areas listed below should be covered; however, the depth of coverage may vary depending upon inspectional findings.

1. Facilities

– cleaning and maintenance

– facility layout and air handling systems for prevention of cross-contamination (e.g. penicillin, beta-lactams, steroids, hormones, cytotoxics, etc.)

– specifically designed areas for the manufacturing operations performed by the firm to prevent contamination or mix-ups

– general air handling systems

– control system for implementing changes in the building

– lighting, potable water, washing and toilet facilities, sewage and refuse disposal

– sanitation of the building, use of rodenticides, fungicides, insecticides, cleaning and sanitizing agents

2. Equipment

– equipment installation and operational qualification where appropriate

– adequacy of equipment design, size, and location

– equipment surfaces should not be reactive, additive, or absorptive

– appropriate use of equipment operations substances, (lubricants, coolants, refrigerants, etc.) contacting products/containers/etc.

– cleaning procedures and cleaning validation

– controls to prevent contamination, particularly with any pesticides or any other toxic materials, or other drug or non-drug chemicals

– qualification, calibration and maintenance of storage equipment, such as refrigerators and freezers for ensuring that standards, raw materials, reagents, etc. are stored at the proper temperatures

– equipment qualification, calibration and maintenance, including computer qualification/ validation and security

– control system for implementing changes in the equipment

– equipment identification practices (where appropriate)

0062009621– documented investigation into any unexpected discrepancy

MATERIALS SYSTEM

For each of the following, the firm should have written and approved procedures and documentation resulting therefrom. The firmís adherence to written procedures should be verified through observation whenever possible. These areas are not limited to finished products, but may also incorporate components and in-process materials. These areas may indicate deficiencies not only in this system but also in other systems that would warrant expansion of coverage. When this system is selected for coverage in addition to the Quality System, all areas listed below should be covered; however, the depth of coverage may vary depending upon inspectional findings.

- training/qualification of personnel
- identification of components, containers, closures
- inventory of components, containers, closures
- storage conditions
- storage under quarantine until tested or examined and released
- representative samples collected, tested or examined using appropriate means
- at least one specific identity test is conducted on each lot of each component
- a visual identification is conducted on each lot of containers and closures
- testing or validation of supplier's test results for components, containers and closures
- rejection of any component, container, closure not meeting acceptance requirements. Investigate fully the firmís procedures for verification of the source of components.
- appropriate retesting/reexamination of components, containers, closures
- first in-first out use of components, containers, closures
- quarantine of rejected materials
- water and process gas supply, design, maintenance, validation and operation
- containers and closures should not be additive, reactive, or absorptive to the drug product
- control system for implementing changes in the materials handling operations
- qualification/validation and security of computerised or automated processes
- finished product distribution records by lot
- documented investigation into any unexpected discrepancy

PRODUCTION SYSTEM

For each of the following, the firm should have written and approved procedures and documentation resulting therefrom. The firmís adherence to written procedures should be verified through observation whenever possible. These areas are not limited to finished products, but may also incorporate components and in-process materials. These areas may indicate deficiencies not only in this system but also in other systems that would warrant expansion of coverage. When this system is selected for coverage in addition to the Quality System, all areas listed below should be covered; however, the depth of coverage may vary depending upon inspectional findings.

- training/qualification of personnel
- control system for implementing changes in processes
- adequate procedure and practice for charge-in of components
- formulation/manufacturing at not less than 100%
- identification of equipment with contents, and where appropriate phase of manufacturing and/or status
- validation and verification of cleaning/sterilization/ depyrogenation of containers and closures
- calculation and documentation of actual yields and percentage of theoretical yields
- contemporaneous and complete batch production documentation

- established time limits for completion of phases of production
- implementation and documentation of in-process controls, tests, and examinations (e.g., pH, adequacy of mix, weight variation, clarity)
- justification and consistency of in-process specifications and drug product final specifications
- prevention of objectionable microorganisms in non-sterile drug products
- adherence to preprocessing procedures (e.g., set-up, line clearance, etc.)
- equipment cleaning and use logs
- master production and control records
- batch production and control records
- process validation, including validation and security of computerised or automated processes
- change control; the need for revalidation evaluated
- documented investigation into any unexpected discrepancy

PACKAGING AND LABELING SYSTEM

For each of the following, the firm should have written and approved procedures and documentation resulting therefrom. The firm's adherence to written procedures should be verified through observation whenever possible. These areas are not limited only to finished products, but may also incorporate components and in-process materials. These areas may indicate deficiencies not only in this system but also in other systems that would warrant expansion of coverage. When this system is selected for coverage in addition to the Quality System, all areas listed below should be covered; however, the depth of coverage may vary depending upon inspectional findings.

- training/qualification of personnel
- acceptance operations for packaging and labeling materials
- control system for implementing changes in packaging and labeling operations
- adequate storage for labels and labeling, both approved and returned after issued
- control of labels which are similar in size, shape, and colour for different products
- finished product cut labels for immediate containers which are similar in appearance without some type of 100 percent electronic or visual verification system or the use of dedicated lines
- gang printing of labels is not done, unless they are differentiated by size, shape, or colour
- control of filled unlabeled containers that are later labeled under multiple private labels
- adequate packaging records that will include specimens of all labels used
- control of issuance of labeling, examination of issued labels and reconciliation of used labels
- examination of the labeled finished product
- adequate inspection (proofing) of incoming labeling
- use of lot numbers, destruction of excess labeling bearing lot/control numbers
- physical/spatial separation between different labeling and packaging lines
- monitoring of printing devices associated with manufacturing lines
- line clearance, inspection and documentation
- adequate expiration dates on the label
- conformance to tamper-evident (TEP) packaging requirements (see 21CFR 211.132 and Compliance Policy Guide, 7132a.17)

- validation of packaging and labeling operations including validation and security of computerised processes
- documented investigation into any unexpected discrepancy

LABORATORY CONTROL SYSTEM

For each of the following, the firm should have written and approved procedures and documentation resulting therefrom. The firmís adherence to written procedures should be verified through observation whenever possible. These areas are not limited only to finished products, but may also incorporate components and in-process materials. These areas may indicate deficiencies not only in this system but also in other systems that would warrant expansion of coverage. When this system is selected for coverage in addition to the Quality System, all areas listed below should be covered; however, the depth of coverage may vary depending upon inspectional findings.

- training/qualification of personnel
- adequacy of staffing for laboratory operations
- adequacy of equipment and facility for intended use
- calibration and maintenance programs for analytical instruments and equipment
- validation and security of computerised or automated processes
- reference standards; source, purity and assay, and tests to establish equivalency to current official reference standards as appropriate
- system suitability checks on chromatographic systems (e.g., GC or HPLC)
- specifications, standards, and representative sampling plans
- adherence to the written methods of analysis
- validation/verification of analytical methods
- control system for implementing changes in laboratory operations
- required testing is performed on the correct samples
- documented investigation into any unexpected discrepancy
- complete analytical records from all tests and summaries of results
- quality and retention of raw data (e.g., chromatograms and spectra)
- correlation of result summaries to raw data; presence of unused data
- adherence to an adequate Out of Specification (OOS) procedure which includes timely completion of the investigation
- adequate reserve samples; documentation of reserve sample examination
- stability testing program, including demonstration of stability indicating capability of the test methods

D. Sampling

Samples of defective product constitute persuasive evidence that significant CGMP problems exist. Physical samples may be an integral part of a CGMP inspection where control deficiencies are observed. Physical samples should be correlated with observed control deficiencies. Consider consulting your servicing laboratory for guidance on quantity and type of samples (in-process or finished) to be collected. Documentary samples may be submitted when the documentation illustrates the deficiencies better than a physical sample. Districts may elect to collect, but not analyze, physical samples, or to collect documentary samples to document CGMP deficiencies. Physical sample analysis is not necessary to document CGMP deficiencies.

When a large number of products have been produced under deficient controls, collect physical and/or documentary samples of products which have the greatest therapeutic significance, narrow range of toxicity, or a low dosage strength. Include samples of products of minimal therapeutic significance only when they illustrate highly significant CGMP deficiencies.

For sampling guidance, refer to IOM, Chapter 4

E.   Inspection Teams

An inspection team (See IOM 502.4) composed of experts from within the District, other Districts, or Headquarters is encouraged when it provides needed expertise and experience.

Contact ORO/Division of Field Investigations if technical assistance is needed (See also FMD 142). Participation of an analyst (chemist or microbiologist) on an inspection team is also encouraged, especially where laboratory issues are extensive or complex. Contact your Drug Servicing Laboratory or ORO/Division of Field Science.

F.   Reporting

The investigator will utilise Subchapter 590 of the IOM for guidance in reporting of inspectional findings. Identify systems covered in the Summary of Findings. Identify and explain in the body of the report the rationale for inspecting the profile classes covered.

Report and discuss in full any adverse findings by systems under separate captions. Add additional information as needed or desired, for example, a description of any significant changes that have occurred since previous inspections.

Reports with specific, specialised information required should be prepared as instructed within the individual assignment/attachment.

PART IV – ANALYTICAL

ANALYZING LABORATORIES

1.   Routine chemical analyses – all Servicing Laboratories except WEAC.

2.   Sterility testing:

Region	Examining Laboratory
NE	NRL
SE	SRL
CE	NRL
SW, PA	SAN-DO

3.   Other microbiological examinations – NRL (for the CE Region), SRL, SAN, and DEN.

Salmonella Serotyping Lab – ARL

4.   Chemical cross-contamination analyses by mass spectrometry (MS) – NRL, SRL, DEN, PRL/NW, and PHI. Non-mass spectrometry laboratories should call one of their own regional MS capable laboratories or Division of Field Science (HFC-140) to determine the most appropriate lab for the determinations to be performed.

5.   Chemical cross-contamination analyses by Nuclear Magnetic Resonance (NMR) spectroscopy NRL. Non-NMR laboratories should call one of their own regional labs equipped with NMR or Division

of Field Science (HFC-140) to determine the most appropriate lab for the determinations to be performed.

6. Dissolution testing – NRL, KAN, SRL, SJN, DET, PHI, DEN, PRL/SW and PRL-NW. Districts without dissolution testing capability should use one of their own regional labs for dissolution testing. Otherwise, call DFS.

7. Antibiotic Analyses:

    ORA Examining Laboratory

    Denver District Lab (HFR-SW260)
    Tetracyclines
    Erythromycins

    Northeast Regional Lab (HFR-NE500)
    Penicillins
    Cephalosporins

    CDER Examining Laboratory
    Office of Testing and Research
    Division of Pharmaceutical Analysis
    (HFD-473)
    All other Antibiotics

7. Bioassays – Division of Testing and Applied Analytical Research, Drug Bioanalysis Branch (HFN-471).

9. Particulate Matter in Injectables: NRL, SRL.

10. Pyrogen/LAL Testing: SRL

ANALYSIS:

1. Samples are to be examined for compliance with applicable specifications as they relate to deficiencies noted during the inspection. Check analyses will be by the official method, or when no official method exists, by other validated procedures. See CPG 7152.01.

2. The presence of cross-contamination must be confirmed by a second method. Spectroscopic methods, such as MS, NMR, UV-Visible, or infrared (IR) are preferred. A second confirmatory method should be employed by different mechanisms than the initial analysis (i.e., ion-pairing vs. conventional reverse phase HPLC).

3. Check Analysis for dissolution rate must be performed by a second dissolution-testing laboratory.

4. Sterility testing methods should be based on current editions of USP and the Sterility Analytical Manual. Other microbiological examinations should be based on appropriate sections of USP and BAM.

PART V – REGULATORY/ADMINISTRATIVE STRATEGY

Inspection findings that demonstrate that a firm is not operating in a state of control may be used as evidence for taking appropriate advisory, administrative and/or judicial actions.

When the management of the firm is unwilling or unable to provide adequate corrective actions in an appropriate time frame, formal agency regulatory actions will be recommended, designed to meet the situation encountered.

When deciding the type of action to recommend, the initial decision should be based on the seriousness of the problem and the most effective way to protect consumers. Outstanding instructions in the Regulatory Procedures Manual (RPM) should be followed.

The endorsement to the inspection report should point out the actions that have been taken or will be taken and when. All deficiencies noted in inspections/audits under this program must be addressed by stating the firm's corrective actions, accomplished or projected, for each as established in the discussion with management at the close of the inspection.

All corrective action approaches in domestic firms are monitored and managed by the District Offices. The approaches may range from shut down of operations, recall of products, conducting testing programs, development of new procedures, modifications of plants and equipment, to simple immediate corrections of conditions. CDER/DMPQ/CMGB/HFD-325 will assist District Offices as requested.

An inspection report that documents that one or more systems is/are out of control should be classified OAI. District Offices may issue Warning Letters per RPM, Chapter 4, to warn firms of violations, to solicit voluntary corrections, and to provide for the initial phase of formal agency regulatory actions.

Issuance of a Warning Letter or taking other regulatory actions pursuant to a surveillance inspection (other than a For Cause Inspection) should result in the classification of all profile classes as unacceptable. Also, the inspection findings will be used as the basis for updating profile classes in FACTS.

FDA laboratory tests which demonstrate effects of absent or inadequate current good manufacturing practice are strong evidence for supporting regulatory actions. Such evidence development should be considered as an inspection progresses and deficiencies are found. However, the lack of violative physical samples is **not** a barrier to pursuing regulatory and/or administrative action provided that CGMP deficiencies have been well documented. Likewise, physical samples found to be in compliance are **not** a barrier to pursuing action under CGMP charges.

Evidence to support significant and/or a trend of deficiencies within a system covered could demonstrate the failure of a system and should result in consideration of the issuance of a Warning Letter or other regulatory action by the District. When deciding the type of action to recommend, the initial decision should be based on the seriousness and/or the frequency of the problem. Examples include the following:

Quality System:

1. Pattern of failure to review/approve procedures.

2. Pattern of failure to document execution of operations as required.

3) Pattern of failure to review documentation.

4) Pattern of failure to conduct investigations and resolve discrepancies/failures/deviations/complaints.

5) Pattern of failure to assess other systems to assure compliance with GMP and SOPs.

Facilities and Equipment

1. Contamination with filth, objectionable microorganisms, toxic chemicals or other drug chemicals, or a reasonable potential for contamination, with demonstrated avenues of contamination, such as airborne or through unclean equipment

2. Pattern of failure to validate cleaning procedures for non-dedicated equipment. Lack of demonstration of effectiveness of cleaning for dedicated equipment.

3. Pattern of failure to document investigation of discrepancies.

4. Pattern of failure to establish/follow a control system for implementing changes in the equipment.

5. Pattern of failure to qualify equipment, including computers.

Materials System

1. Release of materials for use or distribution that do not conform to established specifications.
2. Pattern of failure to conduct one specific identity test for components.
3. Pattern of failure to document investigation of discrepancies.
4. Pattern of failure to establish/follow a control system for implementing changes in the materials handling operations.
5. Lack of validation of water systems as required depending upon the intended use of the water.
6. Lack of validation of computerised processes.

Production System

1. Pattern of failure to establish/follow a control system for implementing changes in the production system operations.
2. Pattern of failure to document investigation of discrepancies
3. Lack of process validation.
4. Lack of validation of computerised processes.
5. Pattern of incomplete or missing batch production records.
6. Pattern of nonconformance to established in-process controls, tests, and/or specifications.

Packaging and Labeling

1. Pattern of failure to establish/follow a control system for implementing changes in the packaging and/or labeling operations.
2. Pattern of failure to document investigation of discrepancies.
3. Lack of validation of computerised processes.
4. Lack of control of packaging and labeling operations that may introduce a potential for mislabeling.
5. Lack of packaging validation.

Laboratory Control System

1. Pattern of failure to establish/follow a control system for implementing changes in the laboratory operations.
2. Pattern of failure to document investigation of discrepancies.
3. Lack of validation of computerised and/or automated processes.
4. Pattern of inadequate sampling practices.
5. Lack of validated analytical methods.
6. Pattern of failure to follow approved analytical procedures.
7. Pattern of failure to follow an adequate OOS procedure.
8. Pattern of failure to retain raw data.
9. Lack of stability indicating methods.
10. Pattern of failure to follow stability programs.

Follow up to a Warning Letter or other significant regulatory action as a result of an Abbreviated Inspection should warrant Full Inspection coverage as defined in this program.

PART VI – REFERENCES, ATTACHMENTS, AND PROGRAM CONTACTS

REFERENCES

1. Federal Food, Drug, and Cosmetic Act, as amended
2. Code of Federal Regulations, Title 21, Parts 210 and 211, as revised, including the General Comments (preamble)
3. Compliance Policy Guides Manual, Chapter 4 Human Drugs
4. Compressed Medical Gases Guideline
5. Guideline on General Principles of Process Validation
6. Guideline on Sterile Drug Products Produced by Aseptic Processing
7. Guide to Inspection of Computerised Systems in Drug Processes
8. Regulatory Procedures Manual, Part 8
9. Inspection Operations Manual (IOM)
10. Guide to Inspections of Dosage Form Drug Manufacturers-CGMPs
11. Guide to Inspections of Lyophilization of Parenterals
12. Guide to Inspections of Pharmaceutical Quality Control Laboratories
13. Guide to Inspections of High Purity Water Systems
14. Guide to Inspections of Validation of Cleaning Processes
15. Guidance to Industry (Draft) Investigation of OOS Test Results
16. 21 CFR Part 11 Electronic Records: Electronic Signatures
17. Compliance Policy Guide7153.17 Enforcement Policy, Part 11 Electronic Records; Electronic Signatures
18. Electronic Records Guide (4/21/98) (Electronic Signatures, 21 CFR Part 11 Answers to Frequently Asked Questions)

Attachments

Attachments to the Drug Process Inspection program may be issued for certain industries, dosage forms, and processes with known problems or unique drug processes. These attachments will contain the guidance needed to perform these specialised inspections.

Some of the attachments to be issued with this program may include reporting requirements specifically designed to obtain industry-wide information on certain practices to permit evaluation of the adequacy of FDA's regulatory efforts.

Attachments and/or reporting requirements will be periodically reviewed and evaluated and deleted from the program when they are no longer needed.

CONTACTS

ORA

ORO/Division of Field Investigations/Drug Group (HFC-132)
Telephone: (301) 827-5653

ORO/Division of Field Science (HFC-140)
Telephone: (301) 827-7605

Center for Drug Evaluation and Research

CGMP Questions

Division of Manufacturing & Product Quality
Chief, Case Management and Guidance Branch (HFD-325)
Telephone: (301) 594-0098
Fax: (301) 594-2202

Product Quality Problems (NDA/ANDA/Compendial Specifications deviations)

Division of Manufacturing & Product Quality
Chief, Case Management & Guidance Branch (HFD-325)
Telephone: (301) 594-0098
Fax: (301) 594-2202

Product Quality Problems (Drug Product Sampling, FAR reporting, ADE reporting)

Team Leader, Post Market Surveillance Team (HFD-336)
Division of Rx Drug Compliance and Surveillance
Telephone: (301) 594-0101
Fax: (301) 594-1146

PART VII – CENTER RESPONSIBILITIES

CENTER FOR DRUG EVALUATION AND RESEARCH

The Division of Manufacturing and Product Quality (DMPQ) HFD-320 will conduct an annual evaluation in order to assess and report on the effectiveness of this program.

Please send any comments on the operation and efficiency and direct any questions regarding application of the program to the Chief, Case Management and Guidance Branch, DMPQ, HFD-325, 301-594-0098, fax 301-594-2202.

3.6   References

1   Brooker, C. *Mosby's Dictionary of Medicine, Nursing and Health Professions*. Elsevier, Edinburgh, 2010

2   FDA (2002) Compliance Program Guidance Manual Program, Document 7356.002, Drug Manufacturing Inspections. Accessed on May 10, 2012 at http://www.fda.gov/downloads/ICECI/ComplianceManuals/ComplianceProgramMan ual/UCM125404.pdf

3   FDA "Pharmaceutical GMPs for the 21st century – a risk based approach", Final Report, US Department for Health and Human Services, US Food and Drug Administration, Rockville, MD, USA, 2004.

4   FDA " Guidance for Industry Quality Systems Approach to Pharmaceutical CGMP Regulations", US Department for Health and Human Services, US Food and Drug Administration, Rockville, MD, USA, 2006.

5   Sandle, T. (2011). 'Risk Management in Microbiology' in Saghee, M.R., Sandle, T. and Tidswell, E. (Eds.) *Microbiology and Sterility Assurance in Pharmaceuticals and Medical Devices*. Business Horizons, India.

6   FDA CDRH (2005) Code of Federal Regulations Title 21

7   FDA (2010) Information Sheet Guidance for IRBs, Clinical Investigators, and Sponsors. FDA Inspections of Clinical Investigators [Online]. Available from: www.fda.gov/downloads/RegulatoryInformation/Ö/UCM126553.pdf (Accessed: 06 May 2012).

8   Hilts, P.J. (2003). Protecting America's Health: The FDA, Business, and One Hundred Years of Regulation. New York: Alfred E. Knopf: USA.

9   FDA (2012). Investigations Operations Manual, Food and Drug Administration, Rockville, MD, USA at: http://www.fda.gov/ICECI/Inspections/IOM/default.htm

10   Elsevier Business Intelligence (2011). The Gold Sheet, "FDA Continues Aggressive Enforcement as Drug GMP Warning Letters Mount", April 01, 2011.

11   Moldenhaurer, J. (2010). *Recent Warning Letters Review for Preparation of an Aseptic Processing Inspection*, Volume 1, PDA: Bethesda, MD, USA

12   US Department of Health and Human Services. Warning Letter. Allure Labs, Inc. 5/24/11. Public Health Service, Food and Drug Administration, San Francisco District, Pacific Region.

13   US Department of Health and Human Services. Warning Letter. Aurobindo Pharma Limited 5/20/11. Public Health Service, *Food and Drug Administration*, Silver Spring MD 20993.

14   US Department of Health and Human Services. Warning Letter. NuSil Technology LLC 3/22/12. Public Health Service, Food and Drug Administration, San Francisco District, Pacific Region.

15 US Department of Health and Human Services. Warning Letter. Wintac Limited 2/23/12. Public Health Service, Food and Drug Administration, Silver Spring MD 20993.

16 US Department of Health and Human Services. Warning Letter. Merck KGaA 12/15/11. Public Health Service, Food and Drug Administration, Silver Spring MD 20993.

17 US Department of Health and Human Services. Warning Letter. Ben Venue Laboratories, Inc. 16-Nov-07. Public Health Service, Food and Drug Administration, Cincinnati District Office , Central Region.

18 US Department of Health and Human Services. Warning Letter. Professional Disposables International 7/5/11. Public Health Service, Food and Drug Administration, Silver Spring MD 20993

19 US Department of Health and Human Services. Warning Letter. Miramar Cosmetics, Inc. 8/18/11. Public Health Service, Food and Drug Administration, Florida District

20 US Department of Health and Human Services. Warning Letter. SmithKline Beecham Limited 10/7/11. Public Health Service, Food and Drug Administration, Silver Spring MD 20993.

21 US Department of Health and Human Services. Warning Letter. Jenahexal Pharm GmbH 10/19/11. Public Health Service, Food and Drug Administration, Silver Spring MD 20993

22 US Department of Health and Human Services. Warning Letter. Novartis International AG 11/18/11. Public Health Service, Food and Drug Administration, Silver Spring MD 20993.

23 US Department of Health and Human Services. Warning Letter. Gulf Pharmaceutical Industries 2/23/12. Public Health Service, Food and Drug Administration, Silver Spring MD 20993.

24 US Department of Health and Human Services. Warning Letter. Bristol Myers Squibb Holdings Pharma., Ltd. 8/30/10. Public Health Service, Food and Drug Administration, Silver Spring MD 20993

25 US Department of Health and Human Services. Warning Letter. Cadila Healthcare 6/21/11. Public Health Service, Food and Drug Administration, Silver Spring MD 20993

26 US Department of Health and Human Services. Warning Letter. Gilead Sciences Inc. 9/21/10. Public Health Service, Food and Drug Administration, Los Angeles District Pacific Region, 19701 Fairchild.

27 US Department of Health and Human Services. Med Watch. Triad Alcohol Prep Pads, Alcohol Swabs, and Alcohol Swabsticks: Recall Due to Potential Microbial Contamination Public Health Service, Food and Drug Administration, Silver Spring MD 20993.

28   US Department of Health and Human Services. Warning Letter Luitpold Pharmaceuticals, Inc. 8/31/11. Public Health Service, Food and Drug Administration, New York District 158-15 Liberty Ave. Jamaica, NY 11433.

29   US Department of Health and Human Services. Warning Letter Sichuan Pharmaceutical Co., Ltd. 9/9/11. Public Health Service, Food and Drug Administration, Silver Spring MD 20993.

30   Lewis T S, (2004) "All Change but Are You in Demonstrable Control?" *Regulatory Affairs Bulletin*, Issue No. 3. 2004), Medical Devices Faraday Partnership

31   Whyte, W. (2001). *Cleanroom Technology: Fundamentals of Design, Testing and Operation*. Wiley, London.

32   Akerlindh, K. (2006). "Regulations Governing Gases Used In Pharmaceutical Production Processes", *Business Briefing: Future of Drug Discovery*, pp1-3

33   Tidswell, E. 'Sterility' in Saghee, M.R., Sandle, T. and Tidswell, E.C. (Eds.) (2010): *Microbiology and Sterility Assurance in Pharmaceuticals and Medical Devices*, New Delhi : Business Horizons, pp589-614

34   Sykes, G. (1970): 'The control of airborne contamination in sterile areas', Aerobiology: Proceedings of the 3rd International Symposium, in Silver, I. H. (ed.), Academic Press, London

35   ISO Standard 14644 – Cleanrooms and associated controlled environments (1999) – Part 1: Classification of air cleanliness; Part 2: Specifications for testing, International Standards Organisation, Geneva, Switzerland.

36   Sandle, T. (2003). 'Selection and use of cleaning and disinfection agents in pharmaceutical manufacturing' in Hodges, N and Hanlon, G. *Industrial Pharmaceutical Microbiology Standards and Controls*, Euromed Communications, England

37   Bloomfield, S. (1990): 'Microbial contamination: spoilage and hazard' in *Guide to microbiological control in pharmaceuticals* edited by Denyer, S. and Baird, R., Ellis Horwood, London.

38   Ljungqvist, B. and Reinmuller, B. (1996): 'Some observations on Environmental Monitoring of Cleanrooms', *European Journal of Parenteral Science*, 1996; 1: 9 -13

39   Food and Drug Administration (2004). Guideline on Sterile Drug Products Produced by Aseptic Processing. *Food and Drug Administration*, Rockville, MD, USA.

40   Sandle, T. 'Practical Approaches to Sterility Testing', in Saghee, M.R., Sandle, T. and Tidswell, E.C. (Eds.) (2010): *Microbiology and Sterility Assurance in Pharmaceuticals and Medical Devices*, New Delhi : Business Horizons, pp173-192

a   The FDA publishes Warning Letters in its Electronic Reading Room available on its website (www.fda.gov).

CHAPTER 4

Preparing for pre-approval inspections

RON JOHNSON,
PRESIDENT OF BECKER & ASSOCIATES CONSULTING, INC. (BECKER CONSULTING), USA

4.1 Introduction

The US Food and Drug Administration's (FDA) approval process for a New Drug Application (NDA), an Abbreviated New Drug Application (ANDA) or a Biologic Licensing Application (BLA) includes an assessment of the manufacturing and support (i.e. testing labs) facilities identified in the submission prior to final approval of the application. This may involve a Pre-Approval Inspection (PAI) of one or more of these facilities. The PAI determines whether the data contained in the submission, and upon which FDA is basing its approval decision, is accurate and credible. Additionally, the PAI assesses the capability of the manufacturer to meet the requirements of the Good Manufacturing Practices (GMP) regulations for pharmaceuticals. The European Medicines Agency (EMA) conducts similar inspections as part of its approval process for the European market.

Preparing for a PAI requires an understanding of the genesis and objectives of FDA's PAI program. Knowing the "why" of the program helps identify the issues and areas that will be the focus of an FDA inspection. Recognizing who within FDA does what in execution of the program helps prioritise and plan preparations. Realizing what FDA instructs its investigators to examine and the "signals" it uses to foretell potential problem areas provides an opportunity to proactively prevent adverse FDA findings. Learning from the mistakes of others as reflected in the observations most frequently cited by FDA investigators helps avoid repeating similar mistakes. Early, proactive preparations will lead to a successful PAI and the timely approval of the associated application. This chapter provides information, insights and suggestions to assure this success.

4.2 Background

The Federal Food, Drug, and Cosmetic Act (FDCA) is the primary law governing how drugs are regulated in the US. This law and many of the numerous amendments that have been made over the years were, unfortunately, initiated in response to tragedies caused by unsafe and/or ineffective drugs reaching patients. FDA enforces the provisions of the FDCA.

The precursor to the FDCA, the Pure Food and Drug Act, was enacted in 1906 in response to widespread reports in the press of unsanitary conditions in the meat packing industry. This law was amended and retitled the FDCA in 1938 in response to the sulfanilamide tragedy. In the mid-1930s, toxic diethylene glycol had been used to manufacture "Elixir of Sulfanilamide" marketed as an anti-infective. The deaths of 107 people resulted. The 1938 amendment to the FDCA required drug manufacturers, for the first time, to prove the safety of their drugs before they were allowed to be marketed.

The next major amendment occurred in 1962 and was in response to the thalidomide tragedy. Thalidomide was marketed in Europe for morning sickness in pregnant women and resulted in hundreds of serious birth defects.[1] While thalidomide was never approved for marketing in the US, this incident served to galvanise public opinion towards more stringent regulation of prescription drugs, including the requirement that manufacturers prove both safety and efficacy.[2] The 1962 amendment authorised FDA to develop GMP regulations for the manufacture of pharmaceuticals and provided FDA with the authority to require compliance with GMPs as an element of the drug approval process. Thereafter in 1963, FDA drafted the first formalised GMPs.[3]

In 1978, the GMP regulations were expanded, revised and codified in Title 21 Code of Federal Regulations Parts 210 and 211. Only minor changes have been made since that time. The GMP regulations establish the minimum standards for the manufacture, processing, packing and testing of drugs marketed in the US. As described by the then FDA Commissioner Donald Kennedy, the GMP regulation is designed "to be general enough to be suitable for essentially all drug products, flexible enough to allow the use of sound judgment and permit innovation, and explicit enough to provide a clear understanding of what is required."[4]

The "generics drug scandal" of the late 1980s provided the impetus for expansion of FDA's authority to conduct PAIs to include ANDA applications for generic drugs. In 1989, FDA learned of widespread corruption in the generic drug industry. Specifically, some generic drug companies were found to be submitting false data in ANDA applications such as bioequivalence data derived from tests in which the innovator drug was used rather than the generic and falsified stability data. It was also discovered that some generic companies had made illegal payments to FDA employees for favourable treatment in review of their ANDAs. Industry and FDA personnel were criminally prosecuted and tainted ANDAs were withdrawn. In 1990, FDA expanded its PAI program to include manufacturing sites identified in ANDAs. In addition to determining compliance with GMPs, PAIs now determined whether data submitted in ANDAs were reliable, accurate and credible. As a result, PAIs extended into product development activities including research and development functions leading to the establishment of product specifications and manufacturing operations.

In 1992, Congress passed the Generic Drug Enforcement Act (GDEA), in order to provide further protection to the public health by granting FDA additional enforcement actions, most notably debarment.[5] Through the GDEA, FDA was authorised and, depending upon the circumstances, required to forbid certain people or drug firms from participating in the drug industry if they were convicted of crimes related to FDA's regulation of drugs. Under this law, an individual convicted of such a crime may not work at a drug firm "in any capacity," and in some cases, could be forbidden from engaging in certain work for a contractor who provides services to a drug firm.[6]

The enactment of the FDA Modernization Act (FDAMA) in 1997 also had implications for the PAI program. FDAMA requires manufacturers of over-the-counter (OTC) drugs to provide FDA investigators access to, and copies of, manufacturing records previously only required of prescription drug makers.[7] Also in 1997, regulations codified in Title 21 CFR Part 11 took effect. Part 11 establishes requirements to assure the security and integrity of electronic records, including electronic signatures.[8]

The PAI has evolved over time becoming an integral element of the FDA approval process for NDAs/ANDAs/BLAs. It has enhanced FDA's ability to rely on submitted data in making scientific decisions related to the safety and effectiveness of drugs marketed in the US It has also provided assurance that manufacturers will produce drugs subject to the applications under conditions compliant with the GMPs.

4.3   Pre-Approval Inspection program

4.3.1   How does FDA decide when a PAI is necessary?

The objectives of FDA's Pre-Approval Inspection program are to determine:

1.   Readiness for commercial manufacturing – Has a quality system been established to assure production of the drug in a GMP-compliant environment?

2.   Conformance to application – Are the representations made in the application consistent with actual conditions at the facility (i.e. are the controls described actually those that will be used)?

3.   Data integrity – Are data contained in the application accurate, credible and reliable? Have all pertinent data been included in the application?

The current program is described in Compliance Program 7346.832.[9] This Compliance Program has been revised periodically, and most recently in May

2010. It is available on FDA's website and should be thoroughly reviewed by any company submitting an NDA/ANDA/BLA, including the original submission, amendments to the Chemistry, Manufacturing, and Controls (CMC) section in a pending original submission, and CMC supplements to approved applications. This internal FDA directive provides instructions to various FDA components in the evaluation of facilities identified in the CMC section.

Evaluation is based upon on-site inspection or in some cases, FDA may have sufficient current and pertinent information to arrive at a scientific decision on site acceptability without conducting a PAI. Thus, if FDA has conduct a recent inspection of the facilities and covered manufacturing operations that are the same or similar to those described in the CMC section of an application, FDA may rely on this information rather than conducting a PAI. Domestic and international PAIs are conducted for generic and innovator drug applications and may cover all facilities associated with a submission including drug component manufacturing such as Active Pharmaceutical Ingredients (APIs), finished drug product manufacturing and control testing laboratories. Current FDA policy exempts manufacturers of excipient component materials from the PAI program even though FDA retains the prerogative to pursue inspection when it sees a good reason.

Several organizational units of the FDA are involved in determining when a facility requires a PAI.

1. Upon receipt of the application, the Center for Drug Evaluation and Research (CDER) identifies the sites named in the CMC section of the application which are proposed as participants in the manufacture of the new product. The Office of Compliance (OC) within CDER evaluates these sites to determine whether a PAI will be necessary. Those applicants and/or named sites meeting the following criteria are considered "Priority" and OC requests the appropriate FDA inspection unit to perform a PAI:

    o   First time applicant/named site including those that have never been inspected or have been inspected only for non-application drugs

    o   Applicant/named site(s) in first ANDA filed for an approved drug

    o   Applicant/named site(s) in an application for finished products containing a New Molecular Entity (NME) other than application supplements

    o   Applicant/named site(s) in application of finished products for which the content assay has a narrow range or the drug is expected to require titrated dosing

- o Applicant/named site(s) in application of products manufactured in a substantially different manufacturing process or dosage form than previously covered at the site(s)

- o Applicant/named site(s) in an application for products with high risk API (i.e. derived from animal tissues), or the intended use has changed (i.e. API previously used in non-sterile product is now intended for a sterile drug product)

- o Applicants with numerous application submissions or product changes expected to pose significant challenge to the state of control of the facility or process

- o Applicant/named site(s) with no qualifying Good Manufacturing Practice (GMP) inspection in the past 2 years (3 years for control laboratories and 4 years for packaging and labeling facilities)

2. Applicant/named site(s) that do not meet the above criteria are considered "Discretionary" and a request for inspection is not sent to the inspection unit. However, in the case of US-based facilities, OC will also request the local FDA inspection unit to provide any relevant local information about the facility and concur that an inspection is not necessary.

3. Even though a Priority inspection request has been made for a US-based facility, the local inspection unit may believe that a PAI is not necessary because it has sufficient information to render a decision on whether the facility should be approved or not. The local FDA office may determine, based upon its experience with the facility, that no PAI is necessary and make a recommendation to approve the application. For example, the local unit may have recently conducted a GMP inspection that found the facility compliant for a related product type and which also met the other objectives outlined in the program for the initial Priority determination. Similarly, based upon information available at the local office, it may recommend that approval of the application be withheld. For example, the local office may be monitoring the facility's efforts to correct unacceptable deficiencies identified in an earlier inspection. Lastly, if it agrees with the request for a PAI, it will conduct an inspection.

4. In the case of facilities located outside the US, CDER OC makes the final determination whether a PAI will be required or not. In some cases, even though the facility meets criteria for a Priority determination, OC may opt not to conduct a PAI based upon the acceptable findings of another recognised regulatory body. It may also determine that a PAI is not necessary, and that a recommendation for approval or withholding of approval can be made based upon current available information as described above.

5. In the case of a Discretionary determination by OC, the local office may still believe a PAI is necessary. In such case, the local inspection office must provide justification to OC for conducting a PAI. Examples that might justify a PAI include:

    - There have been multiple applications submitted in a short period of time for manufacture of finished product at a single establishment;

    - Significant deficiencies were found during the last PAI or the facility has a history of non-compliant PAIs; and

    - Additional potentially adverse information regarding the compliance status of a facility not yet known to OC, such as an expected enforcement action recommendation based upon an ongoing inspection, multiple recalls, or new firm management.

The Pre-Approval Inspection Program embodies a philosophy of "equal voice" meaning that all appropriate expertise should be considered when making decisions to approve or withhold approval of marketing applications. This translates into each of the collaborating FDA units having significant influence on the ultimate agency decision regarding the application. While review and approval of marketing applications was traditionally the exclusive responsibility of CDER, the PAI Program has given the field inspection unit of FDA a strong voice in these decisions. CDER/OC has the authority to overrule a field unit's recommendation to conduct or not conduct a PAI or approve or withhold approval of an application. However, in practice CDER frequently defers to the recommendations of the field inspection unit. This is more true for domestic PAIs than for international PAIs. In the latter case, CDER/OC actually executes many of the responsibilities carried by the domestic field inspection unit.

Understanding what triggers a PAI can help a company anticipate and adequately prepare for such an inspection. If a company or manufacturing site has a recent compliant GMP inspection history, particularly associated with previous PAIs, chances of a PAI for a new submission are reduced. On the other hand, if the site has a history of non-compliant inspections or has a pending Warning Letter, the chances of a PAI are heightened or may even result in a recommendation to withhold the application without a PAI. Companies with such histories should make every effort to resolve all outstanding issues as rapidly as possible and communicate to FDA that corrections have been made. Sometimes a meeting with the FDA offers an opportunity to assure that agreement has been reached on the adequacy of the corrective actions taken.

4.3.2   What happens during a PAI?

The PAI is scheduled once a decision has been made that one is necessary. FDA gives high priority to timely completion of the PAI in order to meet the review time commitments for completing NDA reviews required by the Prescription Drug User Fee Act (PDUFA). There are no similar time frame requirements for ANDAs, however, FDA is currently in active negotiations for legislation that will likely establish them.

For domestic inspections, the local FDA inspection unit will schedule the inspection. The local inspection unit has the discretion whether to notify the facility ahead of time. When notifying a facility of the upcoming inspection, FDA will generally be resistant to any requests for a change in date. Such changes are only considered for unusual circumstances such as the facility being shut down for periodic maintenance.

In the case of foreign facilities, prior notification is traditionally provided because of the lead time necessary for foreign travel arrangements. In these situations, the facility will be notified in writing of the timing of the planned PAI. This notification is generally provided several weeks before the planned inspection. Notification may also include the names of the investigators that will be performing the inspection. FDA foreign inspection trips are generally scheduled for a 2 or 3 week period during which FDA will inspect more than one facility. PAIs of foreign facilities normally take one week and occasionally 2 weeks. The anticipated length of the inspection will be included in the inspection notification to the facility. Because the actual time devoted to facility inspections during the trip may vary, FDA may make last minute changes to the schedule. This can result in the planned date of the inspection moving up or being delayed. In some cases the inspection will be canceled. Thus, foreign facilities should anticipate this possibility and prepare to accommodate such changes.

International inspections have traditionally been conducted by FDA investigators who have been selected from FDA's US-based inspection staff. These investigators tend to be more experienced in pharmaceutical inspections. During the past several years, FDA has also developed a cadre of specialised, more highly trained pharmaceutical investigators as part of a Pharmaceutical Inspectorate. In practice, pharmaceutical inspections including PAI and international inspections could be performed by either of these types of investigators.

FDA has established official offices in several countries over the past several years. These offices have been staffed with managerial and investigational personnel. It is not clear how FDA plans to use investigators stationed at these offices. The agency has communicated that inspections of foreign facilities will continue to be performed by US-based investigators. However, there have been reports of investigators from the foreign FDA office conducting unannounced inspections of facilities within the country. It seems logical that FDA will be relying on in-country investigators to augment its US-based investigators to perform international inspections. Thus, facilities located in countries where FDA has an office should be prepared for inspection by FDA investigators from that office, including unannounced inspections.

The PAI program, more so than FDA's routine GMP inspection program, involves an inspection team. These teams may consist of members of different disciplines (technical or scientific) as warranted by the process and controls proposed in application. Typically the team will consist of a GMP investigator and a laboratory analyst. In some cases, a scientist from CDER will participate. Because of travel expenses, the number of team members for foreign inspections may be limited. In some cases, a single investigator will conduct the inspection, particularly if the investigator has a laboratory background.

Prior to the PAI, the inspection team will have reviewed the CMC section of the application and the Drug Master File (DMF). Normally, this will be augmented by discussions between the team members and the CDER review staff, who may request that specific issues we addressed during the inspection. Communications between the application sponsor and the FDA reviewers can sometimes highlight these issues, allowing the facility opportunity to prepare clarifying or supporting information to be provided during the PAI. It is common even during a foreign PAI that the inspection team will maintain contact with CDER and OC, providing daily updates and/or requesting technical or scientific insight. These communications frequently occur after hours during the inspection and may lead to a change in the inspection plan.

The conduct of the PAI will be directed by the following program objectives (more fully described above):

1.   Readiness for commercial manufacturing.

2.   Conformance to the application.

3.   Data integrity.

4.3.3   Readiness for commercial manufacturing

During the PAI, the inspection team will determine whether the facility has an effective quality system in place to assure sufficient control over the facility and commercial manufacturing operations. If the facility currently commercially manufacturers drugs, FDA expects that its quality system will be mature, robust and fully compliant with the GMP regulations. Even though the drug subject of the application has not been commercially produced, FDA will evaluate the effectiveness of the facility's existing quality system used to produce commercial products. If the facility does not produce any drugs commercially for the US market, FDA will expect that adequate controls are in place to assure the quality of the biobatches/exhibit batches.

In addition, FDA will determine if the facility's quality system has the capability to develop into a fully GMP-compliant system for the planned commercial operations. In either event, FDA will be interested in how the facility manages out-of-specification failures, deviations, changes, and control over raw materials/in-process/finished product, as well as whether there are adequate facilities and equipment and justification to support the manufacturing processes and control measures.

Although, unlike FDA's pre-approval program for medical devices, the PAI Inspection Program for drugs does not require the facility to have completed process validation at the time of the PAI, the inspection team will closely review any process validation documentation available. FDA expects that process validation will be completed before any commercial shipments are made by the facility.

The following signals can be expected to pique the inspection team's interest and precipitate more in-depth investigation. Each of these signals may indicate instability or lack of robustness in the product design, manufacturing process and/or controls:

- Out-of-specification (OOS) results that have not been evaluated with a root cause determination

- Changes made to the manufacturing, control or laboratory practices and methods

- Failures of equipment or facilities (i.e. calibration, qualification, etc.) to be used for commercial production

- Changes in specifications or processing parameters

- Inconsistent API quality

- Unresolved deviations

The following are the deficiencies most commonly cited during GMP inspections. These would also be expected to be cited, as applicable to the specific processes performed, during a PAI, particularly if the facility is commercially producing other drugs.

1. Laboratory Controls

   - Use of test methods that have not been validated

   - System suitability tests not performed

   - Use of secondary reference standards without comparing against official USP standards

   - No investigation of abnormal or missing analytical data

   - Retest of OOS without appropriate investigation

   - Lab data not reviewed for accuracy by second person

2. Equipment Cleaning

   - Inadequate equipment cleaning and validation protocol

   - Cleaning methods not validated

   - Inadequate sampling and testing of equipment surfaces

   - Residue limits for potential contaminants not established

   - Specificity and sensitivity of analytical methods not established

   - No testing for residues of solvents used in API production

3. Process Controls

   - Critical product attributes and critical process parameters not identified and monitored

   - Process yields not established

   - Out of spec API batches blended with batches that have passed specifications

   - Inadequate controls to prevent product contamination and cross contamination

   - Inadequate in-process and end-product testing

4. Process Validation (applicable during a PAI if process validation has been completed)

   - Validation plans not established or not followed

   - Validation protocols do not address equipment, critical process parameters/ranges, sampling, data to collect, number of validation runs, criteria for acceptable results

   - Retrospective validations attempted for an existing manufacturing process that has undergone significant changes in raw materials, equipment, systems, facilities, or production process

- Changes to validated processes not adequately addressed

5. Records/Reports

- Batch records incomplete or do not reflect actual operations
- Manufacturing steps documented before completion
- Lots released before review of batch production records
- Changes to process and equipment are not done by change control system
- Equipment use and maintenance records not maintained
- Product reviews not conducted or are inadequate

6. Water Systems

- Process water not shown to be suitable for intended use
- Specification not established for chemical/microbial quality
- Inadequate investigations and corrective actions following recurring microbiological test results
- Water used to produce sterile products not tested for endotoxins
- Reliance on point of use filters to clean up water while ignoring the system control

7. Stability Programs

- Expiration dates not supported by stability data
- Forced degradation studies not conducted to detect, isolate, identify, and quantify degradants
- Test methods have not been shown to be stability indicating
- Stability samples not stored in containers that approximate the market container, or under conditions specified on the label for the marketed product

8. Buildings/Facilities

- Production facilities not adequately designed to minimise mix-ups and cross contamination
- Pharmaceuticals of varying therapeutic significance, non-pharmaceuticals, and pesticides processed in common areas
- Facilities not provided with air handling and dust control systems to minimise cross contamination

- Production areas not provided with adequate temperature and humidity controls

- Finished product warehouse has not been temperature mapped

9. Component Controls

- Quality Control (QC) sampling of raw materials conducted in open, uncontrolled warehouse

- Components received with a certificate of analysis are not subjected to a specific identity test, nor are the supplier's analytical test results verified at established intervals

- Bulk intermediates not adequately packaged and stored to ensure their suitability for use in further manufacturing

- Time limits and yields not established for critical processing steps

4.3.4   Conformance to application

The inspection team will verify that that the elements of product/process design have not changed since the application was submitted and are unlikely to change upon scale-up and repeated commercial manufacturing. During this phase of the PAI, the inspection team will review and evaluate R&D records to confirm they support information provided in the application. The inspection team will review laboratory notebooks and interview R&D scientists. FDA wants to make sure that the information provided in the application and upon which a decision to approve marketing of the drug is, in fact, reflective of actual conditions in the manufacturing facility and supported by scientific data.

The inspection team will respond to the following signals by conducting more in-depth investigation:

- Biobatch production process is significantly different than that proposed for the commercial product

- Samples submitted, upon FDA request, for FDA analysis are not from the biobatch contained in the application

- Unaccounted API inventory

- Use of an API supplier not referenced in the application

- Analytical records (i.e. chromatograms, lab notebooks, etc.) do not support analytical data provided in the application

- Analytical methods in use at the facility are different from those submitted with the application

- Inability to produce the raw data to support the application or the development report

- Changes in in-process and finished product specifications without scientific justification

- Data reflecting failures of raw materials used in biobatches not included in the application

- No scientific rationale for changes being made between the pivotal and the proposed manufacturing batch record

- Batches that were produced but were not included in the development report

- Lack of characterization of the drug substance or discussion of why some characterization may not be important

- Bioequivalence studies not referenced or contained in the application

4.3.5 Data integrity

The PAI will assess whether the data contained in the application are credible, accurate and reliable. The integrity of data can be compromised in many ways. As discussed above, original or source data may not conform to that referenced or provided in the application. Data can be corrupted by willful, intentional fraudulent activities or as a result of incompetent procedures, practices or employees. The PAI attempts to detect unreliable data and/or practices that may lead to the submission of unreliable data to FDA. If FDA detects fraudulent or false data that have been submitted in an application, it will pursue aggressive enforcement action, including rejection of submitted applications, suspension of review of future applications, withdrawal of previously approved applications, recall, criminal prosecution of companies and individuals, and/or debarment of individuals involved.

During a PAI, the inspection team will be alert to procedures or practices that do or could result in inaccurate or untrustworthy data. The following are signals that alert the inspection team to possible data integrity problems:

1. Poor documentation practices

    - Uncontrolled laboratory notebooks including R&D notebooks

    - Looseleaf lab notebooks, missing pages from bound notebooks

    - Recording completion of a manufacturing step on the batch record before it was performed

- Recording on batch records indicating performance of operations by one employee even though another employee actually performed the operation

- Dated employee signatures/initials on batch record indicating certain activities were performed during production even though facility attendance records indicate that the employee was not present at the facility on the dates recordings were made

- Transposition of date entries indicating activities were performed out of sequence (i.e. a component was used before it was actually received)

- Records indicating that two different products were produced using the same equipment at the same time

2. Data falsification

- Falsifying analytical records to show a passing test result even though the sample failed

- Concealing failed test results and retesting sample until a passing result is obtained

- Substituting acceptable test records for an unrelated sample for a sample from the bio-batch or a stability batch

- Reporting test results which are better than actual passing results (too good) or test results on samples subjected to unrealistic storage conditions (e.g. results on samples held at accelerated conditions for samples said to have been 2-year label condition storage)

- Advance dating of tests – tests done at six-month stability station and reported to be 12-month stability station

- Failing to submit annual reports for stability program testing progress or to avoid reporting failed tests

- Excluding from annual reports failed tests on stability lots to avoid reporting failed tests

- Manipulating/modifying analytical procedure to obtain more favourable results

- Selecting biased samples instead of lot representative samples (i.e. intentionally avoiding a visible problem with film coating, tablet sizes, fill volume, etc.)

3. Inaccurate/untrue information about manufacture of biobatch/stability batch(es)

- Multiple batch(es) /records for the same lot number

- Misrepresentation of the batch size

- Misrepresentation of the container/closure for the sample

- Misrepresentation of the quality of the active ingredient/other components

- Misrepresentation of the source of the active ingredient/other components

- Writing batch production records at a time other than the time of manufacture of a batch

- Excluding from required reports failed in-process test results

- Failure to report as a supplement to an approved application manufacturing process changes that

  – Might affect bioavailability

  – Improve machine processing

  – Correct process failures

It is FDA policy that investigators provide feedback to the facility during the course of the inspection. Each FDA investigator is different and some are more forthcoming while others tend to wait until the end of the inspection to reveal their findings. Some investigators will actually identify findings during the inspection that they plan to put on the FDA-483 (List of Inspectional Observations) which will be left with facility management at the conclusion of the inspection. A request by facility management at the beginning of the inspection for regular feedback may encourage the investigator to be more communicative.

Before the investigator leaves the site, a formal meeting is held with site management during which the investigator will describe his/her findings. These findings may be recorded on the FDA-483 or observations of a minor nature may only be addressed verbally. All findings, whether formally documented on the FDA-483 or not, are important and the company should consider appropriate corrective actions. Senior management should attend the meeting since it is an opportunity to assure understanding of the investigator's findings and offers a final chance for the company to explain it perspective. The company should not be argumentative, but focus on the technical or scientific bases of its position. If the investigator is convinced that the observation is not warranted, he/she will delete it from the FDA-483 by drawing a line through it before leaving the facility.

4.3.6   What happens after the PAI?

If an FDA-483 was issued, a written response to FDA should be made within 15 days. While it seemingly goes without saying, care should be taken to specifically address each individual observation. Unfortunately, on many occasions responses do not directly address the issue FDA has raised. Another common error is to address only the examples FDA has cited rather than addressing the underlying systemic root causes. Timely submitted comprehensive responses can influence FDA's decision as to the approvability of the application. If the observations are significant and extend to ongoing commercial drug production, FDA may seek enforcement action such as a Warning Letter. An effective response to the FDA-483 can also influence FDA's decision to take further action.

Following the PAI, the inspection team will prepare a written narrative report. This report, called the Establishment Inspection Report (EIR), describes the intent of the inspection, compliance history of the site, inspectional findings including those cited on the FDA-483 as well as those verbally discussed, and the facility management's comments, commitments and promises regarding these findings. Importantly, any promised corrective actions will be documented in the EIR for review and consideration by CDER.

The EIR is reviewed at various technical and management levels within the FDA inspection unit. Based upon the findings of the PAI, a decision is made to recommend either to "approve" or "withhold approval" of the application. This recommendation is made to CDER/OC as noted above. If the inspection unit decides to recommend "withhold approval", it may have further discussions with the inspected facility to assure full understanding of the factual circumstances. In either situation, once a decision is made on a recommendation, the inspection unit will send a formal written notification to the inspected site of the recommendation it has made to CDER/OC. This provides another opportunity for the company to attempt to resolve concerns that have led to the "withhold" decision. CDER/OC makes the final decision on approvability of the application based upon the findings of the PAI. However, in view of FDA's "equal voice" philosophy, the inspection unit's recommendation is given significant consideration. In the case of international inspections, CDER/OC makes this determination independently.

FDA has identified the following bases for recommending "withhold approval":

1.   Significant data integrity problems including misrepresented data or other conditions related to the submission batch

2.   Serious GMP concerns with the manufacture of a biobatch or demonstration batch, such as changes to formulation or processing that may cause FDA to question the integrity of the bioequivalence study

3. Significant differences between the process used for pivotal clinical batches and the NDA submission batch

4. Lack of complete manufacturing and control instructions in the master production record or lack of data to support those instructions

5. Lack of capacity to manufacture the drug product or the API

6. Failure to meet application commitments

7. Full scale process validation studies performed prior to the PAI demonstrate that the process is not under control and the facility is not making appropriate changes

8. For products for which full scale summary information is provided in the application, the facility has not demonstrated that the product can be reliably manufactured at commercial scale and meet its critical quality attributes

9. Incomplete or unsuccessful method validation or verification

10. Records for pivotal clinical or submission batches do not clearly identify equipment or processing parameters used

11. Significant failures related to the stability study that raise questions about the stability of the product or API

12. Failure to report adverse findings or failing test data without appropriate justification

4.4 Additional preparation steps

Understanding FDA's PAI Program, its areas of emphasis and common mistakes made by other inspected sites can be used to create a checklist of actions to be taken in anticipation of a PAI. In addition, there are a number of other actions companies should take to assure a successful PAI outcome.

4.4.1 Adoption of quality system principles in research and development

As described above, the PAI will focus on the design and development of the drug subject to the application. This involves review of laboratory notebooks, product development files and interview with R&D scientists. While the GMP regulation does not apply to research activities, adoption of quality system principles in the R&D function will assure that documents and records reviewed during a PAI are complete and accurate, and that design outputs are of high quality. A formal procedure to govern the product development process should provide for quality control checks, verification of design outputs and periodic design reviews. This will assure that the drug is well characterised and the manufacturing processes have been correctly defined. Good documentation practices should be established to assure that records produced during development are accurate and reliable. Documents such as laboratory notebooks and the development files should be controlled with access limited to selected individuals. Transfer of the drug to production should also be directed by a formal procedure and provide for resolution of manufacturing issues.

4.4.2   Inspection preparedness

An obligatory requirement for any facility that is subject to FDA inspection is an established procedure to govern employee actions during an inspection. Such a procedure is useful for FDA inspections including PAIs as well as inspections by other regulatory bodies.

The elements of an inspection management procedure include:

- Assignment of inspection management roles and responsibilities to company employees including:

    o   Inspection Coordinator – senior individual who is the primary interface with FDA investigator(s) and leader of the company's inspection team, and is the authorised spokesperson for the company;

    o   Subject Matter Experts (SMEs) - individuals with deep knowledge of specific operations such as CAPA, design, complaints, etc.;

    o   Scribe(s)–individuals who accompany the FDA investigator and company's inspection team taking notes of conversations, questions asked, and comments/observations made by the investigator; and

    o   Runner(s) – individuals who gather documents that may be requested by SMEs or others.

- Logistics

    o   Pre-identify a location (normally a conference room) where investigator will be escorted upon arrival and where he/she will review documents and interview staff;

    o   Pre-identify a "back room" to be used by the company to assemble, copy, identify (stamp "Confidential"), and create a list of the documents/records provided to the investigator; to brief SMEs before interviews; etc.;

    o   Anticipate documents/records to be reviewed and assure that they are readily available, such as: product development reports, SOPs, change control records, non conformances, failures (OOSs), deviations, field alerts, process validation protocols & reports, system validations (water, gas, steam, computer, etc.), critical process parameters, batch production records, raw material inventories, cleaning validation, environmental monitoring records, complaints, rework/reprocessing records, analytical method validations, stability records, bioequivalence records, etc.;

o   Determine where the investigator will take lunch (i.e. company facility or nearby restaurants) (investigators making international inspections may require recommendations for nearby lodgings).

- Company policies

  o   Clearly articulated company policies in the following areas should be established and described to the FDA investigator at the beginning of the inspection:

    –   Photography – are photographs permitted in the facility? (most companies do not permit photography);

    –   Signing of FDA documents –does the company authorise employees to sign FDA documents such as an affidavit? (most companies do not sign such documents);

    –   Collection of samples – FDA is permitted to collect samples and most companies collect duplicate samples; and

    –   Safety –what safety equipment/precautions does the company require? (i.e., gowning required for controlled areas, medical evaluation, etc.).

- Communication practices

  o   Describe acceptable communication practices and styles to be used when interfacing with the FDA investigator to include:

    –   Professional and courteous demeanor;

    –   Truthfulness;

    –   Non-argumentative/confrontational style;

    –   Responsive to questions, but avoid volunteering information;

    –   Avoid admission of violations;

    –   Avoid commitments for correction except by identified, authorised company individuals.

- Training

  o   All employees having responsibilities established in the inspection procedure must be trained.

Beyond creating a procedure and training, preparations should also include practice in execution. A "walk through" of the inspection procedure should occur to include practice of roles as well as the utility of the logistics. Rehearsal of SME's (and potential backups) ability to be articulate, concise and responsive is critical to a smooth inspection process. This is particularly valuable in preparing R&D scientists who may not have experience in FDA interviews. As described below a mock PAI performed by an external third-party can be an effective vehicle for assessing the company's readiness for a PAI.

4.4.3   Conduct a mock PAI

A mock FDA inspection conducted by an external third-party can serve as a final test of the PAI preparations that have been made. This audit can assess the company's ability to manage an inspection as well as identify any residual quality system or data integrity issues that require correction. The third-party audit should occur with sufficient advance timing, so that any necessary remediation can be made prior to the PAI. The audit should follow FDA's Compliance Program 7346.832. Care should be exercised to select a third-party auditor who is thoroughly familiar with FDA PAI Program, either having previously served as an FDA investigator or a responsible individual in companies that have experienced successful PAIs.

4.5   Comparison of FDA and EMA pre-approval programs

The European Medicines Agency (EMA) has the coordinating role for pre-authorization inspection.[10] However, it is up to the European Economic Area (EEA) individual country to perform the inspection.[11] Specifically, it is EMA's Manufacturing and Quality Compliance Section, within the Compliance and Inspection Sector, which coordinates the GMP inspections of manufacturing sites. The responsibility for carrying out inspections rests with the regulatory authority of the Member State responsible for supervising the manufacturer or importer. Typically, for centralised marketing authorization programs, an inspection will involve several inspectors from different Member States.[12] Overall, an EMA inspection is focused on the manufacturer's adherence to requirements for the manufacturing and documentation process.

The EMA and FDA began a pilot joint inspection program in September 2009. The primary purpose of this program is for the organizations to exchange information within the framework of confidentiality agreements between EMA and FDA and to conduct collaborative inspections. Under the pilot program, as of 2011, seven joint inspections related to three different applications have occurred.[13] Global economic pressures can be expected to encourage greater collaboration and sharing of information between regulatory bodies.

The FDA and EMA inspection processes are compared and contrasted in **Table 1** below.

Table 1: FDA vs. EMA inspection processes[14]

Inspection characteristic	FDA	EMA
Pre-Approval Inspections	Yes. Inspections during the NDA/ANDA/BLA phase are generally "routine," but can be "for cause" or "directed," and can take place during any point in the application approval process.	Yes. There are two types – triggered and routine inspections. Triggered inspections are based on the concerns from the clinical assessors during the review of the dossier.
Pre-Inspection Procedures	International facilities generally always notified in advance. Domestic U.S. facilities notified at discretion of local FDA inspection unit.	EMA sends formal announcement letter and applicant is request to confirm they will be available. Letter will request documents in advance.
Inspection – Information Collection	Focus on design of systems, but to a greater extent focus on records and data. Collection of many documents and notes of employee interviews. Rarely collects samples of product during inspection, but usually will request submission of samples directly to CDER.	Greater focus on system design with copies of data used for report writing, but not with intent for evidence.
Inspection Close Out	At conclusion of inspection, FDA provides written list of objectionable findings and discusses with management. FDA encourages written response to written list of findings. Decision to "approve" or "withhold approval" of application made based upon findings.	Lead inspector highlights findings, but no written summary is left. EU inspections classify each individual observation or finding. Written response from inspected facility is expected in response to inspection report.

4.6   Case studies

4.6.1   Incompetent systems/insufficient staffing compromise data integrity

An FDA PAI found the stability program of an international generic drug company to be out of compliance with GMP regulations and, as a consequence, raised concerns about the integrity of stability data provided in the ANDA. The company's stability program provided for samples to be collected from the manufacturing batch and delivered to the stability laboratory. The laboratory, following an established protocol, identified and labelled the sampled units with the specific stability chamber environment in which they would be stored(i.e. accelerated, room temperature, etc.). After the appropriate period of time in the designated stability chamber, the samples were withdrawn and brought to the laboratory for analysis. The laboratory experienced a temporary shortage of qualified analysts, which contributed to a backlog of unanalyzed samples. This backlog was further exacerbated by a surge in the number of drugs entering the stability program. The laboratory decided to place the unanalyzed samples in cold storage (4°C) until they could be analyzed.

At the time of the PAI, the FDA investigator encountered hundreds of samples in the company's freezer. Because the labelling of the individual samples still bore labelling indicating the stability chamber environment in which they were stored (i.e. accelerated, room temperature, etc.), the FDA investigator concluded that the company was falsifying stability data. For example, finding a sample labelled as an accelerated stability sample in the freezer gave the appearance that the company may be analyzing samples stored under freezer conditions, but reporting them as accelerated stability results. When the investigator discovered this practice, the company was unable to provide documentation showing that the samples in the freezer had previously been properly stored in the designated stability chamber. Moreover, the company had no inventory of the samples in the freezer, or record of when they had been placed there and when they had been removed for analysis. This practice of temporarily storing stability samples in the freezer was not described in the ANDA and the CDER reviewer raised concerns about the impact the temporary freezer storage had even if the sample had been previously stored properly. As a result, FDA withheld approval of the ANDA and required the company to account for all prior stability analyses and revamp its procedures. The company was also required to withdraw several previously approved ANDAs and recall product from the US market.

4.6.2   Unprepared manufacturing facility and ill-trained staff

The batch record for the exhibit batch submitted in an application was reviewed by the FDA investigator prior to the PAI. Upon arrival at this non-US manufacturing facility, the investigator asked to review the attendance records of the employees whose signatures/initials appeared in the batch record. The manufacturing site did not produce any drugs for the commercial market and its quality system was not yet fully developed. Further, because of its status, its security system did not provide for badging in/out of employees, but relied upon a manual system operated by a contracted security firm. The requested attendance documentation provided to the investigator revealed that employees who signed/initialed as performing manufacturing operations in the batch record were actually not present at the facility on some of the dates indicated. The company pointed out that the attendance documentation system was not robust and should not be relied upon by FDA. It produced sworn statements from employees that they were present at the time of production and witnessed the presence of others whose attendance documentation was deficient.

The investigator dug further, ultimately identifying a supervisory employee who admitted that he had initialed a "checked by" entry in the batch record signifying his oversight of the performance of a manufacturing step by another employee even though he was not present at the facility. This supervisor, like many of the other employees involved, was an R&D scientist and was not routinely stationed at this manufacturing facility. When questioned about his rationale for initialing the "checked by" entry, he told the investigator that he did not believe the company's procedures or the GMPs required that he personally witness the performance of the employee he was "checking." Of course, he was wrong on both counts and cemented in the investigator's mind that the entries in the batch record were not reliable. The ANDA was not approved and the company has never attempted to use the facility to commercially produce drugs for the US market.

4.6.3   Omissions in applications can have domino effect

A drug company planned to submit an application for a new drug packaged in seven different sizes. The smallest package size was 15 tablets and the largest was 1000 tablets. During stability testing, the smallest package size failed. The company cancelled the smallest package size and omitted it from the application it ultimately submitted.

During the PAI, the investigator discovered the raw test data for the canceled small package size and realised it was not included in the application. The inspection was expanded into a number of other drugs for which the company had previously obtained approval. The inspection even extended to a second site operated by the company. The expanded inspection revealed that the company had submitted imprecise data in other applications. For example, the company had built a new stability sample storage chamber at the second site. It had moved all of its ongoing and new stability samples into the new chamber. However, it failed to update the form it used for submission of stability data in applications. The form declared that the old stability chamber had been used. This created more uncertainty about the validity of data submitted by the company.

These seemingly innocent errors resulted in numerous communications between the company and FDA and many months of delay in approval of the application.

4.6.4   Inadequately trained staff and inept management

During a PAI of a large drug manufacturer, the investigator became suspicious of spectrograms associated with analyses of different lots of API. They were suspect because they appeared to be identical. The FDA investigator pursued his apprehensions and interviewed the chemists responsible for infrared spectroscopic identity testing. He found that two of the chemists had copied a passing spectrogram and used it in lieu of actually performing the analysis. This was done for multiple lots in multiple shipments of the same material. The spectrogram was copied using a computer so as to appear to be an original document. The chemists explained that they did not believe what they were doing was wrong because "the infrared identity tests on other old lots had always come out OK." Management asserted no knowledge of the practice of the chemists even though supervisory review and approval of each analysis was required by procedure. FDA declined to approve the application under review.

4.7   Conclusion

Development and submission of an NDA/ANDA/BLA is a major financial commitment for most companies and represents a significant aspect of the company's business plan. All the various technical, scientific, and regulatory activities necessary to support a successful application are typically given high priority. These activities do not always include preparation for a Pre-Approval Inspection until after the submission has been made. Understanding the role and intent of the Pre-Approval Inspection in the drug approval process can direct a company's effort in preparing for a PAI. A successful PAI enhances the company's posture with the FDA and avoids the negative financial impact of approval delays.

Preparation for a PAI should begin early in the development process by assuring that quality principles and practices govern R&D activities and extend into the production of biobatches and exhibit batches. Discipline in adherence to established procedures, record keeping and documentation controls assure the scientific rigour and integrity of data contained in the submission. Moreover, it facilitates the ability to produce the data during a PAI. Establishing a procedure that defines roles and responsibilities of company staff during an inspection facilitates an efficient inspection process. Training, and more importantly, rehearsing these roles will assure that required documentation will be readily available for the investigator's review and staff will be prepared to professionally describe its content. Effective inspection management also results in complete and accurate information being provided to the FDA investigator. Beyond rehearsing, periodic tests of the company's ability to successfully perform during a PAI can be achieved through internal audit programs or through the use of an external third-party consultant.

Every company planning to submit a marketing application to the FDA should assess its compliance record with the FDA. Any outstanding deficiencies should promptly be addressed and FDA's concurrence should be sought on the adequacy of any corrections made.

Anticipating and planning for a PAI can prevent unexpected, costly approval delays.

4.8   References

1   Janssen WF. 1981. The Story of the Laws Behind the Labels. *FDA Consumer*. FDA Office of History. US Food and Drug Administration.

2   Immel BK. 2001. A Brief History of the GMPs for Pharmaceutials. *Pharmaceutical Technology*. July 2001: 44-52.

3   Meadows M. 2006. Promoting Safe and Effective Drugs for 100 Years. *FDA Consumer Magazine: The Centennial Edition*. January-February 2006. FDA Office of History. US Food and Drug Administration.

4   43 FR 40577

5   Nordenberg T. 1997. Inside FDA: Barring People from the Drug Industry. *FDA Consumer Magazine*. March 1997. US FDA: Inspections, Compliance, Enforcement, and Criminal Investigations.

6   Fleder JR. 1994. The History, Provisions, and Implementation of the Generic Drug Enforcement Act of 1992. *Food and Drug Law Journal*. 49: 89-108.

7   US House of Representatives. 1997. Food and Drug Administration Modernization Act of 1997: Conference Report 105-399. 105th Congress, 1st Session. November 9, 1997.

8   FDA. 2003. Guidance for Industry: Part 11, Electronic Records; Electronic Signatures – Scope and Application. Pharmaceutical CGMPs. August 2003.

9   FDA. 2010. Chapter 46 – New Drug Evaluation. Program 7346.832. *Compliance Program Guidance Manual*. Form FDA 2438.

10   Wilder B. 2008. Effective Management of Regulatory Good Clinical Practice and Pharmacovigilance Inspections. *Touch Briefings*. http://www.touchbriefings.com/pdf/3227/widler.pdf

11   European Medicines Agency. Coordination of good-manufacturing practice inspections. European Medicines Agency website, last accessed January 6, 2012. http://www.ema.europa.eu/ema/index.jsp?curl=pages/regulation/document_listing/document_listing_000171.jsp&murl=menus/regulations/regulations.jsp&mid=WC0b01ac0580029751&jsenabled=true http://www.ema.europa.eu/docs/en_GB/document_library/Standard_Operating_Procedure_-_SOP/2009/09/WC500003195.pdf

12   EMA and FDA. 2011. Report on the Pilot EMA-FDA GCP Initiative: September 2009 March 2011. 18 July 2011.http://www.fda.gov/downloads/InternationalPrograms/FDABeyondOurBordersForeignOffices/EuropeanUnion/EuropeanUnion/EuropeanCommission/UCM266259.pdf

13   EMA and FDA. 2011. Report on the Pilot EMA-FDA GCP Initiative: September 2009 March 2011. 18 July 2011. http://www.fda.gov/downloads/InternationalPrograms/FDABeyondOurBordersForeignOffices/EuropeanUnion/EuropeanUnion/EuropeanCommission/UCM266259.pdf

14   EMA and FDA. 2011. Report on the Pilot EMA-FDA GCP Initiative: September 2009 March 2011. 18 July 2011. http://www.fda.gov/downloads/InternationalPrograms/FDABeyondOurBordersForeignOffices/EuropeanUnion/EuropeanUnion/EuropeanCommission/UCM266259.pdf

CHAPTER 5

Effectively managing an FDA inspection

JOHN AVELLANET

MANAGING DIRECTOR, CERULEAN ASSOCIATES LLC, USA

5.1   Introduction

Since the 1980s, inspections by the US Food and Drug Administration (FDA) have wrought great anxiety around the world. FDA inspections are no more difficult or stringent than those undertaken for International Standards Organization (ISO) certification or by Australia's Therapeutic Goods Administration or any other regulatory health agency or non-governmental organization. Instead, the difference is based upon impact. If an FDA inspection is failed, the agency may – and has – immediately shut down a company or block its ability to send products to the US. Thus, a failed inspection can quickly cause millions in revenue loss and public humiliation.

While safe and efficacious products are the route to initial market approval under FDA regulations, they are no guarantee, in and of themselves, of continuing marketplace acceptance. The ability to sell products in the US also requires a company to continuously meet evolving FDA requirements and expectations. Thus, a firm must rely upon a quality system supplemented by an effective regulatory intelligence program to maintain continuous compliance. The need to rely upon an effective quality system, braced by an effective records retention program, leaves the firm open to faring poorly on an inspection.

FDA inspectors evaluate a quality system and its supporting documents using both qualitative and quantitative means. Quantitatively, the investigators review a firm's records – completed forms, data generated during production and quality control testing, training records, shipping manifestos, and so on – with questions such as:

- Do the firm's records match the requirements of the regulations and the firm's standard operating procedures (SOPs)?

- Are there missing data or approvals?

- Did all the appropriate personnel attend training?

- Do any non-conformances exceed the investigator's sampling plan level of confidence threshold?

This information serves to verify the judgment rendered during the qualitative aspects of the inspection. During the qualitative portion of the inspection, the investigator seeks to assess whether the actual quality system itself is adequate. Do the SOPs, as they are written, provide the minimum level of process control expected? Is there a records retention program to support the quality system? Do the firm's employees seem to understand their responsibilities under the quality system?

The qualitative aspects of the inspection are the ones that so easily, and so quickly, can put a firm in jeopardy. For if the FDA investigator intuits that all is not right with a company or its controls or its attitude, the inspection can quickly turn negative. In short, for things to go well, firms must appropriately manage the inspection.

5.2 Overview of typical inspections

FDA inspections fall into one of three basic types:

1. Pre-approval

2. Routine

3. "For cause."

Pre-approval inspections are triggered by the submission of a new medicinal product for marketplace approval. For medical device firms, pre-approval inspections are also termed "pre-market approval" (PMA) inspections. In either case, the goal is the same: to assess if the firm is capable of consistently operating in a state-of-control sufficient to regularly produce (and support) a safe and efficacious product. Pre-approval inspections tend to occur within six months of the submission of a new medicinal product application (such as a new drug application, a new biologics application, 510(k), etc.).[1]

Routine inspections are triggered on a calendar basis determined by the risks associated with the products manufactured by a firm. In general, drug firms are to be inspected every two years, device firms less often. This, however, is completely dependent on the product involved and any product approval conditions (i.e. a product that is an entirely new molecular entity may indicate greater risk, and thus necessitate a higher inspectional frequency than a simple generic of a drug whose molecular properties and risks are well understood).

"For cause" inspections are triggered as a result of a specific "cause" such as a product recall, or patient death, or a whistleblower complaint. As a result, "for cause" inspections tend to blind-side company executives and can quickly lead to product seizures and import blocks.

5.3   FDA inspection course

Notice of inspection

Within the US, the FDA does not need to provide ahead of time notice to the firm of an impending inspection. This has given rise to the proverbial "knock on the door" anxiety about FDA inspections. Outside of the US, however, FDA inspections tend to be announced. The FDA will contact the firm through the company's official correspondent two to three months in advance of a window of possible times when the agency would like to come onsite for an inspection.

Upon arrival at the site, the FDA investigator will present his/her credentials (FDA inspectors will send a copy of their passport to the overseas company they wish to inspect) and a copy of the form FDA-482 Notice of Inspection.[2] In addition to his/her photo identification, the investigator may also have a letter from the firm's management, the FDA, or from the host country's regulatory health agency inviting the inspection of the site. Note that Form FDA-482s are not given to non-US based companies, particularly when the FDA investigator is accompanied by the host nation's regulatory health agency auditor. In principle, FDA has no authority to inspect outside US territory without the explicit or implicit approval of the local authority and the local authority always has the right to accompany the inspectors, although this right is often waived.

Opening meeting (investigator's viewpoint)

Following the presentation of credentials, the investigator will look to conduct an introductory or opening meeting. This meeting includes a review of the inspection plan which may include the inspection schedule (including the expected duration), the specific products or operations to be reviewed, an initial set of procedures and/or records to be reviewed, and verification of key elements of the firm's profile. This latter verification reviews many of the basic elements in the first half of a site master file (personnel, reporting structure, site activities, products produced, etc.). For non-US firms, this will also include verification of registration information. During this meeting, the investigator will also discuss various logistical details of the inspection including where to report each day, parking, and so on.

Occasionally, the inspected site will want to demonstrate its capability and give the investigator a slide presentation on the company's history and its products. This is not something to be encouraged unless it is the first time a firm has ever been inspected by the FDA. In such a case, this type of brief overview may be appropriate so long as it is limited to 15 minutes or less, and preferably, less than 10 minutes, especially as the opening meeting typically takes less than one hour.

Facility tour (investigator's viewpoint)

Following the opening meeting, the investigator will likely seek to take a preliminary facility tour or walk through. It is important to realize that this preliminary tour is not a "dry run" for any sort of "real" facility inspection. Far too often, firms see this first investigator walk-through as a visitor or new hire tour. This is not the case for FDA investigators. The tour, from the investigator's viewpoint, allows him/her to be familiar with the site layout and to refine his/her inspection plan. Questions asked during this tour will help determine the level of scrutiny given to various areas of the site and individuals to further interview. Always keep in mind that FDA investigators are at a site specifically to assess its state-of-control. Everything noticed during their time on site is open for questioning and scrutiny.

For instance, if the FDA investigator observes that an outside door is propped open to allow easy access for moving items in and out of a warehouse, this may cause the firm problems if, during the tour, the investigator uncovers that the workers are absent at lunch and the open door allows access to finished stored product. This will result in a non-compliance observation as the company is not maintaining control over its finished medicine. In April 2010, during a simple walk through of a facility, FDA investigators quietly noted that finished product labels, rather than being stored securely, were sitting in open locations and shelves throughout a McNeil Healthcare facility, allowing anyone in the warehouse easy access to the labels.[3] Whether the observation remains minor or turns into a Form FDA-483 observation (or even a Warning Letter or worse) depends on what additional corroborating evidence is gathered during the inspection or if the observation was of an isolated incident with no easily discerned risk to public health or safety.

Closeout meeting (investigator's viewpoint)

Some investigators are more helpful and will end each day of the inspection with a mini-review of the inspectional observations for that day. In general, however, if a firm does not request such a mini-review at the end of each day, the FDA investigator will not offer one. Thus, firms will want to make sure to request a daily summary during the initial inspection opening meeting. If not, the firm may be surprised during the closeout meeting. During any daily debriefing, or even at the closeout meeting, it is important to inform the FDA investigator of any deviation fixes. The firm must have supporting documentation of the fix, such as a revised SOP, a completed supplier qualification dossier, etc. for the investigator to consider not reporting the problem.

The closeout meeting will take place either immediately following the last inspection activities or one day later. The latter is more common if the inspection has been lengthy. The FDA investigator will request a closeout meeting with management individuals having the authority (budgetary, resource allocation, etc.) to fix problems. Remember that FDA enforcement actions are addressed to "management with executive accountability" (i.e. company presidents, chief operating officers, chief executive officers, etc.). At least one of these types of individuals must be in the room for the closeout meeting to be a success.

If there are findings of non-compliance, the FDA investigator will issue a Form FDA-483.[4] More than 60% of companies in the US receive a "483," and more than 80% of the non-US companies receive at least one FDA-483 observation.[5] Recognize that although a 483 observation indicates a non-compliance, it may be a trivial matter or it may be a critical failure. In general, the more 483 observations an investigator has documented, the greater the risk of a Warning Letter or more severe enforcement action (import hold, product seizure, consent decree, etc.). Assuming the 483s are not egregious in nature or do not indicate an immediate public health danger, the general experience is that more than three 483 observations indicates that the firm will receive at least an Untitled Letter, more than twelve 483s indicates at least a Warning Letter, and anything beyond twenty 483 observations of non-compliance enters very dangerous territory indeed.

5.4   Elements of a successful inspection

For an inspection to be successful, not only must the firm be continuously compliant as noted earlier, but a good working relationship should be struck up as rapidly as possible with the investigator. The goal is to encourage the investigator to feel comfortable suggesting remedies for as many of the observations as possible. The more issues that can be fixed as quickly as possible during the inspection, the better the outcome for the company. In such a case, the firm may be one of the few that avoids a written 483.

Using proper terminology

To set the groundwork for a successful inspection, firms need to know and use proper terminology for the FDA and the investigator. The FDA is the "Food and Drug Administration" even though it also regulates medical devices, tobacco, cosmetics, nutritional and dietary supplements, and even lasers. An FDA inspection is conducted by an "investigator," not an auditor or inspector. While these two terminology points may seem minor, many investigators quickly assume company personnel are ignorant of FDA requirements if the company does not use proper terminology when discussing the agency.

Thus, personnel should be trained on this terminology as part of any overview of applicable FDA regulations. For any SOP, company posting, internal memo, etc.

that uses the acronym FDA, make sure to explain the acronym prior to its first usage in the document. Likewise, despite widespread usage of the labels "auditor" and "inspector" and "investigator" interchangeably, company SOPs and forms should refer to the FDA "investigator."

Building rapport

Achieving a success inspection depends in part on quickly establishing a constructive relationship with the investigator. Care must be taken to ensure that communication is open and transparent. This will allow the firm to focus on what it hopes to achieve as a result of the inspection beyond just as few FDA-483 observations as possible.

First, make sure to request a daily debriefing about the investigator's observations and concerns. This will give the firm the opportunity to head off potential trouble as soon as possible and to resolve, just as quickly, as many gaps as possible.

Second, always show the FDA investigator respect, courtesy and professionalism. Just as with any group of people, there are going to be those FDA investigators who may try to bully a firm or push the boundaries of propriety. However, as long as company personnel remain respectful and professional, communication lines can remain open. Remember that the FDA investigator is a representative of the FDA, therefore make sure that senior management is present early on in the inspection, preferably in the opening meeting. This shows the investigator that the company respects FDA's time and effort.

Third, be very careful – and preferably, do not even mention – the names of any former FDA personnel who have served as consultants to the company. Since 2003, FDA has grown weary of firms trying to claim that a former FDAer helped them and thus their systems are compliant. As an FDA official noted in a 2011 conference, "Think of all your various former employees. How many of those would you rely upon to provide someone a true insight into your company's goals, expectations, operations and priorities? Hiring former FDA employees as consultants does not, in any way, provide the current FDA with assurance of your compliance status. If anything, it may generate a question in the mind of the investigator: Was this former FDAer's advice based on how the agency was a decade ago or based on today's FDA? The former will likely cause you significant heartburn during the inspection."[6] Several FDA officials have noted an additional risk by name-dropping former FDA employees as consultants – the former FDAer could have been a disliked supervisor or colleague of the investigator.[7]

Fourth, make sure that any company spokesman, or the company individual appointed to take notes on the inspector's activities and observations (see below,

"Hosting the inspection"), can comport him or herself appropriately. Do not select someone who is openly critical of the agency or government regulations in general. It is important to ensure that the investigator's judgments of the firm are not unduly influenced by the personalities of those with whom he/she regularly interacts during the inspection.

Disagreeing with the investigator

The investigator's role is to provide an objective, independent opinion on the activities and records under review. This means that in order to successfully disagree with the investigator, a firm must provide facts and documented proof that support any differing point of view. Like any large organization, FDA has a large degree of variability in the different levels of its personnel's expertise. Investigators are usually prepared to explore differences, especially if a firm's processes and/or systems are cutting edge and the investigator has not seen their like before.

Most disagreements fall into one of two categories:

1. The investigator's assessment is off-base because not all the facts were evident; or

2. The investigator is essentially accurate but it's a matter of opinion as to the severity of the situation.

In the first case, the only remedy is to gather the facts and arguments to sway the investigator to revise her/her opinion. Years ago, a medical device firm had to place several of their finished products temporarily in with raw materials and quarantined storage. As luck would have it, this was the same day the investigator did a review of the warehouse. The finished product was immediately found and the investigator shared the deviation at the end of the day summary meeting. The following day, the finished product were moved back to finished goods storage and the firm gathered its evidence to show that this was a one-time, temporary and controlled situation. The firm did not re-present its SOP on finished goods storage (the investigator already had a copy of that SOP) and instead presented the following facts and supporting documents:

*   The floors in the finished storage area were being refinished section by section – this was planned months in advance and a month's old project plan/schedule was presented as evidence;

*   The manufacturing line had overproduced several lots worth of product as the manufacturing line was next on the project plan to have its floors

refinished and would thus be shut down for several days (again, the project plan showed this information);

- The additional production had filled the finished storage area at a time when sections of it were being cordoned off to allow refinishing of the floor (the packed to capacity storage room provided ample proof of this);

- The quality control engineer, also in charge of releasing product to distribution, had decided to move a small amount of finished product to the raw materials/quarantined area to use that as a "swap" area for any overfill (this was documented in an email sent to warehouse personnel several weeks beforehand – along with a reminder to all personnel not to handle or move such items during the two day period they would be in quarantine storage);

- The quality control engineer had carefully recorded all serial numbers of the devices temporarily located in the "swap" area (the log was provided as proof of this);

- The quality control engineer had documented (in the email to his supervisor) that he would conduct visual examinations of those finished devices in the raw materials/quarantined area to ensure their sterility seals were still intact and the serial numbers were the same before they were placed back into finished product storage; and

- The quality control engineer had documented his visual review in the log along with the results before the devices were moved back into finished product storage.

The result was a change in the investigator's opinion. The facts and supporting proof clearly demonstrated that the storage location was temporary, planned and controlled.

The second type of disagreement – wherein the interpretation of the significance of the situation is in dispute – typically occurs between the investigator and the firm's management. Unfortunately, these are much more difficult to resolve as it can quickly become a battle of wills. A good investigator (or auditor for that matter) will be able to present specific criteria to support and explain his/her opinion. FDA trains its investigators on the regulations and on multiple regulatory harmonized guidelines from the International Conference on Harmonization (ICH) and the Global Harmonization Task Force (GHTF). These guidance documents may form the basis for the investigator's judgment of a deviation's significance and will make any pushback from company personnel more difficult.

The only means to productively resolve this type of difference is to avoid the battles of criteria and wills. Instead focus on what the investigator wants the company to do to resolve the non-conformance. Two opportunities may emerge from this solutions-focused discussion. First, the firm may already have other controls that render the ones that the investigator found lacking to be moot. Imagine a situation wherein finished product was stored in an open warehouse environment with other materials. Normally this would be a clear deviation; however, if only quality control release personnel were given access to the warehouse then management could legitimately argue that both the physical boundaries and the procedural controls were functional for the intent of the regulations. Second, by focusing on what the investigator wants the firm to do to resolve the issue, the firm's personnel may be able to present other guidance documents and controls to support their argument that the deviation is less severe than the investigator believes.

Ensuring management visibility

One means to head off disagreements with the investigator is to ensure that management is involved early and often with the inspection. During the opening meeting, executive management will want to make sure they have a firm grasp of the investigator's scope, objectives, approach and evaluation criteria. This is the first moment for management to raise any concerns prior to the inspection getting underway.

Throughout the inspection, as the firm continues with its day-to-day activities, management should be visibly present, especially if senior executives have a habit of "management by walking around." Now is the time to make sure such walk-throughs occur. Conversely, be cautious about encouraging management that never makes an appearance to suddenly show up on the factory floor. FDA investigators are trained to observe people's reactions closely and will undoubtedly pick up on the unease which will come from a company president walking into a manufacturing bay when workers have never seen him/her before at the site, much less on the production floor.

If the investigator is willing to schedule daily inspection debriefings, a member of management should be present. This will show the investigator that the company takes its commitments seriously to safe and efficacious products, and to compliance with the FDA. Having management only put in appearances at the beginning and/or the end of the inspection is a good way to demonstrate that management does not actively support and oversee implementation of an effective quality system or other compliance program. Likewise, a firm can go too far and appear defensive if the management team travels in packs and is frequently accompanied by a member of the legal staff. To have a successful inspection, management must be present, visible, and engaged, not wary and standoffish.

Conducting timely fixes

When there are problems – and every firm has at least one or two issues that will be found by the investigator – what is the firm's reaction? Being defensive can quickly escalate a minor issue to one that will be documented on the Form FDA-483. Assuming the management team is not immediately defensive, does management take a vested interest in solving the problem and preventing its reoccurrence? Investigators do not want to see just their one observed gap fixed. When it comes to compliance, consider an investigator's observation as a symptom. FDA's experience over the past 40 years has been that inspectional observations are the visible surface of underlying systemic problems. Thus, while correcting the observed deviation is clearly a necessity, it is only one step in implementing a timely fix.

Instead, a firm needs to view the investigator's observation, whether minor or major, as the trigger point for an investigation to determine the full scope of the deviation. For instance, in the event of batch record discrepancies, the firm should investigate if there are any other batches implicated by this deviation. Then, the firm needs to determine not only how to resolve the problem but how to prevent similar issues in the future. In order to successfully resolve a problem, the firm needs to show the FDA the records that prove the problem was fixed and the records associated with preventive controls such as including revised procedures, additional automated controls, new training, more frequent auditing, etc.

Many FDA Warning Letters cite a firm for failure to support a claimed resolution to an inspection observation. Claiming that an FDA-483 observation has been resolved by implementation of a revised control, such as a revised SOP, without having records to prove the revised control is operating effectively will not assuage FDA's concerns. This is a key reason for immediately undertaking timely fixes during the inspection. Firms need to rapidly start generating records of revised control effectiveness. These records can then be shown to the investigator if the inspection is still on-going or can be part of a firm's response to the FDA following the inspection, especially if the ultimate resolution to a systemic problem will take significant time.

5.5   Hosting the inspection

In the article "Rapid deployment tips to prepare for an inspection quickly," the author describes seven steps to take when time is of the essence.[8] For international inspections, as previously noted, a sudden "knock upon the door" is extremely unlikely and the firm will have ample time to review its procedures for hosting an FDA inspection. No matter the amount of forewarning, the day must come for the investigator to arrive.

As soon as the investigator arrives (note that most FDA inspections are conducted by two investigators), the investigator should be treated as first time visitors. Have the investigator sign-in, explain who they are, identify who their contact at the firm is, and so on. Inspections occur during normal business hours so the firm should factor this into its expectations and training of security guards, receptionists, etc. The receptionist or security guard can then notify the firm's designated "inspection host." Typically, the inspection host is a senior regulatory affairs, quality department or compliance team employee; if a firm has only one product, the employee who serves as the official correspondent with the FDA may be the initial host until the firm's management decides otherwise.

Opening meeting (firm's viewpoint)

Once the host arrives on the scene, he/she should verify the investigator's credentials and notice of inspection. The firm's host should then escort the investigator to the meeting room or other area that has been designated for the inspection's opening meeting. Exchanging business cards and enquiring about refreshments prior to the start of the meeting will allow time for the firm's key personnel, including at least one member of senior management, to make their way to the meeting. It is important to understand that an FDA inspection is a critical moment for the company. Normal day-to-day concerns may need to be set aside or at least downgraded in priority. Presenting an impression of laxness or misplaced priorities to the monitor of US public's health and safety (i.e. the FDA investigator) is not a good way to begin the inspection. Management's presence at the opening meeting helps to show commitment to day-to-day oversight of both the regulated functions of the company and the attendant quality system controls.

As the investigator discusses his/her logical needs, the host should verify that a designated work area exists for the inspection team. Preferably this needs to be an area that can be closed off for privacy. Not only may this area be used for interviews, but confidential documents may be kept within it during the course of the inspection. Prior to escorting the investigator to the work area, the host should conduct a quick walk-through of the area to ensure it is prepared to receive the investigator. Are there any confidential or other internal company documents left lying about from a previous meeting? Is the work area clean? If the firm has a guest network or internet access that bypasses the company network, the host should make sure this guest access is operating appropriately. A nice touch is to include a basic map of the facility from the standpoint of the investigator's work area showing the nearest bathrooms, the location of the break room, any photocopiers, and so on; also make sure to note the host's office and phone number (and the phone number of the investigator work area). Just this little bit of preparation can help show the investigator that the firm takes the time to get the details correct.

As part of the logistics discussion in the opening meeting, make sure to clarify to the investigator any ground rules required of access to the specific areas such as a cleanroom or a manufacturing floor. If special equipment is required (i.e. a hard hat, hearing protection, lab coat, etc.), make sure spares in good condition are available. If the firm normally conducts some level of safety and/or sanitary training for onsite contractors and/or long-term visitors, make sure the investigator receives this training. Do not forget that once the investigator has shown his/her credentials, the inspection has technically begun and everything an investigator notices (such as a lack of training for him/her) is fodder for the inspection report. If the investigator refuses to cooperate with the training or protective clothing, the host should contact the firm's legal counsel to discuss the matter and determine appropriate next steps. Some investigators initially refuse such training just to gauge the firm's response.

Following the opening meeting, the host should communicate the onsite presence of FDA to all site personnel. By waiting until after the opening meeting, the host can provide specific information to company personnel:

*   How long the inspection is expected to take;

*   Number and names of the investigators;

*   Where the inspection team will be located; and

*   The inspection's anticipated scope.

Firms may also want to briefly issue reminders to personnel such as an expectation that investigator questions are to be answered promptly and efficiently, that day-to-day activities and schedules should continue as normal unless specifically instructed otherwise, and that the firm's designated host will be able to answer any questions or concerns. It may be wise to remind personnel that an answer of "I don't know" is always preferable to a guess or even making something up. Human nature being what it is, many people have a hard time admitting ignorance on a particular topic, but it is best to be honest rather than state something that turns out to be incorrect. The latter will always raise suspicions in a skeptical investigator.

Facility tour (firm's viewpoint)

The facility tour that typically follows the opening meeting is a good time to show the investigator his/her working area (including the basic facility map if so provided), the host's office, and other logistical details. For each area within the facility that requires protective gear, make sure to not only point that out but, whenever applicable, provide the gear nonchalantly if, for instance, the tour will go through a noisy factory floor.

Use the tour time to have the investigator meet any of the key personnel unable to attend the opening meeting, including any senior management. Be cautious about trying to schedule a "make-up" opening meeting for members of senior management who were unable to attend the early meeting. Trying to schedule a make-up session sends the wrong signal – that FDA, especially in its role as guardian of the safety of medicines in the US – has little better to do than sit around waiting for when a meeting is convenient for the firm's senior management.

Providing access

Following the tour, the FDA inspection team will begin its first document review. The firm should log each document requested by the investigator. Sometimes, the logging is simple and straightforward (e.g. "Phase II clinical trial initiative approval form"). However, the vast majority of records requested will not be single forms but large sets of records such as vendor dossiers, batch records, training records, and so on. In these cases, log each set as discretely as possible (e.g. "vendor dossier for API manufacturer, vendor dossier for formulation manufacturer") without going into each individual document within the set.

If the investigator asks for a copy of a record (or record set), especially if he/she is planning on leaving the premises with the document, the firm should make a duplicate copy and keep it in a file. Label the file "FDA Inspection (inspection dates)." Noting the dates on the label will help appropriately retain the file once the inspection is complete. Any notes taken during the inspection, meeting minutes, document logs, etc. should also go into the file. If the investigator requests a physical sample (of a product, chemical, component, etc.), make sure to log this and to take a duplicate sample. Some companies have gone so far as to duplicate the tests FDA plans to run on the sample, but this is usually not necessary.

One often overlooked aspect is to ensure that confidential documents are marked as such. Typically, records relating to proprietary manufacturing processes, formulation specifications, design control and so on (i.e. related to intellectual property) are deemed confidential by the firm. The host will want to ensure this notice of confidentiality is clear on all such documents. In addition, for firms that prohibit photography onsite, make sure this rule is clearly explained to the inspection team, preferably during the opening meeting.

Another easily overlooked point of risk can be found within a firm's SOP on its procedures or quality document control, usually within one phrase along the lines of "Copies of SOPs shall be marked as 'unofficial' copies and initialled by the individual who made the copies." Make sure to follow this process precisely, even when printing off copies of SOPs maintained electronically. As soon as an investigator notices that such a small rule or step was not followed, he/she will wonder what other SOPs or policies are not adequately followed.

Escorting the investigator

Beyond simply documents, the FDA inspection team will want to inspect various areas of the site more closely than during the initial facility tour. The investigator will also want to meet and interview different individuals and watch processes in action. It is imperative that the firm escort the investigator outside the designated inspection work area. The escort needs to be familiar with, although not necessarily expert in, each of the activities at the site. The escort should feel comfortable introducing the investigator to specific subject matter experts in the company, even if that means, in the worst case, interrupting a meeting.

A key responsibility of the escort is to take notes on what the investigator observes, notes about the interviews conducted, notes of any feedback discussions with the investigator. The firm must have its own set of notes for three reasons:

1. To allow early warning of observed non-conformances and deviations;
2. To provide specific information to support an alternative viewpoint in the event a disagreement with the investigator arises;
3. To allow a summary, from the company's perspective, of each day's activities.

Tracking the inspection's progress

At the end of each day, the escort(s) gather with the official company host to review the progress of the inspection so far. This is not the time for arguments over responsibility or accountability for any deviations found. Rather, the goal is to obtain a quick summary of the day's activities and prepare for the next day of the inspection. If the investigator is willing to conduct regular debriefings of the inspection, the company's review session should follow immediately thereafter.

During the company review session, individuals should share impressions, discuss potential corrective actions to take, and determine if everything is available for the next day's inspectional activities. Following this review session, the inspection host may want to update senior management on the inspection's progress. The agenda for the daily review session should cover the following points:

*   Non-compliance findings (if any) so far
*   Interview summaries

- Investigator discussion summaries

- Other notes of relevance

- Activities for the next day

- Quick fixes to be implemented immediately

- Potential longer-term gaps to be resolved

If the firm has a consultant who previously conducted a mock FDA audit and/or to whom they expect to look to for advice on responding to the formal Form FDA-483, the firm may want the consultant to participate in these review sessions; such a discussion can easily happen virtually, via teleconference, video-conference or online.

As part of its internal debriefings, the firm will want to start to create an observation-closure matrix (see **Table One**). Keep the matrix simple to allow for rapid review and analysis. The first column should be a short summary of the investigator's non-conformance observation. The second column is for the name of the individual accountable for resolving the issue. Since the FDA wants to see a company's management team taking accountability for compliance, the individual listed in this column needs to be someone with the budgetary and resource authority to fix the problem (i.e. a director, senior manager, etc.) and not necessarily the person who will perform the actual activity. The third column indicates the current status of the closure. And in the fourth column, list the specific proof – the records – generated. Remember, without the proof, the activity did not occur. Thus, in order to show the FDA that a deviation has been resolved, the firm must have data to support the claim of resolution. Make sure to use the observation-closure matrix during any debriefings with the FDA inspection team (including the final closeout meeting). This will help show that the firm takes compliance seriously and is actively engaged in accountable gap closure.

Table 1: An observation-closure matrix.

Inspectional Observation	Owner	Status	Proof (records)

Closing out the inspection (firm's viewpoint)

As the inspection winds down, the firm will want to schedule the closeout meeting so that as many members of senior management can be present. If the host can influence the meeting schedule, rather than having the meeting on the last day of the inspection, it may be worth everyone's time to have the meeting the following day. This allows the FDA inspection team to synthesize its observations and come to consensus on the ones that will end up in the Form FDA-483. For the firm, a pause for even a single evening can allow management enough time to quickly assess the inspection overall, estimate the most likely critical observations, and determine questions to raise during the closeout meeting.

The closeout meeting is the last opportunity the firm has to make a positive impression on the FDA inspection team and understand its expectations of compliance. Therefore, seek clarity on the investigator's viewpoint. Once the FDA has identified and documented specific compliance deviations within a firm, the company is obligated to solve the problems.

Listening to the investigator requires active listening skills and clarifying questions. One of the best techniques for clarifying questions is to use the "If we did.../would that...?" approach. For instance, at one firm's inspection closeout meeting, the FDA inspection team identified that the firm's corrective and preventative action (CAPA) procedures were strong on fixing the immediate problem but weak on root cause analysis and preventative controls. During the closeout meeting, the management asked, "So, if we revised our CAPA process to have a quality department sign-off on the root cause analysis and revised our internal quality audit procedure to not only cover the specific fix from the CAPA but at least four or five other related items to ensure the overall area was operating in a state-of-control, would that meet your concerns?" This allowed the investigator to say either "yes/no/almost" and then to explain the logic underlying his/her original observation. That can then easily segue into a discussion about best practices and how to achieve a resolution on the original observation, all without anyone getting defensive.

While the "If we did.../would that...?" technique also has the obvious benefit of drawing out the investigator's recommendation to fix the immediate problem and overall systemic concerns, it provides a more subtle benefit: showcasing to the investigator that the firm takes the deviation seriously and is already thinking/planning how to ensure an optimal return to compliance within business confines.

To give both the FDA district office the complete picture of current activities being taken to resolve deviations, the firm's management will want to ask the

FDA investigator to annotate his/her observations with the various actions the company has taken and/or is beginning to take to resolve the issues. These annotations can help FDA officials who are reviewing the inspection report to quickly determine whether or not the company "gets it" and is committed to doing the right thing.[9] If the investigator is amenable, he/she may be willing to remove those items that were fixed during the inspection to his/her satisfaction.

Throughout the closeout meeting, continually check for understanding on the FDA investigator's observations even if the observation is seemingly obvious. As noted earlier, the FDA uses the individual observations as symptoms of larger systemic, state-of-control operational deficiencies. It may be helpful to rely upon the observation-closure matrix discussed earlier to refine any planned corrective actions.

Finally, keep in mind that the inspection does not completely end until the FDA inspection team has left the site. Non-conformances noticed after the closeout meeting may end up on the Establishment Inspection Report (EIR), and contribute to a decision by the agency to take strong enforcement action. Thus, within 30 days of the inspection completion, the firm should file a Freedom of Information Act request to obtain a copy of the final EIR from the FDA.

5.6   SOP on inspection handling

The firm's SOP on handling regulatory agency inspections should be focused on two points:

1.   Ensuring broad applicability
2.   Activities for company personnel

Ensuring broad applicability is important in an international marketplace. Executives cannot afford to look only as far as their own region. Rather, draft the SOP using a generalized title such as "Handling Regulatory Health Agency Inspections" or "Handling Regulatory Inspections and Third-Party Audits." Depending on the level of detail in a firm's SOPs, the procedure should ideally be crafted such that it is applicable to any audit or inspection by an outside entity, whether a regulator, independent consultant (performing a mock FDA audit), certification agency, or a potential business partner. By ensuring as broad an application as possible, the firm will minimize the number of inspection-related SOPs to maintain while maximizing its flexibility in handling different types of inspections and audits.

Second, the SOP should center on activities to be undertaken by the firm's personnel when it comes to dealing with outside auditors and inspectors. Far too often, inspection handling SOPs are little more than descriptions of how an inspection proceeds and what company personnel should not do. It is best to leave this type of information to either a training session or to a policy on employee conduct. Instead, keep the SOP focused on specific responsibilities and process steps such as creation and maintenance of an observation-closure matrix, verifying FDA investigator credentials, reviewing provided documents to ensure they are properly marked, and so on.

Finally, firms should not train all company personnel on this type of SOP. Ideally, only those individuals who will have specific tasks under the procedure will be trained, leaving company management and other personnel to be trained on an overall policy on conduct during an inspection and a training session on FDA regulations and enforcement.[10] Make sure to describe the SOP in any policy training, however. It is important for quality and regulatory affairs departments to ensure that the firm's management has a level of comfort in the knowledge that any FDA inspection will be handled in a manner designed to limit risk, ensure compliance, and keep the company in the best light.

5.7 Final thoughts

There is an old saying in academia: the best doctoral dissertation is a done dissertation. Certainly the same can be said of almost any significant undertaking: eventually it will fade into memory. As much as an FDA inspection is an anxious, critical time for a company, the inspection will end. And by following the pattern and tactics in this chapter, executives can easily end an FDA inspection with heads held high and the inspection behind them.

To summarize, this chapter reviewed the following primary advice for successfully navigating an FDA inspection:

- Prioritize time with the FDA investigator in order to make a consistently positive impression;

- Be transparent to company personnel about the inspection and location of the investigators;

- Make sure that at least one member of senior management is consistently present in both the opening and closing meetings;

- Use the initial tour to showcase specific, visible controls such as protective clothing, specific training required, security access restrictions, etc.;

- Keep a log of all records and record sets provided to the inspection team (store this log in a dedicated inspection file to be maintained as per the firm's records retention policies);

- Always escort any FDA investigator and make sure to take notes on what he/she observes and who he/she interviews;

- Disagree only with the investigator when facts are available to support a different point of view;

- Undertake daily internal debriefings and start populating the observation-closure matrix;

- For all claims of gap closure, ensure that proper records (log files, batches, documents, etc.) exist to prove the resolution is effective;

- Use the closeout meeting to obtain clarity on observations and underlying FDA concerns;

- Make sure to obtain a copy of the final EIR to close out the inspection file;

- Ensure any inspection handling SOP adheres to lean compliance tenets of broad applicability, business flexibility and tasks accomplishable by individuals.

Throughout the chapter, smaller insights have been added to provide the best possible handling of any inspection or audit, much less an FDA inspection. Finally, keep in mind that the notes taken throughout the inspection can be combined with the early creation and use of an observation-closure matrix to help streamline and simplify responding to the inspection.

Are you ready?

5.8   References

1   University of Michigan Medical School, *The Inspector Cometh: Inspection Process for Clinical Trials*, a presentation at University of Michigan, http://www.wlap.org/file-archive/cacr/CRE-Talk_1.ppt, accessed on 9/12/2011.

2   Current versions of Form FDA 482 (and other inspectional related forms) can be found online at the fda.gov website: http://www.fda.gov/ICECI/Inspections/IOM/ucm127372.htm, accessed on 9/12/2011.

3   April Hollis, "McNeil Fort Washington Inspection Finds QA Responsibility Issues," *Drug GMP Report*, 2010; August, accessed online on 9/12/2011 at http://www.fdanews.com/newsletter/article?articleId=129138&issueId=13917.

4   Ibid.

5   Michael Anisfeld, "How to Guarantee You Get a Warning Letter from FDA!" *Journal of Validation Technology*, 2008; Spring: p. 11.

6   Kimberly Trautman, *Purchasing Controls – Background and Examples*. Speech to the FDAnews Supplier Quality Congress, 10 August 2011.

7   Ronald Johnson, "How to Handle an FDA Inspection" in How to Work with the FDA, 2nd edition, edited by Wayne L. Pines, Washington, D.C.: FDLI, 2003.

8   John Avellanet, "Rapid Deployment Tips to Prepare for an Inspection Quickly," *PDA Letter*, 2010; September: 18-22 (a copy of the article is available online at http://www.ceruleanllc.com/resources/published-articles-case-studies/, accessed online 9/19/2011.

9   Timothy Ulatowski, *Avoiding Enforcement Actions*. Speech to the FDLI Enforcement and Litigation Conference, 6 February 2007.

10   For an excellent discussion of the differences between a policy and an SOP and a work or task instruction, see Larry Peabody, Sending Clear Signals in Written Directions, 3rd edition, Lacy, Washington: Writing Services, 2010.

CHAPTER 6

Guide for successful EU inspection management

Siegfried Schmitt and Nabila Nazir
PAREXEL Consulting, UK

6.1 Introduction

Other authors have described in great detail the inspection process by regulatory authorities such as the United States Food and Drug Administration (US FDA) and have thereby covered to some degree the process as it applies in the European Union (EU). However, unlike the United States of America, Europe is a conglomerate of individual member states who are only loosely linked through common regulations and whose governments are at liberty to interpret European law when implementing it into their legislation. The result is a much more varied and individual approach to inspections than experienced elsewhere. This chapter covers the key legislation in the EU, the major agencies and the specifics of an inspection by EU inspectors.

The European Union is a unique economic and political partnership between 27 European countries, the Member States[1]. The European Economic Area (EEA) consists of the European Union plus Iceland, Liechtenstein and Norway. The regulatory authorities in the Member States are called Competent Authorities[2]. The Heads of Medicines Agencies[3] is a network of the Heads of 44 national Competent Authorities from the 27 EU Member States and the 3 countries from the EEA. Most Member States have separate agencies for human and veterinary drugs.

The European Commission (EC) is one of the main institutions of the European Union. It represents and upholds the interests of the EU as a whole. It drafts proposals for new European laws. It manages the day-to-day business of implementing EU policies and spending EU funds. It is the governing body for the European Union. The EC, located in Brussels, Belgium, issues Directives, which are implemented by each Member State into their national legislation. It created the European Medicines Agency (EMA) to implement and enforce Directives in 1995. The interaction of the various institutions under the differing legislative systems impacts on the inspection process.

To aid the pharmaceutical manufacturer to prepare for a regulatory inspection the chapter is divided into the following sections:

- Part A: European Pharmaceutical Legislation – an Overview
- Part B: EU GMP Guidance: EudraLex Volume 4

- Part C: Overview of EU GMP inspections

- Part D: Hosting and Managing the Inspection

- Part E: Managing the Exit Meeting

- Part F: The EU GMP database and information sharing practices

- Part G: Risk-based re-inspection schedules

6.2   Part A: European pharmaceutical legislation – an overview

In the European Union, the European Commission's remit is: To guarantee the highest possible level of public health and to secure the availability of medicinal products to citizens across the European Union; all medicinal products for human use have to be authorised either at Member State or Community level before they can be placed on the EU market. Special rules exist for the authorisation of medicinal products for paediatric use, orphan medicines, traditional herbal medicines, vaccines and clinical trials. Furthermore, to ensure that medicinal products are consistently produced and controlled against the quality standards appropriate to their intended use, the European Union has set quality standards known as 'Good Manufacturing Practice' (GMP). Compliance with these principles and guidelines is mandatory within the European Economic Area.

In addition, once a medicinal product has been authorised in the Community and placed on the market, its safety is monitored throughout its entire lifespan to ensure that, in case of adverse reactions that present an unacceptable level of risk under normal conditions of use, it is rapidly withdrawn from the market. This is done through the EU system of pharmacovigilance.

In order to help the European Union ensure the highest possible level of public health protection, in 1994 the EU established the European Medicines Agency (EMA) with the main task of coordinating the scientific evaluation of the quality, safety and efficacy of medicinal products which undergo an authorisation procedure, and providing scientific advice of the highest possible quality[4].

The EU legal framework for medicinal products for human use is intended to ensure a high level of public health protection and to promote the functioning of the internal market, with measures which moreover encourage innovation. It is based on the principle that the placing on the market of medicinal products is made subject to the granting of a Marketing Authorisation (MA) by the competent authorities.

A large body of legislation has developed around this principle, with the progressive harmonisation of requirements for the granting of marketing authorisations since the 1960s, implemented across the European Economic Area (EEA).

The requirements and procedures for the marketing authorisation for medicinal products for human use, as well as the rules for the constant supervision of products after they have been authorised, are primarily laid down in Directive 2001/83/EC and in Regulation (EC) No 726/2004. These texts additionally lay down harmonised provisions in related areas such as the manufacturing, wholesaling or advertising of medicinal products for human use. In addition, various rules have been adopted to address the particularities of certain types of medicinal products and promote research in specific areas: orphan medicinal products (Regulation (EC) No 141/2000), medicinal products for children (Regulation (EC) No 1901/2006) and advanced therapy medicinal products (Regulation (EC) No 1394/2007)[5].

All Community legislation in the area of medicinal products for human use is contained in volume 1[6] of "The Rules Governing Medicinal Products in the European Union".

To facilitate the interpretation of the legislation and its uniform application across the EU, numerous guidelines of regulatory and scientific nature have additionally been adopted:

A detailed explanation of the marketing authorisation procedures and other regulatory guidance intended for applicants is contained in volume 2 (Notice to Applicants)[7].

Scientific guidance on the quality, safety and efficacy of medicinal products is provided in volume 3[8].

Specific guidance on the legal requirements concerning good manufacturing practices, pharmacovigilance and clinical trials is laid down in volumes 4[9], 9[10] and 10[11], respectively.

- For access to EU pharmaceutical legislation consult EUDRALEX[12]
- For a summary of EU pharmaceutical legislation consult SCADPlus[13].

Marketing Authorisation procedures

The European system offers several routes for the authorisation of medicinal products:

- The **centralised procedure**, which is compulsory for products derived from biotechnology, for orphan medicinal products and for medicinal products for human use which contain an active substance authorised in the Community after 20 May 2004 (date of entry into force of Regulation (EC) No 726/2004) and which are intended for the treatment of AIDS, cancer, neurodegenerative disorders or diabetes. The centralised procedure is also mandatory for veterinary medicinal products intended primarily for use as performance enhancers in order to promote growth or to increase yields from treated animals.

  Applications for the centralised procedure are made directly to the European Medicines Agency (EMA) and lead to the granting of a European marketing authorisation by the Commission which is binding in all Member States.

- The **mutual recognition procedure**, which is applicable to the majority of conventional medicinal products, is based on the principle of recognition of an already existing national marketing authorisation by one or more Member States.

- The **decentralised procedure**, which was introduced with the legislative review of 2004, is also applicable to the majority of conventional medicinal products. Through this procedure an application for the marketing authorisation of a medicinal product is submitted simultaneously in several Member States, one of them being chosen as the "Reference Member State". At the end of the procedure national marketing authorisations are granted in the reference and in the concerned Member States.

Purely national authorisations are still available for medicinal products to be marketed in one Member State only.

Special rules exist for the authorisation of medicinal products for paediatric use, orphan drugs, traditional herbal medicinal products, vaccines and clinical trials[14].

The centralised procedure

The 'centralised procedure'[15] for authorising medicinal products is laid down in Regulation (EC) No 726/2004.

The centralised procedure, which came into operation in 1995, allows applicants to obtain a marketing authorisation that is valid throughout the EU. It is compulsory for medicinal products manufactured using biotechnological processes, for orphan medicinal products and for human products containing a new active substance which was not authorised in the Community before 20 May 2004 (date of entry into force of Regulation (EC) No 726/2004) and which are intended for the treatment of AIDS, cancer, neurodegenerative disorder or diabetes. The centralised procedure is also mandatory for veterinary medicinal products intended primarily for use as performance enhancers in order to promote growth of treated animals or to increase yields from treated animals.

The centralised procedure is optional for any other products containing new active substances not authorised in the Community before 20 May 2004 or for products which constitute a significant therapeutic, scientific or technical innovation or for which a Community authorisation is in the interests of patients or animal health at Community level.

When a company wishes to place on the market a medicinal product that is eligible for the centralised procedure, it sends an application directly to the European Medicines Agency, to be assessed by the Committee for Medicinal Products for Human Use (CHMP) or the Committee for Medicinal Products for Veterinary Use (CVMP).

The procedure results in a Commission decision, which is valid in all EU Member States. Centrally-authorised products may be marketed in all Member States.

Applications from persons or companies seeking 'orphan medicinal product designation' for products they intend to develop for the diagnosis, prevention or treatment of life-threatening or very serious conditions that affect not more than 5 in 10,000 persons in the European Union are reviewed by the Committee for Orphan Medicinal Products (COMP).

The Committee on Herbal Medicinal Products (HMPC) is responsible for establishing a list of herbal substances, preparations and combinations thereof for use in traditional herbal medicinal products. It is also responsible for establishing Community herbal monographs.

Once the pharmaceutical company has submitted its application for a marketing authorisation, full copies of the marketing authorisation application file are sent to a rapporteur and a co-rapporteur designated by the competent EMA scientific committee. They co-ordinate the EMA's assessment of the medicinal product and prepare draft reports.

Once the draft reports are prepared (other experts might be called upon for this purpose), they are sent to the CHMP or CVMP, whose comments or objections are communicated to the applicant. The rapporteur is therefore the privileged interlocutor of the applicant and continues to play this role, even after the marketing authorisation has been granted.

The rapporteur and co-rapporteur then assess the applicant's replies, submit them for discussion to the CHMP or CVMP and, taking into account the conclusions of this debate, prepare a final assessment report. Once the evaluation is completed, the CHMP gives a favourable or unfavourable opinion as to whether to grant the authorisation. When the opinion is favourable, it shall include the draft summary of the product's characteristics, the package leaflet and the texts proposed for the various packaging materials.

The time limit for the evaluation procedure is 210 days (the application formalities and the detailed procedure are described in the Notice to Applicants which is available in the EudraLex volumes).

The EMA then has 15 days to forward its opinion to the Commission. This is the start of the second phase of the procedure: the decision-making process. The Agency sends to the Commission its opinion and assessment report, together with annexes containing:

- the summary of product characteristics (Annex 1);

- the particulars of the manufacturing authorisation holder responsible for batch release, the particulars of and the manufacturer of the biological active substance and the conditions of the marketing authorisation (Annex 2);

- the labelling and the package leaflet (Annex 3).

The annexes are translated into the 22 other official languages of the EU.

During the decision-making process, the Commission services verify that the marketing authorisation complies with Union law.

The Commission has fifteen days to prepare a draft decision. The medicinal product is assigned a Community registration number, which will be placed on its packaging if the marketing authorisation is granted. During this period, various Commission directorates-general are consulted on the draft marketing authorisation decision.

The draft decision is then sent to the Standing Committee on Medicinal Products for Human Use, or the Standing Committee on Veterinary Medicinal Products (Member States have one representative each in both of these committees) for their opinions.

Member States have 15 days to return their linguistic comments and 22 days for scientific and technical ones. This procedure is conducted in writing but if a duly justified objection is raised by one or more Member States, the committee holds a plenary meeting to discuss it.

When the opinion is favourable, the draft decision is adopted the empowerment procedure.

The Commission's Secretariat-General then notifies the Commission Decision to the marketing authorisation holder. The decision is then published in the Community register.

Marketing authorisations are valid for five years. Applications for renewal must be made to the EMA at least six months before this five-year period expires.

The mutual recognition procedure

Basic arrangements for implementing the mutual recognition procedure[16] laid down in Directive 2001/83/EC have been made in all Member States. To be eligible for this procedure, a medicinal product must have already received a marketing authorisation in one Member State.

Since 1 January 1998, the mutual recognition procedure is compulsory for all medicinal products to be marketed in a Member State other than that in which they were first authorised. Any national marketing authorisation granted by an EU Member State's national authority can be used to support an application for its mutual recognition by other Member States.

The mutual recognition procedure is based on the principle of the mutual recognition by EU Member States of their respective national marketing authorisations. An application for mutual recognition may be addressed to one or more Member States. The applications submitted must be identical and all Member States must be notified of them. As soon as one Member State decides to evaluate the medicinal product (at which point it becomes the "Reference Member State"), it notifies this decision to other Member States (which then become the "Concerned Member States"), to whom applications have also been submitted. Concerned Member States will then suspend their own evaluations, and await the Reference Member State's decision on the product.

This evaluation procedure undertaken by the Reference Member State may take up to 210 days, and ends with the granting of a marketing authorisation in that Member State. It can also occur that a marketing authorisation had already been granted by the Reference Member State. In such a case, it shall update the existing assessment report in 90 days. As soon as the assessment is completed, copies of this report are sent to all Member States, together with the approved summary of product characteristics (SPC), labelling and package leaflet. The Concerned Member States then have 90 days to recognise the decision of the Reference Member State and the SPC, labelling and package leaflet as approved by it. National marketing authorisations shall be granted within 30 days after acknowledgement of the agreement.

Should any Member State refuse to recognise the original national authorisation, on the grounds of potential serious risk to public health, the issue will be referred to the coordination group. Within a timeframe of 60 days, Member States shall, within the coordination group, make all efforts to reach a consensus. In case this fails, the procedure is submitted to the appropriate EMA scientific committee (CHMP or CVMP, as appropriate), for arbitration. The opinion of the EMA Committee is then forwarded to the Commission, for the start of the decision making process. As in the centralised procedure, this process entails consulting various Commission Directorates General and the Standing Committee on Human Medicinal Products or the Standing Committee on Veterinary Medicinal Products, as appropriate

The Decentralised Procedure

The decentralised procedure[17] was introduced by Directive 2004/27/EC. As the mutual recognition procedure, it is also based on recognition by national authorities of a first assessment performed by one Member State. The difference lies in that it applies to medicinal products which have not received a marketing authorisation at the time of application.

An identical application for marketing authorisation is submitted simultaneously to the competent authorities of the Reference Member State and of the Concerned Member States. At the end of the procedure, the draft assessment report, SPC, labelling and package leaflet, as proposed by the Reference Member State, are approved.

The subsequent steps are identical to the mutual recognition procedure.

All European agencies, following the example of EMA, which charges fees in accordance with Regulation (EC) No 726/2004, which states in paragraph (21): The Agency's budget should be composed of fees paid by the private sector and contributions paid out of the Community budget to implement Community policies.

These fees vary depending on the type of application. Fees levied by national competent authorities vary widely and are subject to change at short notice. It has been noticed that in the current climate of austerity, agencies are keen to maximise income from fees as they have to cope with otherwise reduced budgets and resources.

The European Medicines Agency (EMA)

The European Medicines Agency[18] is a decentralised agency of the European Union, located in London. The mission of the European Medicines Agency is to foster scientific excellence in the evaluation and supervision of medicines, for the benefit of public and animal health. The Agency is responsible for the scientific evaluation of medicines developed by pharmaceutical companies for use in the European Union. The EMA does not perform inspections. The Agency has a co-ordinating role for inspections whilst the responsibility for carrying them out rests with the Competent Authority under whose responsibility the manufacturer falls.

EMA has published a "Compilation of Community Procedures on Inspections and Exchange of information" document on 5th July 2011[19], which details the procedures related to Good Manufacturing Practice (GMP) and Good Distribution Practice (GDP).

Active Pharmaceutical Ingredients (API) legislation

The United Kingdom (UK) Medicines and Healthcare products Regulatory Agency (MHRA) took the lead in making clear expectations for API (drug substance) manufacturers. In 2010 they published their expectations on their website[20]:

Marketing and manufacturing authorisation holders are reminded that as all sites named on approved marketing authorisations (MAs) can be used for the activity they have been registered for without further variation or provision of evidence of Good Manufacturing Practice (GMP) compliance to the MHRA, it is important that all named sites are fully maintained as approved suppliers and thus available for use.

Through its inspection programme, the MHRA has encountered numerous examples of API manufacturing sites that are named on marketing authorisations but are not currently used and not fully maintained as approved suppliers. In some cases such API sites may never have supplied API for marketed product. This can also apply to other sites, e.g. drug product manufacturing, importation or batch release sites.

Therefore in order to clarify MHRA expectations, the following guidance is provided:

- All API manufacturing sites named on UK marketing authorisations (product licences) must be actively maintained as approved suppliers (in line with EU GMP expectations).

- The MHRA regards all API manufacturers that are registered on UK marketing authorisations to be approved to supply without further notice and with immediate effect; sites without proven documented GMP compliance status cannot be regarded as an effective back up supply source of starting material.

- As a minimum, and in order to comply with EU directives on the use of GMP compliant APIs, the supplier approval process must be supported by evidence of effective GMP compliance of the API manufacturing site(s). As per European Medicine Agency (EMA) guidance it is expected that this will be confirmed via audit of the API manufacturing site by or on behalf of the relevant manufacturing authorisation holder. Audits should have been conducted at intervals not exceeding three years, by persons with appropriate training and experience, to confirm the current GMP status of the site.

- Any API manufacturer listed on a UK marketing authorisation that has not been maintained as an approved supplier does not meet the EU requirements for starting materials to have been manufactured in compliance with EU GMP requirements.

- The MHRA can inspect API sites named on marketing authorisations at any time where GMP compliance status is unknown or suspected to be deficient. Confirmation of non-compliance can result in regulatory action being taken by the MHRA.

- Companies not wishing to maintain back up API manufacturers to GMP requirements should remove them from marketing authorisations by submission of a Type 1A variation change code A.7.

- The MHRA will continue to monitor compliance with this guidance through its inspection programme of MA holders and as necessary by inspection of marketing authorisation holders.

In addition to reviewing the approval status of API manufacturers named on marketing authorisations, holders are also requested to remove any other sites that are no longer used/maintained, e.g. drug product manufacturing, importation or batch release sites, etc. by use of a Type 1A variation, change code A.7.

Drug product legislation

GMP inspections of drug product manufacturers are mandated in the EU, and in some cases the frequency of such inspections is fixed in the laws of member states (e.g. three years in the UK). In order to achieve consistency between the various European agencies, a standard inspection report format has been defined[21].

Very specific to EU legislation is the definition of the Qualified Person (QP). The QP is the only authorised person to release product onto the market. Each batch is required to undergo certification by a QP within the European Union. The QP has to provide a written statement that the batch has been manufactured in accordance with GMP and the Marketing Authorisation[22]. QPs must be listed on the Marketing Authorisation Application and the national competent authorities can accept or refuse listed QPs. The EU member states have highly differing rules on who can become a QP and what prerequisites must be met. Only a QP listed on the Marketing Authorisation can release a batch of the drug product.

6.3 Part B: EU GMP Guidance: EudraLex Volume 4

The EU legislation compilation named EudraLex can be found at http://ec.europa.eu/health/documents/EudraLex/index_en.htm. Volume 4 of this compilation covers the guidelines for good manufacturing practices for medicinal products for human and veterinary use. This section is by no means static and readers are advised to check regularly for changes and updates to this section. In 2011 a three-part structure was introduced, whereby there is some ambiguity with regards to the newly created Part III. Though the authorities maintain that this part does not introduce new GMP requirements and that some of the content is purely voluntary, there is some doubt in industry that this is effectively the case. Although EMA regularly consults with industry before publishing new or revised guidelines, they are not bound to follow such a process.

The structure of EudraLex is as follows:

- Introduction and references to the governing European Commission directives
- Volume 4 – Guidelines for good manufacturing practices for medicinal products for human and veterinary use
    o Chapter 1 Quality Management (revision February 2008)
    o Chapter 2 Personnel
    o Chapter 3 Premise and Equipment
    o Chapter 4 Documentation (Revision January 2011) – Coming into operation by 30 June 2011

- o   Chapter 5 Production
- o   Chapter 6 Quality Control
- o   Chapter 7 Contract Manufacture and Analysis
- o   Chapter 8 Complaints and Product Recall
- o   Chapter 9 Self Inspection
- • Part II – Basic Requirements for Active Substances used as Starting Materials
- • Part III – GMP related documents
  - o   Site Master File
  - o   Q9 Quality Risk Management
  - o   Q10 Note for Guidance on Pharmaceutical Quality System
  - o   MRA Batch Certificate
- • Annexes
  - o   Annex 1Manufacture of Sterile Medicinal Products
  - o   Annex 2 Manufacture of Biological Medicinal Products for Human Use
  - o   Annex 3 Manufacture of Radiopharmaceuticals
  - o   Annex 4 Manufacture of Veterinary Medicinal Products other than Immunological Veterinary Medicinal Products
  - o   Annex 5 Manufacture of Immunological Veterinary Medicinal Products
  - o   Annex 6 Manufacture of Medicinal Gases
  - o   Annex 7 Manufacture of Herbal Medicinal Products
  - o   Annex 8 Sampling of Starting and Packaging Materials
  - o   Annex 9 Manufacture of Liquids, Creams and Ointments
  - o   Annex 10 Manufacture of Pressurised Metered Dose Aerosol Preparations for Inhalation
  - o   Annex 11 Computerised Systems (revision January 2011)
  - o   Annex 12 Use of Ionising Radiation in the Manufacture of Medicinal Products
  - o   Annex 13 Manufacture of Investigational Medicinal Products
  - o   Annex 14 Manufacture of Products derived from Human Blood or Human Plasma Deadline for coming into operation: 30 November 2011
  - o   Annex 15 Qualification and validation

- o   Annex 16 Certification by a Qualified person and Batch Release

- o   Annex 17 Parametric Release

Annex 18 is missing as this has since become Part II.

- o   Annex 19 Reference and Retention Samples

- •   Glossary

- •   Other documents related to GMP

  - o   Compilation of Community Procedures on Inspections and Exchange of Information updated to include new EU formats and procedures

  - o   Guidelines on Good Distribution Practice of Medicinal Products for Human Use (94/C 63/03)

These guidelines provide the basis for inspector queries.

PIC/S Guidance

The Pharmaceutical Inspection Convention and Pharmaceutical Inspection Co-operation Scheme (jointly referred to as PIC/S) are two international instruments between countries and pharmaceutical inspection authorities, which provide together an active and constructive co-operation in the field of GMP[23]. Most of the EEA member state regulatory authorities are members of PIC/S. PIC/S publishes guidance for inspectors and also provides training to these. Though not binding, these guidance documents are widely used by EU inspectors and inspectorates. It is advisable to familiarise oneself with these before an inspection.

Individual member state regulatory authorities may also publish guidance documents, which may find wider use by other agencies. Examples are the German Zentralstelle der Länder für Gesundheitsschutz bei Arzneimitteln und Medizinprodukten[24], whose aide memoires are of exceptional quality (albeit in German), and the Swedish Medical Products Agency (MPA)[25], which has published a proposal for guidelines regarding classification of software-based information systems used in health care.

6.4   Part C: Overview of EU GMP inspections

Aim of the inspection: Inspections of a firm's manufacturing operation are essential to evaluate commercial manufacturing capability, adequacy of production and control procedures, suitability of equipment and facilities, and effectiveness of the quality system in assuring the overall state of control. Pre-approval inspections include the added evaluation of authenticity of submitted data and link to dossier.

The role of the European Medicines Agency is to co-ordinate inspections to verify compliance with the principles of:

- Good manufacturing practice (GMP)[26]

- Good clinical practice (GCP)[27]

- Good laboratory practice (GLP)[28]

Inspections also verify compliance with other aspects of the supervision of authorised medicinal products in use in the European Union. These are described in Regulation (EC) 726/2004[29].

The EMA itself does not have inspectors and inspections are performed by inspectors from the Member States competent authorities. Similarly, the EMA does not have an enforcement role, this also lies with the national authorities.

Inspection requests from committees

The Agency is responsible for co-ordinating any inspection requested by the Committee for Medicinal Products for Human Use (CHMP)[30] or Committee for Medicinal Products for Veterinary Use (CVMP)[31] in connection with the assessment of marketing-authorisation applications or matters referred to these committees.

These inspections may cover GCP, GLP, GMP, pharmacovigilance (PhV)[32] or vaccine antigen master file (VAMF)[33] or plasma master file (PMF)[34] certification. They may be necessary to verify specific aspects of the clinical or laboratory testing or manufacture and control of the product or to ensure compliance with GMP, GCP, GLP or pharmacovigilance quality assurance systems.

Inspector meetings

The Agency organises and chairs regular meetings of GCP and GMP inspectors from the European Economic Area (EEA). These meetings:

- develop harmonisation of inspection-related procedures and guidance documents;

- manage the Agency's process analytical technology (PAT) team. This is a forum for dialogue and understanding between the Quality Working Party[35] and the GMP/GDP Inspectors Working Group[36] with the aim of reviewing the implications of PAT, which has been set up under its responsibility.

Sampling and testing

The Agency implements a sampling and testing programme[37] aimed at supervising the quality of centrally authorised medicines available on the European market.

EMA Regulatory and procedural guidance on inspections can be found on the EMA website (www.ema.europe.eu) in the Regulatory, Human Medicines, Inspections section. As the contents of the EMA website changes often, links provided in this chapter were only correct at the time of printing and may have changed since.

There are two types of inspections:

General GMP inspections (also termed regular, periodic, planned or routine) should be carried out before the authorisation referred to in Article 40 of Directive 2001/83/EC and Article 44 of Directive 2001/82/EC respectively, is granted and periodically afterwards as required to assess compliance with the terms and conditions of the manufacturing authorisation. This kind of inspection may also be necessary for a significant variation of the manufacturing authorisation and if there is a history of non-compliance. This includes follow-up inspections to monitor the corrective actions required following the previous inspection.

On-site assessment of quality control laboratories is normally part of a GMP inspection.

Product or process related inspections (also termed pre-authorisation, pre-marketing, special, problem orientated) focus on the compliance of the manufacturer to the terms and conditions of the marketing authorisation and on the manufacture and documentation related to the product. It is also indicated when complaints and product recalls may concern one product or group of products or processing procedures (e.g. sterilisation, labelling, etc.).

Contract QC laboratories are according to Article 20(b) of Directive 2001/83/EC or Article 24(b)of Directive 2001/82/EC or Article 13.1 of Directive 2001/20/EC subject to these inspections.

Like other regulatory agencies, the EU authorities perform inspections of manufacturers in third countries. The EMA has detailed that process in "Outline of a Procedure for Co-ordinating the Verification of the GMP Status of Manufacturers in Third Countries"[38]

Inspections are performed by the Competent Authorities (CA) (i.e. the national agencies). Depending on the structure of these CAs, they will perform all or specific inspections (e.g. of drugs for human or for veterinary use). In some EU member states, the inspecting authorities are only acting in a specified region (e.g. Bundesland in Germany) or are bound by other restrictions.

Inspection notification

Though in principle, inspections do not have to be announced, the European agencies do normally inform the company about the intention to inspect and a mutual agreeable date is arranged. The Marketing Authorisation Holder or applicant is required to cover all of the inspectors' expenses and make the necessary travel arrangements. The fees are significant and differ from one authority to the other. An example can be found on the MHRA website http://www.mhra.gov.uk/Howweregulate/Medicines/Licensingofmedicines/Feesfo rmedicinesbloodestablishmentsandbloodbanks/Inspectionfees/index.htm.

The notification can be either via formal letter or by email. The notification will also typically contain a list of the areas the inspection will cover and detail the documents the inspectors wish to receive prior to the inspection. Unlike the US FDA inspectors, EU inspectors do not need to present formal credentials. In fact, in many cases the local inspectors and the company may have established a working relationship over many years and thus know each other quite well.

Closeout meeting

Unlike the US FDA, the European inspectors do not issue a written summary of objectionable findings (i.e. no equivalent to form FDA-483). Instead the findings will be discussed verbally and a report will be issued at a later date.

Inspector qualifications

Typically, inspectors in the EU have had many years (typically 5 – 10 years) of industry experience before starting work as an inspector. They are likely to have a Ph.D. degree in a scientific discipline, may have Qualified Person status, have a proven track record in auditing and are often multi-lingual. This hands-on expertise makes them much more likely to understand alternative approaches to compliance, but on the other hand also allows them to readily identify non-compliances from observing actual operations. For that reason EU inspectors often spend more time on walk-abouts than non-EU agency inspectors. The focus of these inspections is on the operations, less on the systems, and specifically how senior management assures compliance with the regulations. Indeed, senior management awareness and compliance governance is a critical element in many an inspection, which has resulted in critical inspection observations.

6.5   Part D: Hosting and Managing the Inspection[†]

Preparing for an Inspection

Agency inspections are a familiar part of the regulatory environment. A company can be inspected as part of a routine or a triggered inspection by any regulatory agency. Whatever the reason for an inspection, it is essential that all employees are well prepared. An inspection provides the company with an opportunity to present to the agencies that quality is inbuilt into processes and procedures; that quality is embedded in the company's culture of doing business, and does not come as an afterthought.

Before an inspection has been announced, it is important for companies to have in place Standard Operating Procedures (SOPs) defining how an agency inspection should be prepared for and handled.

When preparing such an SOP or guidance document for inspections, the following points should be addressed:

- Receiving inspectors
    - o   Ensure that their credentials have been verified
    - o   Assign at least one staff member who will be the main point of contact and will be responsible for staying with the inspectors at all times
- The duties of key individuals
    - o   This should include everyone from the reception staff to members of the inspection teams
- Set up of the inspection rooms
    - o   It is advisable to have a meeting room assigned to the inspectors and the inspection for the duration of the visit. This room should be close enough to any amenities (restrooms, etc.) but also, if possible, separate from the rest of the office/working space
    - o   A document or control room should also be set up. This room should be staffed at all times throughout the inspection with people who are able to provide and access any additional documents required
- General employee preparation and identification
    - o   Although until the inspection is announced it will not be possible to determine the scope and areas to be covered, it is worth documenting

† This Part is based mainly on an article published in RAPS Focus in 2011[39]

the general kinds of roles that will need to be assigned during an inspection, for example:

- Facilitator(s) – the main contact throughout the inspections

- Scribes and runners

- Subject matter experts

- Document/control room staff

- Focus on employee preparation. As mentioned previously, it is key that each employee understands what the quality processes are within the company/department, that they understand their job role and that any training is current and up to date.

- Mock inspections/internal audits

  o As a matter of good practice, it is advisable for the department to undergo an internal 'mock' inspection and/or internal audit on a regular basis. This is a good way to identify any gaps in the department's procedures and also a good training method for staff who may not have participated in an inspection before.

The importance of the opening meeting

European inspectors invariably prefer to begin any inspection with an opening meeting during which the purpose and the expectations of the inspection are outlined and discussed.

It is advisable to have a meeting room assigned to the inspectors and the inspection for the duration of the visit. This room should be close enough to any amenities (restrooms, etc.) but also, if possible, separate from the rest of the office/working space.

A document or control room should also be set up. This room should be staffed at all times throughout the inspection with people who are able to provide and access any additional documents required.

On the day of inspection, once the inspectors have identified themselves, the reception should notify the facilitator/primary contact. The facilitator should introduce him/herself to the inspectors and take them to the assigned meeting room. Only personnel assigned for the welcome presentation should be present in the meeting room and introductions should be done. It is important to note that the all meetings and discussions with the inspectors should be professional and respectful; do not underestimate the need to be polite.

At this stage the agency will usually open the meeting with the lead inspector outlining the purpose and the scope of the meeting and confirm the inspection plan. The inspection plan will operate as the agenda for the inspection with time allotted for review of documents, interviews with personnel and any inspections of systems and/or facilities.

The opening meeting will, to an extent, set the tone for the inspection. The company should use the meeting as an opportunity to make a good first impression on the inspectors with regards to its preparedness and professionalism.

What to include and what to exclude

At the opening meeting, it is important to ensure that there is a clear plan in place in terms of what will be presented to the inspectors. If appropriate, a presentation including topics such as a brief overview of the company and what the company understands to the scope of the inspection for example, should be prepared. Prior to an inspection, some agencies will request certain documents to be made available to them as part of the inspection process. In this case, it is appropriate to have copies of the documents requested available. Where requests for documents have been made, it is common sense to have these ready at the time of inspection rather than trying to arrange for them whilst the inspectors are on site. This also provides a good opportunity for ensuring that document archive systems are in a good working order; if there are issues or concerns, these can then be resolved prior to the inspection.

If specific documents have not been requested prior to the inspection, it may be appropriate to have certain documents on hand in order to be prepared. For example, for a manufacturing site inspection, having the Site Master File printed and copies available might be advisable. The Site Master File is a must have in Europe.

It also helps to understand specific concerns that some agencies are known to have, e.g. supply chain control. It is not always possible to predict the 'hot topics' for certain agencies, but checking with colleagues based in the country that is inspecting or even informal conversations with fellow industry colleagues, can provide some helpful pointers.

As with all other aspects of the inspection, being prepared is important; having any documents that were or are likely to be requested, ready and copies available for distribution, makes a good first impression.

Roles and duration

It is advisable to have a core team set up who will be involved throughout the duration of the inspection. This core teams does not necessarily include all of the people likely to be interviewed by the inspectors. The inspectors can request to interview any employee, however, based on the scope of the inspection, it should be possible to identify the employees most likely to be interviewed.

The core inspection team should consider including the following roles:

- Primary Contact – this should be the main contact person, who is responsible for leading all aspects of the inspection from an internal perspective. The lead should coordinate all activities, ensure that there are sufficient staff levels and also that appropriate staff members are involved

- Administrator(s) – there should be adequate support with regards to all administrative activities, for example photocopying, general document management. The administrator(s) should be positioned in the document management / coordination room and should be available throughout the inspection in order to provide any documents as and when required

- Note taker – there should be someone responsible for taking minutes and notes throughout the interviews and inspection. Ideally there should be a minimum of two people assigned, this will ensure that all notes are captured and should there be any discrepancies or misunderstandings, these can be identified and resolved during the inspection

- Personnel assigned to accompany inspectors – the inspectors should be accompanied at all times. Depending on the number of inspectors and the size of the site, it may be necessary to have more than one person assigned. It is possible that the inspectors may choose the split up to inspect or tour different parts of the department/site

- Management – it is advisable to include management personnel at the start and end of the inspection days to cover the briefing and debriefing sessions.

These roles should be well defined; everyone should be aware of their responsibilities and familiar with the role. The assignment of employees to these roles will of course depend on the scope of the inspection. Depending on the area(s) to be inspected, the right person for each job will vary. Once the scope of the inspection is known, the team should be quickly assembled and trained. Mock inspections in the lead up to the actual inspection will help the team to understand and perfect their roles.

Following the inspection, the primary contact should remain involved in the co-ordination and follow-up of post inspection activities. Although they may not be the correct individual to respond to all of the requests, they should nevertheless be involved as they will have been present throughout the inspection and will have an oversight of the process overall.

Until all activities pertaining to the inspection are complete and there are no follow up actions pending from the agency, the primary contact should continue to meet with the management personnel and all other individuals responsible for follow up on a regular basis. This is to ensure that all actions are completed as well as ensuring there is adequate follow-up internally.

Once all activities are complete and the inspection is finally closed out, an internal final debrief is recommended. The aim of this debrief should be to highlight any internal processes which should be amended; an inspection offers a company the opportunity for an external view of familiar internal procedures. This fresh view should allow a company to really consider if they working in the best possible way. Even if certain processes or procedures pass inspection, the inspection process should encourage a review to ensure that those processes and procedures still represent the best way of working.

Answering questions and providing information

The kind of questions that will arise will vary depending on the nature of the inspection, however there are some example questions listed below which may be expected. These questions are general in nature and are in no way intended to be exhaustive, however these are the most basic questions that each employee, irrespective of their role, would be expected to answer comfortably:

- What are your responsibilities?
- What procedures do you follow / what standards to you work to?
- How were you trained?
- How is training documented?
- How are you informed of relevant regulation changes?

It is quite interesting to observe these questions being answered without prior practicing or second thought. For example, an inspector will have looked at the organisational chart before arriving on site and will have an idea of the person's role and responsibilities. Often, the issue is that the job description and the actual scope of the role may have drifted apart, or that the person answering is not entirely sure of their responsibilities (e.g. if only started work a few weeks prior to the inspection). All too often those being questioned by an inspector tend to go on and on and on, rather than take a minute to reflect on the actual question.

More specific questions can be determined depending on the scope of the planned inspection.

Companies should be aware that inspectors may ask questions they already know the answers to. This is not a trap, their purpose is to ascertain that information provided to them prior to the inspection is consistent with the processes/activities under review.

Mind your language

In some companies slang or careless use of terminology can annoy inspectors. Especially EU inspectors are wary of companies where the term quality is overused, or as one well-known MHRA inspector put it: "The number of times the word quality is used is inversely proportional to the actual quality of the organisation".

In many an organisation, some of the key documents (e.g. mission statements) contain language that is incomprehensible even to the learned reader. If you cannot say it in plain English or if it cannot be understood by all the statement is directed to, it misses the point and becomes a liability. Senior management is most liable to fall into this trap.

Senior Management

EU inspectors are keen to understand to what degree senior management considers themselves responsible for quality and compliance. An inspection at a well known contract manufacturer was cut short when management retorted that they hired the best of the best and thus everything must be totally in order. They admitted not understanding the concepts and principles of compliance. The inspectors immediately left and the company's products were barred from import into the EU, which killed several clinical trials at a high cost to the companies involved, and even worse to the patients.

Preparing all employees to answer the questions

One of the main challenges during an inspection is staff behaviour. Employees must be encouraged to listen to the questions and allow the inspector to finish before providing an answer. Although employees may be eager to demonstrate their knowledge or confidence on the topic being discussed, it is important that only the information requested is provided and nothing more. If the answer to a question is not known, the employee should respond honestly by saying they do not know, but that they will try to find the answer. It is important that the employee is comfortable answering in this way and that they do not feel pressured to answer off topic or make up an answer out of nervousness. Inspectors may be silent for lengths of time ("pregnant pause") and this should not lead to interviewees to solicit unrequested information.

Inspectors are usually well versed in using various questioning techniques, such as asking open questions, which provides the person answering with an opportunity to show their expertise and experience. Often, these opening questions are then followed up with more probing, specific questions, which helps to verify that the information given before is indeed correct.

Some questions are aimed at getting acceptance of an issue, such as the need to revise or amend a SOP. Commitments must be followed up by actions. Other questions may be purely hypothetical, with the aim of understanding a person's capability to react to a situation that may not have been foreseen. Though difficult to describe, inspectors are quite capable of asking questions that invoke emotions, such as empathy or outrage, which may be difficult to handle by some staff.

In principle, it is fine to ask clarifying questions to assure a question by the inspector was correctly understood. However, to do this as a response to every question will risk the inspection come to a premature close.

Body language is incredibly important and everyone should be made aware of this. A really useful tool in assessing this is to use video footage of inspection situations. It is amazing what one can learn about oneself. There is much cultural bias in body language as every seasoned traveller will be able to tell you, however not everyone may be aware of this.

Are you actually answering?

All too often, interviewees have preconceptions and preformulated responses rather than having listened properly to the question asked by the inspector. Not actually listening and understanding the question is as damaging as giving an irrelevant answer. Inspectors may get really worried when they get all the right answers to their questions: are they missing something or are they asking the wrong questions? Never mind, sometimes everything is actually in order – it is known to happen.

It is essential that employees are aware of their job roles and responsibilities, have been appropriately trained and have the relevant experience and expertise for their job. Remember, the agency may choose to interview any person within the department as they see fit; so it is important that staff are fully familiar with and competent in all processes and procedures relevant to them.

It is often useful to employ the services of an independent (consulting) company to assist with a mock inspection. As the person in the role of the inspector will be unknown to the interviewees, their behaviour is likely to be more realistic than when talking to another member of staff. Furthermore, the independent adviser will be unfamiliar with any company jargon or the processes; being therefore in a much better position to judge if the answers provided are comprehensive and comprehensible.

The human factor

It could easily get forgotten that inspectors are people like you and me. They may not like to travel, despise the local food, be plagued by jet-lag and may have had a bad day altogether. So ignoring them upon arrival, bombarding them with irrelevant information and making their lives difficult is really not conducive to a successful inspection.

Creature comforts are cheap and easy to provide, so why not do it? Is it a cup of tea the inspector prefers to a cup of coffee? Is the inspector diabetic and needs to eat at specified times? Are they hard of hearing and do they prefer to sit on their own? An inspection is not only stressful for the auditee, also for the auditors.

Following completion of the mock inspection, a report should be prepared highlighting any areas of concern which may require further training and preparation. The inspection team should disseminate the finding of this report to all appropriate personnel involved and ensure that any action items identified are addressed prior to the actual inspection.

What the inspector can inspect and what is off limits

The inspectors should be accompanied by at least one employee at all times and this includes to the washrooms. If a tour of the facility is planned, the most appropriate route around the site should be identified. It should be direct and allow the inspector to get an overview of the facility. As with all other aspects of the inspection, preparation is key. Make sure that employees are aware of the tour, a clear desk policy should be in place, computers should be set to screensavers when unattended and discretion should be employed during any discussions which may be overheard. The inspectors cannot start rifling through desks or cabinets, if they do, it is perfectly acceptable to ask them to stop (politely, of course) and to let them know if they would like to see a particular document, you can provide it.

Inspectors should not generally be shown financial information, audit reports or personnel records, beyond standard training records, company curricula vitae and job descriptions. If for any reason, this type of information is requested, the inspectors should be informed of the need for a formal written request which would then require approval from your legal department.

An inspection always has a scope, even a 'surprise' inspection is organised with an agenda in mind and this is important to remember. The inspectors are not at your company for a 'free for all' fact finding mission; they have clear objectives and aims and their requests should be limited to the aims of the inspection.

Openness and honesty vs. formality

How does a company deal with a visitor on site? Hopefully the answer for most is by being welcoming, friendly and co-operative. Inspectors, while clearly having a purview that could have serious implications for the company, should be regarding in much the same way.

Employees should aim to be polite, respectful and above all, confident. If you have prepared for the inspection and know you have robust procedures in place, then you should be confident about presenting your company in this light.

As discussed above about answering questions honestly, if you don't know the answer, it's acceptable to say as much. Guessing the answer or, even worse, lying about the answer will be viewed in a much more negative light that not knowing will.

Regarding formality, as with other visitors, it's important to be respectful and polite. In the same spirit of politeness, do not argue with the inspector or engage in an unconstructive argument. Red flags for inspectors include phrases such as 'that's impossible' or 'that would never happen here'; confidence is good, overconfidence and bluster is not.

In an increasingly globalised world, we often feel connected with and even familiar with other cultures and customs, but this should not encourage over familiarity. Ensuring you are respectful of the culture of your inspectors makes sense. Cracking jokes for example, while in most social situations can help lighten a situation, should not be attempted here. An offended inspector is not an ideal outcome.

Working with the inspector

The inspectors, irrespective of the agency or country they represent, should be considered experts in the field they are responsible for. Despite this, you, as the company representative, are the expert in your company processes/procedures and your products. There should be a mutual respect between the company and the inspectors; it is an opportunity for the inspectors to learn more about the company and for the company to show their expertise as well as taking on board improvements suggested by the inspectors.

The key to working with the inspector is in the sentence itself; the company should work with the inspectors. This means, being helpful and respectful and not trying to obscure or obstruct the inspectors. This does not mean that the inspectors can have free reign, they should remain within the scope of their inspection, and anything that seems to be out of scope can be (politely) questioned.

What to provide

All copies of documents provided to the inspectors should be checked, copies kept on file and traced as to their whereabouts, as should any samples (e.g. Package Insert Leaflets) provided. The facilitator should ensure that all appropriate procedures for obtaining documents and/or samples are followed.

Hospitality: what is acceptable/expected

The issue of hospitality can vary according to the country/agency inspecting. Many companies will have clear guidelines on hospitality which will stipulate that gifts to government officials are not permitted. Whilst this should be clear and common sense should prevail with respect to knowing not to offer or provide inspectors with any gifts, there can be a grey area with regards to things such as dinners, hotel accommodation and entertainment.

Well established agencies such as the EMA have clearly defined requirements which will be adhered to. Difficulties can arise with less well established agencies for which the requirements for the inspectors are not as well defined.

What is unacceptable in one region or country with regards to hospitality may be expected from a cultural stand point in another. It is a tricky area and can be difficult to navigate; therefore it is advisable to ensure that your legal and or compliance group are kept included in any arrangements so that they can ensure that all appropriate guidelines and laws are respected and maintained.

Dealing with issues as they arise

Imagine the scenario: You have met with the inspectors, you have exchanged niceties and are now at the introductory meeting. You have a presentation lined up which will show your company in the best possible light, you have prepared various hand-outs; all that needs to happen now is for the person who will be presenting, to start. The presentation goes well, the inspectors have some questions. To your dismay, the presenter reacts badly to the questions: the answers are defensive and hostile, some questions are answered in a manner that is arrogant and worse still, displays ignorance. What do you do?

What is done cannot be undone, you cannot re-do the presentation with a more suitable host, but you can resolve to ensure as soon as this portion is done, the person causing issues is no longer part of the internal inspection team. You could wait for this portion of the meeting to end, politely suggest a short comfort break and take the opportunity to speak to the person responsible and re-assign the role. If there are people who are proving to be unsuitable for any aspect of the inspection or with regards to the role assigned to them, change them. It is much better to identify an issue and resolve it than hoping it will resolve itself and that it is better to battle on in the name of continuity. A lot is made of first impressions and whilst they are important, the ability to tackle problems as they arise and the impression that makes, should not be underestimated.

During the inspection, if you are asked for particular documents and they cannot be located, let the inspectors know how long they are expected to wait for them. Donot leave them hanging with vague promises; it is always better to be open and honest; anything else looks as though you have something to hide.

Similarly, if the inspectors want to meet with specific people, if they are in the office/site, make sure it happens. If they want to meet the CEO, get the CEO. Again, being evasive about people's availability makes it look as though there is something to hide. Prior to an inspection, everyone in the organisation should have been prepared about the possibility of being interviewed and this should apply equally to senior management as well.

Unless they have been asked to attend by the inspectors, it is usually wise to ensure that legal personnel should not be included as part of the internal inspection team. Without wishing to offend esteemed legal colleagues, during an inspection, that tends to be primarily technical, the presence of lawyers might send out a signal that something is not quite right. There should not be anything that needs to be hidden anyway, so having a lawyer present indicates that there may be things which a company would prefer to keep under wraps. Even if your legal colleague does not speak, think about the message you are sending to the inspectors.

At the end of each of the inspection days, when the inspectors leave, whilst the temptation to throw your hands up in the air, heave a huge sigh of relief and leave as quickly as possible can be tempting, the internal inspection team should stay behind to discuss the day's events. What went well? What went badly? What is the plan for the next day? Is everyone prepared? It is so important to understand and manage the happenings of the day. If there were any discrepancies or things that came up that did not make sense, if discussed the day they happen, they can be put into the agenda for the next day. This will again, allow for issues to be dealt with as they arise.

6.6 Part E: Managing the exit meeting

What to expect, how to respond

At the end of the inspection, most or all agencies will have a close out meeting at which the findings of the inspection are reported verbally and the next steps in terms of issuance of the written report and timelines will be communicated. Some agencies may choose not to provide a verbal account of their findings; in this case, it is acceptable to request information on the next steps and expected timelines for receipt of the written report.

Once the inspection is complete, the facilitator shall ensure a de-briefing session is held with the internal inspection team and any observations and findings communicated to senior management.

The agency will send out an inspection report, which will include all information on their observations, any findings resulting from the inspection and any actions which need to be addressed. The agency will also usually provide the company with a timeframe for them to respond to the report; in cases where no timeframe is identified, good practice would indicate responding to the agency within 15 working days from receipt of the report. It is essential that the response to the inspectorate includes all the evidence and not merely references to documents. For example, if page 4 of the Marketing Authorisation is referenced, a copy of the page will be attached to the response. Though timelines must be achievable, they should be as tight as possible. After all, you are remediating / rectifying deficiencies.

Clearing up misunderstandings: The Post-Inspection Process

Once the inspectors have gone, the company should instigate a full de-brief to review what was provided to the inspectors and how the inspection went.

It is possible during this process that certain discrepancies or misunderstandings will come to light. Perhaps a document that was requested was never actually provided, or inadvertently, the wrong information was provided to a specific response.

This is not the end of the world. Mistakes happen and even the best organised companies will sometimes make mistakes. The key is to rectify these errors as quickly as possible. You do not have to wait for an official letter from the agency, in fact, if identified early, it is better to rectify these errors prior to the official report.

When corresponding, as with all other aspects of the inspection, be open and honest about any mistakes made, politely inform the agency about what happened and provide them with the correct information.

The sin is not usually in the error but in the failure to amend it.

The post-inspection letter – how to interpret it and how to respond

The internal inspection team must action any findings and ensure that a plan is put into place identifying the action owner and the timeframe for the response. A project plan should be initiated with regular checks to ensure that all actions are dealt with and that there is time allotted for internal reviews and checks of any proposed responses.

Should there be any changes or delays in responding to the agency, these should be communicated as soon as possible and discussed with the agency. Regular communication with the agency detailing progress against plan is a must, even though the authority may chose not to respond to such correspondence.

How to learn and improve from the experience

Some of the key points in getting a department inspection ready can be summarised as follows:

- be prepared
- be prepared
- be prepared!

The importance of being prepared, of having everything running smoothly and of having well briefed staff, cannot be underestimated. This is your opportunity to demonstrate key aspects of your organisation; by presenting a calm front the message you give out to the agency is that you have everything in proper order and are ready for them.

Of course, all of these tips and plans can only be effective if quality is inbuilt into the organisation. Preparation will help a good company to get through an inspection but it will not mask the shortcomings of a poor company.

Whilst undoubtedly stressful and involving a lot of work for the companies, an inspection is a valuable opportunity to engage with the agency, to demonstrate your strengths as a company and to show your commitment to your customers, the patients. Even the findings and observations that result from the inspection offer you the chance, as an organisation, to improve. As an adage, an inspector should never find anything you did not already know about.

Possible negative actions resulting from an inspection

A listing of such agency actions are listed for example on the MHRA website[40].

For Good Manufacturing Practice (GMP)/Good Distribution Practice (GDP)/Blood Establishment Authorisation (BEA) these include:

- refusal to grant a licence or a variation

- proposal to suspend the licence for a stated period

- notification of immediate suspension of the licence for a stated period (no longer than three months)

- proposal to revoke the licence

- action to remove a Qualified Person/Responsible Person QP/RP from the licence

- issue a Cease and Desist order in relation to a Blood Establishment Authorisation

- issue a warning letter to the company/individual

- request a written justification for actions of a QP/RP

- referral of a QP to his/her professional body

- increased inspection frequency

- request the company/individual attend a meeting at the Agency

- refer to the Enforcement Group for further consideration

For Good Clinical Practice (GCP)/Good Pharmacovigilance Practice (GPvP) these include:

- issue an infringement notice in relation to a clinical trial
- suspend or revoke a clinical trial authorisation
- further follow-up inspections, or triggered inspections at related organisations (e.g. issues in GCP may trigger a GMP inspection)
- referral to CHMP (Committee for Medicinal Products for Human Use) for consideration for against a marketing authorisation (e.g. suspended, varied or revoked)
- liaison and coordinated action with EMA (European Medicines Agency) and other Member States regarding concerns
- refer the case to the EMA for consideration of the use of the EU Infringement Regulation (which could result in a fine)
- request a written justification for action of a QPPV (Qualified Person responsible for pharmacovigilance)
- request the company/individual attend a meeting at the Agency
- refer to the Enforcement Group for further consideration

In the case of inspections in third countries:

- a refusal to name a site on a marketing authorisation
- a recommendation that a site be removed from a marketing authorisation
- the issuing of a GMP non-compliance statement
- in the case of an adverse (voluntary or triggered) active pharmaceutical ingredients (API) inspection, this could result in the removal of the API site from the marketing authorisation
- in the case of an adverse (voluntary or triggered) investigational medicinal products (IMP) inspection, this could result in the suspension of a clinical trial

In all cases, an action could result in the withdrawal of product (API, IMP medicinal product, etc.) from the market

Follow-up actions may include:

- a re-inspection to ensure corrective actions implemented
- request for regular updates on the corrective action plan

- the issue of a short-dated GMP certificate

- recommended increase of inspection frequency

- continued monitoring of the company by IAG via inspectorate updates

- if serious and persistent non-compliance continues, referral for consideration of criminal prosecution.

Additional details can be found in the EMA publication "Compilation of Community Procedures on Inspections and Exchange of information" document[41].

6.7 Part F: The EU GMP database, the EDMQ Supplier Certification Database and information sharing practices

EudraGMP database

On 4th August 2009 the EMA published a press release titled "EudraGMP 2.0 gives public access to information about good manufacturing practice (GMP)". Following the launch of a new version, the EudraGMP database is now providing public access to information about manufacturing, importation authorisations and Good Manufacturing Practice (GMP) certificates. EudraGMP can be accessed using the following URL http://eudragmp.emea.europa.eu or via the EMA website at http://www.ema.europa.eu/ema/index.jsp?curl=pages/regulation/general/eudra_gmp_database.jsp&mid=WC0b01ac058006e06e&jsenabled=true

The EudraGMP database was initially launched in April 2007 in order to facilitate the exchange of information on compliance with good manufacturing practice (GMP) among the competent regulatory authorities within the European medicines network. The database contains information on manufacturing and importation authorisations issued by the national competent authorities within the network, i.e. the EU Member States and Iceland, Liechtenstein and Norway. It also contains information on GMP certificates, which the competent authorities issue following each GMP inspection conducted either within the European Economic Area or in third countries.

Version 2.0 of the database will now also contain Non-Compliance Statements. These statements will be issued in cases where the reporting inspection service is of the opinion that a manufacturer's non-compliance with GMP is so severe that regulatory action is required to remove a potential risk to public or animal health.

This database is still being built and there is no guarantee that all information is as yet in the database and/or freely accessible by the public. With time, the database will contain all:

- manufacturing and import authorisations (MIAs) for medicinal products issued to pharmaceutical companies within the EEA

- GMP certificates for all GMP inspections conducted within the EEA

- GMP certificates for all GMP inspections conducted in third countries, including API inspections conducted by an authority in the EEA

The European legislation does not require mandatory routine GMP inspections for active substance manufacturers. Therefore, the absence of a GMP certificate for a manufacturer of active substances in EudraGMP does not automatically mean that the manufacturer does not comply with GMP.

It is expected that EudraGMP will provide information on around 10,000 manufacturers and importers in the EEA, and that each year more than 3,000 new GMP certificates will need to be entered into the database.

EEA competent authorities currently have full read/write access to EudraGMP. Discussions are ongoing about providing MRA partners access to EudraGMP and giving authorities outside the EEA access for regulatory purposes[42].

European Directorate for the Quality of Medicines (EDQM) Supplier Certification Database

The EDQM's mission is to contribute to the basic human right of access to good quality medicines and healthcare and to promote and protect human and animal health by[43]:

- establishing and providing official standards which apply to the manufacture and quality control of medicines in all signatory States of the "Convention on the Elaboration of a European Pharmacopoeia" and beyond;

- ensuring the application of these official standards to substances used in the production of medicines;

- co-ordinating a network of Official Medicines Control Laboratories (OMCL) to collaborate and share expertise among Member States and to effectively use limited resources;

- proposing ethical, safety and quality standards:

  – for the collection, preparation, storage, distribution and appropriate use of blood components in blood transfusion;

  – for the transplantation of organs, tissues and cells;

- collaborating with national, European and international organisations in efforts to combat counterfeiting of medical products and similar crimes;

- providing policies and model approaches for the safe use of medicines in Europe, including guidelines on pharmaceutical care; and by

- establishing standards and co-ordinating controls for cosmetics and food packaging.

The procedure for 'Certification of Suitability to the monographs of the European Pharmacopoeia' was established in 1994 and was in the beginning restricted to controlling the chemical purity of pharmaceutical substances. In 1999, the procedure was extended to include products with a risk of transmissible spongiform encephalopathy (TSE), thus enabling their certification on the basis of the European Pharmacopoeia general chapter 5.2.8 'Minimising the risk of transmitting animal spongiform encephalopathy agents via medicinal products' and of the new monograph on 'Products with risk of transmitting agents of animal spongiform encephalopathies (1483)'. The general chapter 5.2.8 is a verbatim reproduction of the guidance issued by the EMA's Committee for Medicinal Products for Human Use (CHMP), previously known as CPMP and by their Committee for Medicinal Products for Veterinary Use (CVMP). The procedure was further revised to allow for the control of herbal drugs and herbal drug preparations.

In 1999, the EDQM initiated an inspection programme for manufacturing sites, covered by an application for certificate(s) of suitability to the monographs of the European Pharmacopoeia (CEPs). CEPs are recognised by the signatories of the Convention on the Elaboration of a European Pharmacopoeia, i.e. all Member States and the European Union. They are also recognised by other countries, e.g. Canada, Australia, New Zealand, Tunisia and Morocco.

The database can be accessed through this link https://extranet.edqm.eu/publications/recherches_CEP.shtml.

Collaboration between FDA and EMA

The framework for collaboration was established through[44]:

EMA/EC – FDA Bilateral Meetings

EMA/EC – FDA Confidentiality Arrangements

- Signed in September 2003
- Implementation plan agreed in 2004 (H) + 2008 (V)
- Extended for 5 years in September 2005
- Extended indefinitely in September 2010

Transatlantic Administrative Simplification

- Action plan agreed June 2008

An update on these activities was published by the EMA in June 2011 in a report titled "Interactions between the European Medicines Agency and U.S. Food and Drug Administration September 2009-September 2010"[45]

A further update was published by EMA on 2nd August 2011, as a press release titled: European Medicines Agency and its international partners complete successful inspection pilots; International collaboration on drug quality and safety to continue[46].

One pilot programme carried out between EMA and FDA focussed on good clinical practice (GCP). Under the joint GCP inspection pilot, the two agencies exchanged more than 250 documents relating to 54 different medicines and organised 13 collaborative inspections of clinical trials. According to EMA, 'this lays the foundation for a more efficient use of limited resources, improved inspectional coverage and better understanding of each agency's inspection procedures.' The joint inspections could also help to protect participants in clinical trials and better ensure the integrity of data submitted as the basis for drug approvals.

The other pilot programme focused on active pharmaceutical ingredients (APIs) and was carried out jointly by EMA, France, Germany, Ireland, Italy, UK, the European Directorate for the Quality of Medicines and HealthCare, FDA and Australia's TGA. The participants shared their API surveillance lists over a 24-month period and found 97 sites common to all three regions, resulting in the exchange of nearly 100 inspection reports and nine joint inspections[47].

The agencies announced that this has been a 'positive experience' and have therefore agreed to continue with their collaboration on inspections, taking into account the experiences and lessons learned during the pilot phases.

These are not the only collaborations being carried out between EMA and FDA. The agencies launched a three-year pilot programme on 1 April 2011 that will allow parallel evaluation of quality elements, known as Quality by Design, of selected applications that are submitted to both agencies at the same time[48].

6.8 Part G: Risk-based re-inspection schedules

The UK MHRA has been the leading agency in establishing the risk-based inspection concept. This was triggered by the Hampton Review on "Reducing administrative burdens: effective inspection and enforcement"[49] published in 2005. This has resulted in a sophisticated risk-based inspection programme for good practice inspections[50].

The EMA publication "Compilation of Community Procedures on Inspections and Exchange of information" document[51] specifically calls for the establishment of risk management as part of the inspection process:

11.3.1 The pharmaceutical inspectorate should implement risk management for assigning resources and prioritising tasks and activities to carry out its obligations (e.g. planning of inspections).

11.3.2 The pharmaceutical inspectorate should also implement risk approach in the conducting of inspection.

The same document is specific with regards to re-inspection frequencies:

6.1   In general, authorities with supervisory responsibility for a third country manufacturing site should ensure that it is re-inspected by an EEA authority or MRA partner authority, under the terms of an MRA, between every two to three years.

6.2   Where inspection reports and information exchange based on inspections conducted more than three years ago are available, as there is evidence of acceptable GMP standards, it should not be necessary to withhold any application or variation pending the results of a new inspection unless information is available from other sources suggesting that this status may have changed. Steps should nevertheless be taken to obtain an updated report.

6.3   Inspection reports, and information exchange based on inspections or distant assessments conducted more than five years ago, from whatever source, should not normally be taken into consideration.

This risk-based approach is summarised in this document as shown in **Table 1**:

Table 1. Inspection frequency

Category	Description*	Inspection interval
Compliance Factor I Poor Compliance	The last inspection revealed critical and/or more than/equal to six (≥6) major deficiencies	1 year
Compliance Factor II Acceptable Compliance	The last two inspections revealed no critical and less than six (<6) major deficiencies	2 years
Compliance Factor III Good Compliance	The last two inspections revealed no critical and major deficiencies	3 years

* Deficiencideficiencieses are categorised according to the Definition of Significant Deficiencies laid down in the GMP Inspection report – Community format.

These approaches will be widely adopted by the various agencies, as they, just as the rest of the economy, are under pressure to achieve more with fewer resources. Doing the right things better is the common motto.

6.9   Conclusion

EU inspectors come from a diverse range of agencies and cultural backgrounds, which means there is more of a variety in the approach to inspections compared with, for example, US FDA inspectors. Nonetheless, there is a lot of consensus and co-operation between the agencies on how to conduct inspections, classify findings and come to conclusions on the audited company's compliance status. Companies will find that the EU inspectors' way of inspecting is leaning more towards providing guidance rather than being purely investigative. Though these inspections are as tough as any other, auditees will usually find inspectors to be highly competent and friendly. They're human after all.

6.10   References

1.   http://europa.eu/about-eu/countries/index_en.htm

2.   http://www.ema.europa.eu/ema/index.jsp?curl=pages/medicines/general/general_content_000155.jsp&mid=WC0b01ac0580036d63

3.   http://www.hma.eu/242.html

4.   http://ec.europa.eu/health/human-use/index_en.htm

5.   http://ec.europa.eu/health/human-use/legal-framework/index_en.htm

6.   http://ec.europa.eu/health/documents/EudraLex/vol-1/index_en.htm

7.   http://ec.europa.eu/health/documents/EudraLex/vol-2/index_en.htm

8.   http://ec.europa.eu/health/documents/EudraLex/vol-3/index_en.htm

9.   http://ec.europa.eu/health/documents/EudraLex/vol-4/index_en.htm

10. http://ec.europa.eu/health/documents/EudraLex/vol-9/index_en.htm

11. http://ec.europa.eu/health/documents/EudraLex/vol-10

12. http://ec.europa.eu/health/documents/EudraLex/index_en.htm

13. http://europa.eu/legislation_summaries/internal_market/single_market_for_goods/
     pharmaceutical_and_cosmetic_products/index_en.htm

14. http://ec.europa.eu/health/authorisation-procedures_en.htm

15. http://ec.europa.eu/health/authorisation-procedures-centralised_en.htm

16. http://ec.europa.eu/health/authorisation-procedures-mutual-recognition_en.htm

17. http://ec.europa.eu/health/authorisation-procedures-decentralised_en.htm

18. www.ema.europe.eu

19. http://www.ema.europa.eu/docs/en_GB/document_library/Regulatory_and_
     procedural_guideline/2009/10/WC500004706.pdf

20. http://www.mhra.gov.uk/Howweregulate/Medicines/Medicinesregulatorynews/
     CON088198

21. http://www.ema.europa.eu/docs/en_GB/document_library/Regulatory_and_
     procedural_guideline/2011/07/WC500108770.pdf

22. http://www.imb.ie/images/uploaded/documents/GMP%20info%20day%20
     presentations/15_Role%20of%20the%20QP%20in%20MA%20Compliance_Breda
     %20Gleeson.pdf

23. www.picscheme.org

24. www.zlg.de

25. www.lakemedelsverket.se

26. http://www.ema.europa.eu/ema/index.jsp?curl=pages/regulation/document_listing/
     document_listing_000154.jsp&murl=menus/regulations/regulations.jsp&mid=WC0b
     01ac0580027088

27. http://www.ema.europa.eu/ema/index.jsp?curl=pages/regulation/general/general_
     content_000072.jsp&murl=menus/regulations/regulations.jsp&mid=WC0b01ac0580
     0268ad

28. http://www.ema.europa.eu/ema/index.jsp?curl=pages/regulation/general/general_
     content_000158.jsp&murl=menus/regulations/regulations.jsp&mid=WC0b01ac0580
     0268ae

29. http://ec.europa.eu/health/files/EudraLex/vol-1/reg_2004_726_cons/reg_2004_726_cons_en.pdf

30. http://www.ema.europa.eu/ema/index.jsp?curl=pages/about_us/general/general_content_000094.jsp&murl=menus/about_us/about_us.jsp&mid=WC0b01ac0580028c79

31. http://www.ema.europa.eu/ema/index.jsp?curl=pages/about_us/general/general_content_000262.jsp&murl=menus/about_us/about_us.jsp&mid=WC0b01ac0580028dd8

32. http://www.ema.europa.eu/ema/index.jsp?curl=pages/regulation/general/general_content_000145.jsp&murl=menus/regulations/regulations.jsp&mid=WC0b01ac058002708a

33. http://www.ema.europa.eu/ema/index.jsp?curl=pages/regulation/general/general_content_000165.jsp&murl=menus/regulations/regulations.jsp&mid=WC0b01ac058002708c

34. http://www.ema.europa.eu/ema/index.jsp?curl=pages/regulation/general/general_content_000163.jsp&murl=menus/regulations/regulations.jsp&mid=WC0b01ac058002708b

35. http://www.ema.europa.eu/ema/index.jsp?curl=pages/contacts/CHMP/people_listing_000016.jsp&murl=menus/about_us/about_us.jsp&mid=WC0b01ac0580028d31

36. http://www.ema.europa.eu/ema/index.jsp?curl=pages/regulation/document_listing/document_listing_000161.jsp&murl=menus/regulations/regulations.jsp&mid=WC0b01ac05800296c9

37. http://www.ema.europa.eu/ema/index.jsp?curl=pages/regulation/document_listing/document_listing_000174.jsp&murl=menus/regulations/regulations.jsp&mid=WC0b01ac058002708d

38. http://www.ema.europa.eu/docs/en_GB/document_library/Regulatory_and_procedural_guideline/2009/10/WC500004706.pdf

39. Nabila Nazir and Siegfried Schmitt, "What Happens When the Regulatory Professional is Inspected?", Regulatory Focus, October 2011, 22 – 28 – www.raps.org

40. http://www.mhra.gov.uk/PrintPreview/DefaultSplashPP/CON126020?ResultCount=1

41. http://www.ema.europa.eu/docs/en_GB/document_library/Regulatory_and_procedural_guideline/2009/10/WC500004706.pdf

42. http://laegemiddelstyrelsen.dk/en/topics/authorisation-and-supervision/company-authorisations/news/public-access-to-emas-community-database-eudragmp

43. http://www.edqm.eu/en/Vision-Mission-Values-604.html

44. http://www.topra.org/sites/default/files/2._Boone.pdf

45. http://www.fda.gov/downloads/InternationalPrograms/FDABeyondOurBorders
ForeignOffices/EuropeanUnion/EuropeanUnion/EuropeanCommission/UCM261565.
pdf

46. http://www.ema.europa.eu/docs/en_GB/document_library/Press_release/2011/08/
WC500109704.pdf

47. http://www.ema.europa.eu/docs/en_GB/document_library/Report/2011/07/
WC500108655.pdf

48. http://www.ppme.eu/About-PPME/News/Pilot-programmes-between-EMA-FDA-
and-TGA-a-success

49. www.hm-treasury.gov.uk/hampton

50. http://www.mhra.gov.uk/Howweregulate/Medicines/Inspectionandstandards/Risk-
basedInspectionProgrammeforgoodpracticeinspections/index.htm

51. http://www.ema.europa.eu/docs/en_GB/document_library/Regulatory_and_
procedural_guideline/2009/10/WC500004706.pdf

CHAPTER 7

Regulatory requirements for GMP inspection in Japan

Yoshikazu Hayashi
International Liaison Officer, Pharmaceuticals and Medical Devices Agency (PMDA), Japan

7.1 Marketing Authorization License and Marketing Approval system in Japan

In Japan, pharmaceutical products are regulated in line with rules established by the Pharmaceutical Affairs Law (PAL). The Marketing Authorization License and Marketing Approval system under this law provides the legal basis for the GMP inspection activity by PMDA (Pharmaceuticals and Medical Devices Agency). The PAL has, roughly speaking, two objectives as follows:

1. By imposing necessary regulations, to assure quality, efficacy and safety of pharmaceuticals, quasi drugs*, cosmetics and medical devices. This objective enables you to prevent substandard ineffective products and/or inferior, unhealthy and harmful products to be marketed.

2. It aims to improve public health and hygiene by taking necessary measures to encourage and promote the research and development of innovative pharmaceuticals and medical devices. In this way, the PAL has a function to prevent negative aspects of pharmaceuticals and to develop and promote good aspects of them. Marketing Authorization License and Product Marketing Approval which systematically supports achieving this aim are laid down in the Law with its flowchart as shown in **Figure 1**.

To market pharmaceutical products the company first has to be granted a Marketing Authorization License from the Prefectural Governor. The companies who are granted an Authorization License to market pharmaceutical products are required to assume consistent supervisory responsibility for their whole lifecycle, from APIs to the pharmacovigilance of the authorised and marketed products. Aside from this, they have to acquire an Establishment License with respect to each manufacturing site or factory, which is a facility requirement. This license is granted for each manufacturing facility after its buildings and facilities are checked. It has to be renewed every five-years. Pharmaceutical products should only be manufactured in the licensed facility.

* Quasi drugs are the products ranked between drugs and cosmetics, which include hair dye, permanent wave solution, medicated cosmetics and medicated tooth paste, etc.

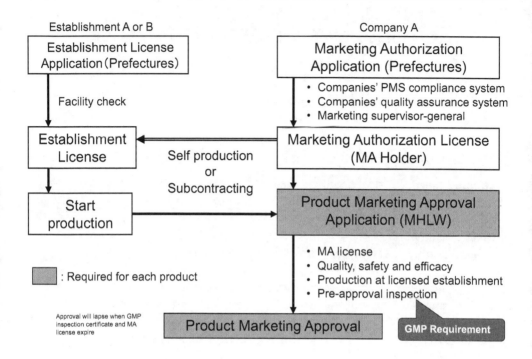

Figure 1: Flowchart of License and Approval

In addition, in order to market pharmaceutical products, the company need to receive a Product Marketing Approval from the Minister for Health, Labour and Welfare for each product. The review of the Product Marketing Approval Application is based on a dossier submitted by the company itemising manufacturing method, quality characterization, results of non-clinical and clinical studies, etc. and the results of conducting a GMP inspection for the manufacturing site. This ensures quality, efficacy and safety of pharmaceutical products.

Once granted a Marketing Authorisation License, the company has to employ a "marketing supervisor-general" responsible for managing and supervising both quality assurance and post-marketing safety measures. He/she has final responsibility for the products on the market as well as the need to prevent any harm to public health if something does go wrong with a marketed products. Requirements for the marketing supervisor-general are that he/she should be a pharmacist specialising in pharmaceuticals, so that he/she can make the necessary judgement and decision about measures for preventing any harm and injury to public health. The standard for the quality control system is laid down in GQP (Good Quality Practice), while that for a post-marketing safety control system is provided in GVP (Good Vigilance Practice).

After the Marketing Authorisation License and the Establishment License (or in the case of an overseas manufacturer, an Accreditation of Foreign Manufacturer) are granted, the Marketing Authorisation License Holder may start production of pharmaceutical products even before the Product Marketing Approval is granted.

After the Marketing Authorisation License is granted, the License Holder files a Marketing Approval Application. In the review of Marketing Application, quality, efficacy and safety of pharmaceutical products are examined. And, as a part of the approval review process, a GMP inspection is conducted, which covers the standards for the buildings and facilities required for the manufacturing site as well as the standards for manufacturing control and quality control. Compliance with GMP is a requirement for granting the Marketing Approval, which is necessary for each product.

Figure 2, shows the flowchart of New Drug Approval from the Application to the granting of the Product Marketing Approval. The Marketing Approval Application is received by PMDA, where it is first examined by "Conformity Audits" to check that the contents of the dossier conform to the required standards. After this the PMDA conducts "Scientific Reviews" by experts with medical, pharmacological, veterinary and bio-statistic backgrounds, who evaluate a new product to ensure that its quality, efficacy and safety meet scientific and

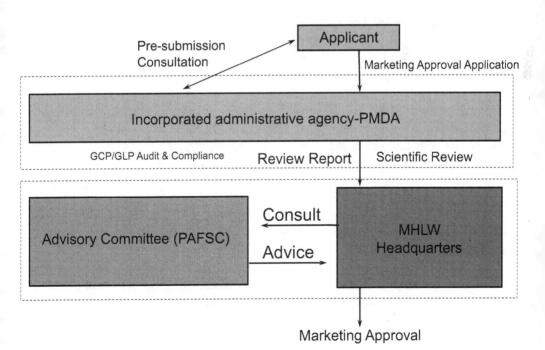

Figure 2: Flowchart for New Drug Approval

technological standards. If necessary, opinions of external experts can be obtained. As a part of its service, PMDA also provides "Scientific Advice" for applicants.

After the review is completed, PMDA submits a report to the Minister who makes the final decision on approval after hearing opinions of the PAFSC (Pharmaceutical Affairs and Food Sanitation Council) which consists of university professors, researchers and other experts with a relevant scientific background. The average review period is set at 12 months.

Under the PAL, the Marketing Authorization License Holder shall assume, as explained above, total responsibility for marketing pharmaceutical products. This means that for quality assurance the Marketing Authorization License Holder has to supervise and manage all actions taken by the relevant manufacturers and has to ensure proper release to the market based on the GQP; if the manufacturers are located overseas, the Marketing Authorization License Holder has to verify that each manufacturer complies with the GMP requirements. For the purpose of quality assurance, the Marketing Authorization License Holder is required to make the necessary arrangements with manufacturers of APIs, etc. and to check and oversee the manufacturers periodically by collecting relevant information from them.

7.2   Japanese Pharmacopoeia (JP)

The Japanese Pharmacopoeia (JP) is an official document which provides the standards required to ensure the quality of medicines in Japan according to the progress of science and technology and medical demands at the time. It has a role in clarifying the criteria for quality assurance of pharmaceutical products which are essential for public health and medical treatment. In pursuit of this goal, it defines the standards for test procedures and acceptance criteria to ensure overall quality of pharmaceutical products. It is established and published by the Minister for Health, Labour and Welfare, based on the provision of Art. 41-1 of the PAL after hearing an opinion of the Committee on JP, PAFSC.

It is laid down in the PAL that the JP shall be subject to a complete revision at least every 10 years and such revisions have actually taken place every 5 years since its 9th revision in April 1976. In addition to this 5-year revision, a partial revision is made as necessary, taking into account the recent progress of science and technology and in the interests of the international harmonization by the PDA (Pharmacopoeial Discussion Group) and the ICH (International Conference on Harmonisation of Technical Requirements for Registration of Pharmaceuticals for Human Use).

As the JP has been prepared with the input from many professionals in the area of pharmaceuticals, it has the characteristics of an official standard, and as such, it should be widely used by all parties concerned. It fulfills accountability to the public through providing information on the quality of pharmaceutical products. It also contributes to smooth and effective pharmaceutical administration on the quality of pharmaceutical products as well as promoting and maintaining the international consistency of the technical requirements.

In July 2006, the Committee on JP recommended the basic principles for the preparation of the 16th edition of JP, elaborating the roles and characteristics of the JP, the measures taken for the revision and the due date of its implementation (1st April 2011). The Committee established five basic principles of the JP. 1. To include all medicines which are important to health care and medical treatment; 2. To make qualitative improvements by introducing the latest science and technology; 3. To promote internationalization; 4. To make prompt partial revisions as necessary for promoting smooth administrative operation; and 5. To ensure transparency in the revision as well as to disseminate the JP to the public. Further information in English on the JP can be found on the PMDA website[3].

When you employ an API listed in the JP (Note: In the following sections the JP, Drug master file system and the GMP inspection are explained using the API as an example) and excerpt specifications provided in the JP as API specifications in the Marketing Approval Application, you are requested to make the specifications conform to those in the JP. Specifications, including test procedures and acceptance criteria provided in monographs for each API are, in some cases, not necessarily identical among JP, USP and Ph Eur. However, you are advised to make sure that specifications comply with the JP when you receive the GMP inspection for an overseas manufacturing site by PMDA.

7.3   Drug master file system

Figure 3 shows the flowchart of the marketing approval review and the master file registration. In the case of overseas API manufacturer, to begin with, an application for Master File registration is submitted to PMDA via its co-opted in-country caretaker. Then, a Product Marketing Approval Application citing the Master File registration number is submitted to PMDA by the Marketing Authorization License Holder. In the review of an application, PMDA makes inquiries about the pharmaceutical product to the applicant and asks the Master File registrant about registered APIs. By this means, in case an overseas API manufacturer does not want to disclose registered information, e.g. manufacturing method, to the Marketing Approval Applicant, he/she can register the confidential information to PMDA via its co-opted in-country caretaker.

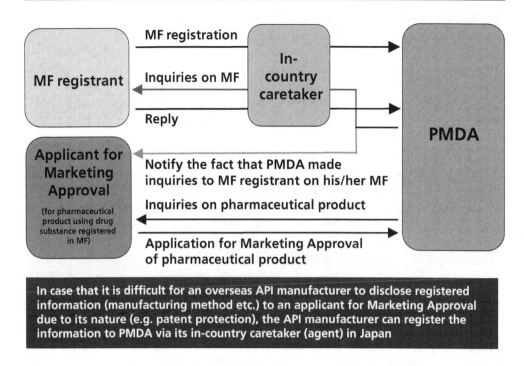

Figure 3: Approval review of pharmaceutical products using drug master files (MFs)

In this way, the Master File system allows you to submit information on the quality and manufacturing method of APIs to be used for pharmaceutical products without involving other competitors. Registration is optional. Registered APIs can be used with the confidential information being protected. The Master File system adopted in Japan is mostly the same as that in Europe and the US, though, you still need to take heed of the following points.

1   *Who can apply for Master File registration?* A manufacturer of APIs in Japan or overseas. An overseas API manufacturer who applies for registration should assign an in-country caretaker for APIs. The in-country caretaker performs the administrator's work in Japan for the overseas manufacturer. The overseas API manufacturer should assign an in-country caretaker before applying for Master File registration.

2   *What is an in-country caretaker's role?* His/her base of operation should be located in Japan. He/she should act as an agent for a Master File registrant. When he/she submits Master File documents, the application and the Quality Overall Summary should be in Japanese. PMDA accepts, however, the data from a CTD Module 3 in English. In the review process, the in-

country caretaker should act as a contact person for the inquiries from PMDA and also should be engaged in relevant administrative procedures and management of the File after registration. For a smooth operation, overseas registrants are advised to keep in close communication with their in-country caretaker.

3   *How to describe manufacturing methods in the Master File.* For chemical APIs, for example, you should describe manufacturing processes, in principle, for at least two reaction processes. In that case, starting materials should be determined in accordance with the concept shown in ICH Q7. The reason why you should avoid just one reaction process control is because in such cases, for example, impurities included in the starting material may affect the quality of APIs. As for residual solvents, in general they should be controlled through a Specified Manufacturing Process or Specification & Testing methods. The classes of solvents used should be considered.

If the target values and/or set values of standard batch sizes or the process parameters are highly significant or highly critical of the quality of APIs, they should be enclosed in brackets. When you change these things in the future, you will need a "Partial Change Approval" which usually becomes a complex procedure. On the other hand, if these values have only low significance for the quality of APIs, they should be enclosed in another type of bracket as a "Minor Change Notification". The values other than target and set values with low significance for the quality of APIs, should be enclosed with quotation marks and is also a "Minor Change Notification" matter. In brief, you should describe critical manufacturing processes in detail. For further information, please consult PFSB/ELD Notification[1]. **Figure 4** shows an example of a manufacturing process description using brackets for the values enclosed in brackets for which the "Partial Change Approval" is required and the values enclosed in different brackets or quotation marks for which only the "minor change notification" is needed.

4   *Notes for changing information registered in the Drug Master File.* When manufacturing methods in the File are partially changed, a registrant is required to obtain an approval for partial change or to submit a Minor Change Notification according to the nature of the change as laid down in the notification. Hence, the registrant, and the in-country caretaker, should not fail to communicate any change to the registered information to the Product Marketing Approval Applicant and to the Marketing Approval License Holder of the relevant pharmaceutical product. The above points are also elaborated on PMDA's English website[2].

Step 1 (Critical process)

Mix 2-(1-triphenylmethyl-1*H*-tetrazole-5-yl)-4'-bromomethylbiphenyl [1] 「(21.6 kg)」 , 2-formyl-5-[(1*E*,3*E*)-1,3-pentadienyl]-1*H*-imidazole [2] 「(6.9 kg)」 , potassium carbonate 「(11.8 kg)」 , and dimethylformaldehyde 「(60 L)」 at 「25°C for 24 hours」 . Add sodium borohydride 「(3.2 kg)」 , and mix further at 「25°C for 24 hours」 . Filter the reaction mixture, and remove the insoluble matter. Concentrate the filtrate under vacuum. Add water 「(50 L)」 to the residue, and extract it with ethyl acetate 「(50 L)」 . Wash the organic layer with water 「(50 L)」 and "10%" saline solution 「(30 L)」 . Concentrate the organic layer under vacuum until it is reduced by approximately one-half. Stir the residue at 「5°C for 3 hours」 . Centrifuge the precipitated crystals and wash them with ethyl acetate 「(10L)」 . Dry the crystals under vacuum

> **Scope of minor change notification**
> → 「 」, " "

hours, and obtain 1-[2'-(1-trityl-1*H*-tetrazole-5-yl)-4-biphenylmethyl]- 5-[(1*E*,3*E*)-1,3-pentadienyl]-2-hydroxymethylimidazole [3].

Step 2

Mix the [3] approximately 「(22 kg)」 obtained in Step 1, "10%" hydrochloric acid 「(200 L)」 , and tetrahydrofuran 「(400 L)」 at 「25°C for 4 hours」 . Add "10%" sodium hydroxide aqueous solution 「(200 L)」 to the reaction mixture. Concentrate the mixed liquid under vacuum. Add water 「(100 L)」 to the residue. Filter it and remove insoluble matter. Adjust the pH of the filtrate to pH 3±0.5 with "35%" hydrochloric acid. Centrifuge the precipitated crystals, and wash them with water. Dry the crystals under vacuum at 《40°C》 , and obtain crude crystals of 1-[2'-(1*H*-tetrazole-5-yl)biphenyl-4-yl]methyl]- 5-[(1*E*,3*E*)-1,3-pentadienyl] [4].

> **Scope of partial change approval application**
> → 《 》

Step 3 ....

Figure 4: Example of description (a part of the process)

7.4   GMP inspections by PMDA (on-site inspection and desktop inspection)

The PAL, as explained, is the basis for ensuring and controlling quality, safety and efficacy of pharmaceutical products in Japan. The Law went through a big amendment , when new GMP standards were enforced from April 2005: (1) Detailed descriptions of manufacturing in the column of "Method of Manufacture" were required, (2) As for the post-approval change in the description of "Method of Manufacture", the applicant was required to specify each parameter as a major or minor change. Correspondingly, the applicant has to submit a partial change application or a minor change notification as already explained in the part of DMF and (3) A Drug Master File system was introduced.

On the other hand, in GMP regulation, (A) Pre-approval GMP inspection has become a prior condition of the Marketing Approval, (B) Inspection of overseas manufacturers has begun and (C) each product now has to be inspected every 5 years. For these changes, a 5-year transitional period, which was already finished at the end of March 2010, was granted to the products already approved at the time the new GMP standards was put in force. These products were, during this period, therefore exempted from the obligation of (1), (2), (3) and (C). If any change is made in the description of "Method of Manufacture" of the pharmaceutical products approved and marketed before the revision of the law, it can be notified as a minor change if it is "just" adding a detailed description without making any change in actual manufacturing method. These minor change notifications submitted are not reviewed by PMDA's review division at the time that notifications are submitted. Instead, they should be subject to a 5-year regular (i.e. periodical) inspection and adequacy of the minor change notification is checked in the GMP inspection. If it is found that the change which is supposed to be a major one is submitted as a minor change notification, the submission itself becomes invalid and may result in an enforcement action.

PMDA conducts a variety of inspections including Pre-approval Inspections and Periodical (or Regular) Compliance Inspections, both of which are based on applications, and Inspections of Buildings, Facilities and Equipments for domestic and overseas manufacturers. There are also other on-site inspections caused by recall, etc. and an ad-hoc inspection not based on an application. **Figure 5** gives a snapshot of the participants' relationship in the GMP inspection. This flowchart is valid for the Periodical Inspection and also the Pre-approval Inspections. GMP is laid down in the Ministerial Ordinance on Standards for Manufacturing Control and Quality Control for Drugs and Quasi-drugs (MHLW Ministerial Ordinance No 179, 2004)[4].

By their nature, Inspections by PMDA are categorised into two types. One is on-site inspection and the other is desktop inspection. The latter is also known as "document inspection" or "paper inspection". Due to a resource constraint in

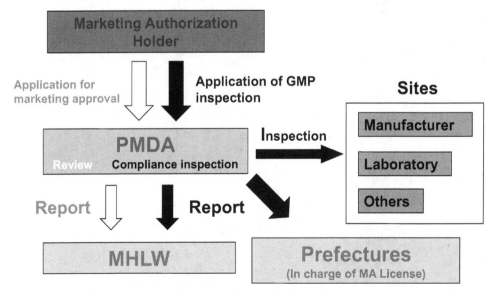

Figure 5: Flowchart of GMP Inspection

PMDA, desktop inspections are carried out for low-risk product and low-risk sites. The decision for an on-site or desktop inspection is made based on the outcome of the risk assessment procedure. In the risk-assessment procedure, the following are taken into account when choosing either on-site or desktop inspection: information about manufacturing site, new products or already marketed products, dosage form, manufacturing process, inspection history by overseas authorities and outcome of past PMDA inspections.

7.5   Workflow of inspections by PMDA

Figure 6 shows a standard inspection schedule by PMDA. It normally takes 3 days. In the morning of Day 1, following the opening session, an overview of the company and site is given by the manufacturer. And then, from morning to afternoon, a tour of the site takes place including warehouse, manufacturing

Figure 6: On-site inspection schedule

Inspection Schedule (standard case)

Day 1: Morning

1.   Opening meeting
2.   Company and site overview (presentation by manufacturer)
3.   Plant tour (1)
   • Warehouse

Day 1: Afternoon

   • Manufacturing area, packaging area, in-process laboratory
   • Support system
      – Water system
      – Air handling system
      – Waste treatment system
4.   Plant tour (2)
   • QC laboratory
      – Sample receiving area
      – Testing area
      – Biological tests
      – Stability test area
      – Retain sample area

Day 2: Morning

5.   Document check (1)

  (1) Quality system

  - GMP organization, job description of key personnel
  - Document control
  - Standard codes, SOPs
    Manufacturing control
    Quality control
    Manufacturing hygiene control
    Product master file

Day 2: Afternoon

  - Handling of Information on Quality, etc.
  - Recall
  - Deviation control
  - Change control
  - Training
  - Release procedure
  - Self inspection
  - Control of contract manufacture, supplier control
  - Written agreement

Day 3: Morning

6.   Document check (2)

  (2) Buildings and facilities

  - Daily check of buildings, equipment and utilities
  - Preventative maintenance

  (3) Control of raw materials, packaging materials

  - Material receipt, quarantine
  - Sampling procedure, storage of released material, use for manufacture
  - Compliance to the Japanese standard for the raw materials of animal origin

Figure 6: On-site inspection (contd.)

Day 3: Afternoon
(4) Manufacture • Product standard code (Master formula) manufacturing instruction, batch record • Validation • Reprocess, rework (5) Packaging and labelling (6) Test and inspection 7. Summary by inspectors 8. Closing meeting • Comments by inspectors • Discussion

area, supporting system and QC laboratory. On the 2nd day, a document check is made on the quality system including document control, deviation control, change control, training, release procedure and written agreement with the Japanese Marketing Authorization License Holder. And on the 3rd day, the document check continues on buildings and facilities, control of raw materials, manufacture, packaging and test, concluded by a summary by inspectors and discussion with them.

Following the inspection, a notice on the observations pointed out during the inspection is sent to the manufacturer. It is written in Japanese and sent via the Marketing Authorization License Holder within about three weeks after inspection. The manufacturer is required to respond to it in Japanese within about three weeks, with a specific deadline being shown in each notice.

PMDA normally requests the following for the written response by the manufacturer: (1) To submit photos or drawings if it is easier to demonstrate corrective actions, (2) To submit a summary report if some tests or validation works are carried out, (3) If SOPs are revised, to submit a copy of the revised part. It is sufficient to submit just the revised part and (4) If it takes some time before completing the corrective action, submit a schedule of action which includes specific actions to be taken and expected time to report.

A judgment of compliance with the GMP requirements is made by PMDA. Observations in the inspection are dealt with, depending on its criticality, as Rank A (Conformance), B (Minor observation), C (Moderate observation) and D (Critical observation). When all the requirements are duly complied with (including a case that an immediate correction measure is taken on site), it is assessed as Rank A. If it apparently conflicts with the GMP requirements, it is assessed as Rank D. An assessment of Rank B is given to a case where correction is required for the sake of completeness in the implementation of the GMP requirements, though it has almost no effect on the quality of a manufacturing item. And an assessment of Rank C is provided if any correction is required in view of the implementation of the GMP requirements, since one cannot deny its effects on the quality of a manufacturing item. An overall outcome of the inspections is then made based on the above judgements on a site by site basis with respect to each manufacturing item as the following categories: Compliance, Almost compliance, Correction required and Non-compliance.

7.6   Documents required for the desktop inspection

Another important aspect is "desktop inspection". Since accessible information is limited, the main part of the inspection is the check of discrepancy between the Application file or Master file and actual manufacturing process in order to ensure the quality of pharmaceutical products. From this point of view, PMDA requires documents reflecting the actual and detailed manufacturing process at the site. This is the reason why PMDA requests batch records and other manufacturing related documents.

The list of documents requested for the review is shown in **Figure 7**[5]. For No.6 "Method of manufacture", a copy of the relevant part of technical standard or SOP is submitted, which includes the items shown in the lower half of **Figure 8**. The copy of the technical standard or SOP should specifically be a master batch manufacturing instruction, a master batch record, a master test and inspection record, or a manufacturing and testing SOP at the site.

If an original copy of a GMP certificate issued by MRA (Mutual Recognition of Agreement) or MoU (Memorandum of Understanding) country** is submitted, you do not have to submit documents Nos. 2-11. Current MRA with the EU, however, does NOT cover APIs. If the 12 documents are not submitted, PMDA will request the staff of the manufacturing site to bring these documents to them. Otherwise, PMDA will go to the on-site inspection.

Documents submitted should be written in Japanese. If a major part of the documents are written in a foreign language other than English, not all but at least an overview must be submitted in Japanese or in English. When a site master

1. Outline of the site and the product(s)

2. Layout of the manufacturing site

3. Floor plan of facilities (premises, equipment and HVAC system etc.)

4. GMP organization and quality assurance organization

5. List of GMP documents

6. **Method of manufacture**

7. Validation status or annual validation program

8. Routine production status

9. Batch release procedure

10. Deviation managing procedure and a list of actual record including summary

11. Change control procedure and a list of actual record including summary

12. Document indicating compliance status to the standard for biological ingredients

13. (Site Master File, if available)

Figure 7: List of documents requested for review (1)

6. **Method of manufacture**
   - A copy of the relevant part of technical standard or SOP for the product subject to the inspection
     (*e.g., master batch manufacturing instruction, master batch record, master test and inspection record, or manufacturing and testing SOP actually implemented and operated at the site*)

   ① Manufacturing flowchart and detailed description of the manufacturing process
   ② In-process control value and control limits
   ③ Specifications and test methods of the intermediate and product
   ④ Specifications and test methods of the raw materials

Figure 8: List of documents requested for review (2)

file which is in Japanese or in English in accordance with the form specified by PIC/S (Pharmaceutical Inspection Convention and Pharmaceutical Inspection Co-operation Scheme) is submitted, since it already includes the information on No. 2 to 5 and 9 to 12, you do not have to submit these documents.

** Japan has the MRA on GMP with the EU. The scope of the MRA is the pharmaceutical products "except" sterile pharmaceuticals, biopharmaceuticals and APIs. European countries subject to the MRA are Belgium, Denmark, Germany, Greece, Spain, France, Ireland, Italy, Luxemburg, the Netherlands, Austria, Portugal, Finland, Sweden and the UK. Japan also exchanges a MoU with Germany, Sweden, Switzerland and Austria.

7.7   Common issues experienced in overseas inspections

Common issues that the PMDA has experienced with regard to overseas on-site as well as desktop inspections are: (1) It is commonly noted that there is insufficient supervision and management by Japanese Marketing Authorisation License Holders of relevant manufacturers, (2) In some Master File in-country caretakers, lack of knowledge about Japan's pharmaceutical regulation is observed. They sometimes fail to provide necessary information in a timely manner to overseas manufacturers and Master file registration holders, which makes it difficult for overseas companies to take appropriate actions. (3) It seems that all of these problems come from the lack of good communication among Marketing Authorisation License Holders, Overseas manufacturers and Master File in-country caretakers.

7.8   References

1   PFSB/ELD Notification No. 0210001, Feb 10, 2005
    http://www.pmda.go.jp/english/service/pdf/guideline_application-2.pdf

2   PMDA's English website
    http://www.pmda.go.jp/english/service/gmp.html

3   PMDA's English website on Japanese Pharmacopoeia (JP)
    http://www.pmda.go.jp/english/pharmacopoeia/index.html

4   The Ministerial Ordinance on Standards for Manufacturing Control and Quality Control for Drugs and Quasi-drugs (MHLW Ministerial Ordinance No 179, 2004)
    http://www.pmda.go.jp/english/service/pdf/ministerial/050909betsu2.pdf

5   The complete list is also posted on PMDA's English website which is accessible through the following
    URL.http://www.pmda.go.jp/operations/shonin/info/iyaku/file/080901-e-2.pdf

CHAPTER 8

Preparing and management of international inspections

Tim Sandle, Bio Products Laboratory, UK
Andreas Brutsche, Novartis, Switzerland

8.1   Introduction

Regulatory inspections are part of the overall drug quality assurance system. The objective of inspecting pharmaceutical manufacturing facilities is either to enforce Good Manufacturing Practice (GMP) compliance or to provide authorization for the manufacture of specific pharmaceutical products, usually in relation to an application for marketing authorization. A further aspect of pharmaceutical regulatory inspections is for monitoring the quality of pharmaceutical products in distribution channels, from the point of manufacture to delivery to the recipient, as a means of eliminating the hazard posed by the infiltration of counterfeit medicines[1].

The regulation of drugs encompasses a variety of areas. These areas include:

*   Licensing

*   Inspection of manufacturing facilities and distribution channels

*   Product assessment and registration

*   Adverse drug reaction (ADR) monitoring

*   Quality control

*   Control of drug promotion and advertising

*   Control of clinical drug trials (evaluation of safety and efficacy data from animal and clinical trials)

Each of these functions targets a different aspect of pharmaceutical activity[2].

Regulatory inspections are either 'national', that is inspections undertaken by the inspectorate based in the country in which a pharmaceutical company is located, or 'international', that is a regulatory agency from a different country to the one in which the company is located conducts the inspection.

Inspection programmes by international regulators are orientated towards those global establishments that are identified by the regulator as having the greatest public health risk potential if they experience a manufacturing defect. Thus 'risk based inspections' are especially applicable to international regulatory inspections.

International inspections, like national inspections, are very important and failure to 'pass' an inspection can lead to pharmaceutical products being debarred from sale within the territory covered by the international regulatory agency. Thus, if inspections go wrong, they can result in negative outcomes. Such outcomes include:

- Inspection observation (in the case the FDA, a Form-483)

- Warning Letter (or European equivalent)

- Product Recall

- Facility Lockdown

- Non approval for Facility or Product License

- Injunction

- Debarment of individuals to operate in the pharmaceutical industry

- Civil penalties, fines or judgements

- Criminal penalties, fines, imprisonment

Although the approach, standards and guideline used by different inspectorate bodies differs internationally, there is some commonality in approach under the auspice of the International Conference on Harmonisation and the Quality Risk Management guidance established by ICH Q9[3]. This risk-centric approach applies both to industry and regulators. For inspection, key factors need to be included to establish the quality risk-based approach as those linked to the product (e.g. aseptic, toxic and paediatric) as well as the inspection history of the site and the company[4].

Furthermore, where possible, international regulatory agencies are attempting to minimise the number of inspections made outside of their own territories through developing mutual ways of working and establishing a harmonised approach to inspections. This movement is generally supported by the pharmaceutical industry. This strengthening of international collaboration in GMP inspections is with the view to one day develop a regulatory inspection framework approach at a global level (e.g. Mutual Recognition Agreements (MRA) between national/regional authorities; confidentiality agreements for sharing inspection reports; joint GMP inspections; acceptance of GMP-certificates based on legal formats (e.g. WHO Certificate of Pharmaceutical Product [(CPP)])[5]; and recognition of harmonised approaches for GMP (e.g. WHO, the

Pharmaceutical Inspection Convention and Pharmaceutical Inspection Co-operation Scheme (jointly referred to as PIC/S), ICH Q7, Q9 and Q10)[6].

For example, in 2011 under the Good Clinical Practice (GCP) initiative the FDA (Food and Drug Administration) and EMA (European Medicines Agency)[a] exchanged over 250 documents concerning 54 different medications. Together, the two agencies organised 13 clinical trial inspections. Nonetheless, despite such initiatives, international regulatory inspections continue and may continue for some time.

One of the prominent bodies for the promotion of pan-national regulatory collaboration is the Pharmaceutical Inspection Convention and Pharmaceutical Inspection Co-operation Scheme (PIC/S). Together these comprise two international instruments for use between countries and pharmaceutical inspection authorities. The PIC/S is intended to be an instrument to improve co-operation in the field of Good Manufacturing Practices between regulatory authorities and the pharmaceutical industry.

This chapter focuses on the preparation and management of international inspections. This is in relation to receiving inspections from inspectorate groups which are not resident to the home country of the company being inspected. In focusing on these more specific aspects, the chapter does not set out to provide an overall plan for managing regulatory inspections in general, moreover the emphasis is upon issues applicable to international regulatory inspections.

If there is one key learning point from the chapter that the authors wish the reader to take away it is preparation. Preparation is the key to successfully passing a regulatory inspection. For this, preparations should start as early as possible. This gives the company enough time to audit their own systems, to identify gaps and to resolve the issues. In addition it provides sufficient time to prepare and train the team on site.

For the purpose of outlining approaches to international regulatory inspections, the chapter is divided into three main areas:

1. Pre-inspection preparation activities

2. Inspection activities

3. Post-inspection activities

In doing so, the chapter additionally offers some general advice for preparing for, managing and responding to regulatory inspections. Much of this is based on the authors' own experiences.

8.2   International regulatory agencies and the key differences

Before examining each of the three steps involved in preparing for, dealing with, and responding to inspections, it is important to note that there are a multitude of different international regulatory agencies. Some of these agencies are longer established and more influential than others. Here the prime examples are the FDA, the agencies of the EMA, and in Japan. In the European Union (EU), regulatory activities are carried out by the European Medicines Agency (EMA) and the local regulatory agencies in the individual EU member states.

In the USA, the laws governing medicines are found in the Food Drug and Cosmetic Act [21 U.S.C. 301, *et seq.*] as well as other statutes. In addition to laws, the FDA has the power to create regulations, which are published in the Federal Register and in the Code of Federal Regulations (CFR). The FDA also issues Guidelines, which specify the agency's current thinking and preferences on laws and regulations. While non-binding, these Guidelines are generally followed by the industry and are often used by inspectors[7].

The European authorisation system for pharmaceuticals was established in 1993 (Regulation (EEC) No. 2309/93, OJ No. L 214 of 24/8/1993)[8]. This legislation means that, within the EU, the EMA evaluates and supervises the approval and use of medicinal products that are centrally approved in the EU (a process whereby a single application leads to an EU-wide approval). In addition, products that are not centrally approved are evaluated and supervised either by individual national regulatory authorities, or by a group of member states' regulatory authorities, with one of the member states acting as the "reference member state" and leading the process (known as the decentralised or mutual recognition procedures)[9].

EU legislation is arguably more complicated than in the U.S., because it derives from treaties and agreements between the different sovereign countries (Member States of the European Union). This includes laws that are specified in Regulations (laws directly applicable to all Member States, without the need for additional implementation at the national level); and Directives (which bind Member States to implement the contents of the Directive into their national laws within a certain time period). There are also Guidance documents which, although not law themselves, should be complied with. In addition, scientific opinions on applications are elaborated by an institutionalised network of experts from member state regulatory agencies within the Committee for Human Medicinal Products (CHMP)[b].

The general responsibilities of regulatory agencies are[10]:

*   Reviewing clinical research and consulting with experts to decide whether or not to approve a new medicine,

- Providing safety information on approved medicines,

- Overseeing the safety information generated on a product during its lifetime and any changes that need to be made to the safety information as a result,

- Inspecting facilities where medicines are tested or manufactured, and

- Carrying out safety reviews of a product when necessary.

Pharmaceutical companies must report safety information about their medicines to applicable regulatory agencies like the FDA and other relevant agencies (or health authorities) worldwide[11]. Whilst all agencies have the same objectives and obligations to safeguard public health when assessing the safety, quality, and efficacy of medicines, there are differences with regard to their organisation, scope of activities and remit[12].

The full list of national and international regulatory agencies (at the time of writing this chapter) is as follows:

International

- International Conference on Harmonisation (ICH)

- United Nations Health Care Organization (UNHCO)

- World Health Organization (WHO)

- World Trade Organization (WTO)

Of these, the ICH (International Conference on Harmonisation of Technical Requirements for Registration of Pharmaceuticals for Human Use) sets a number of global standards which have been adopted by regulatory authorities, notably the EMA and FDA, and which includes the ICH quality management standards discussed above. For the developing world, the WHO is enormously influential and sets out a range of GMP standards[13]. Of the others, UNHCO is an advisory body on health care best practice; whereas the WTO is not an inspectorate body, it does, however, affect the rules pertaining to trade and Good Distribution Practice (GDP)[14].

- Argentina: Ministry of Health - National Administration of Drugs, Food & Medical Technology (ANMAT)

- Armenia: Drug and Medical Technology Centre, Ministry of Health

- Australia: Therapeutic Goods Administration (TGA)

- Austria: Bundesministerium für Gesundhe

- Bahrain: Ministry of Health
- Bangladesh: Ministry of Health
- Belgium: Federal Public Service (FPS) Health, Food Chain Safety and Environment
- Belize: Ministry of Health
- Bolivia: Ministry of Health and Sports
- Botswana: Ministry of Health
- Brazil: Ministry of Health- National Health Surveillance Agency (Anvisa)
- Brunei: Ministry of Health
- Bulgaria: Bulgarian Drug Agency (BDA)
- Canada: Health Canada
- Chile: Ministry of Health
- China: State Food and Drug Administration (SFDA)
- Colombia: Ministry of Health-National Institute of Food and Drug Monitoring (INVIMA)
- Costa Rica: Ministry of Health
- Croatia: Ministry of Health and Social Welfare
- Cuba: Ministry of Public Health
- Czech Republic: State Institute for Drug Control
- Denmark: Danish Medicines Agency
- Ecuador: Ministry of Public Health
- Egypt: Ministry of Health and Population
- El Salvador: Ministry of Health
- Estonia: State Agency of Medicines
- Europe: EU Legislation – Eudralex, European Directorate for the Quality of Medicines and Healthcare (EDQM), European Medicines Agency (EMA), Heads of Medicines Agencies (HMA)
- Fiji: Ministry of Health
- Finland: Finnish Medicines Agency

- France: Agence Française de Sécurité Sanitaire des Produits de Santé, Ministry of Health
- Georgia: Ministry of Labour, Health and Social Affairs of Georgia
- Germany: Federal Institute for Drugs and Medical Devices (BfArM), Ministry of Health, Paul-Ehrlich-Instituts in Langen (PEI)
- Greece: National Organization for Medicines (EOF)
- Guam: Department of Public Health and Social Services
- Guatemala: Ministry of Health and Welfare
- Guyana: Ministry of Health
- Hong Kong: Department of Health: Pharmaceutical Services
- Hungary: National Institute for Pharmacy
- Iceland: Icelandic Medicines Agency, Ministry of Health & Social Security
- India: Central Drug Standard Control Organization (CDSCO), Government of India Directory of Health and Family Welfare, Indian Council of Medical Research (ICMR) , Ministry of Health and Family Welfare
- Indonesia: Ministry of Health
- Ireland: Irish Medicines Board
- Israel: Ministry of Health
- Italy: Italian Pharmaceutical Agency, Ministry of Health
- Japan: Ministry of Health and Welfare , National Institute of Infectious Diseases, National Institute of Health Sciences
- Jamaica: Ministry of Health
- Jordan: Ministry of Health
- Kenya: Ministry of Health
- Latvia: State Agency of Medicines
- Lebanon: Ministry of Public Health
- Lithuania: State Medicines Control Agency
- Luxembourg: Ministry of Health
- Malaysia: National Pharmaceutical Control Bureau
- Maldives: Ministry of Health and Family

- Malta: Medicines Authority, Ministry of Health
- Mauritius: Ministry of Health and Quality of Life
- Mexico: Ministry of Health
- Morocco: Ministry of Health
- Namibia: Ministry of Health and Social Services
- Nepal: Ministry of Health and Population
- Netherlands: Medicines Evaluation Board
- New Zealand: Medsafe - Medicines and Medical Devices Safety Authority, New Zealand Ministry of Health
- Nicaragua: Ministry of Health
- Nigeria: National Agency for Food and Drug Administration and Control (NAFDAC)
- Norway: Ministry of Health and Care Services, Norwegian Medicines Agency
- Pakistan: Drug Control Organisation, Ministry of Health
- Palestine: Ministry of Health
- Panama: Ministry of Health
- Papua New Guinea: Department of Health
- Paraguay: Ministry of Health
- Peru: Ministry of Health
- Philippines: Department of Health, Philippine Council for Health Research and Development (PCHRD)
- Poland: Ministry of Health & Social Welfare
- Portugal: The National Institute of Pharmacy and Medicines (Infarmed)
- Romania: Ministry of Health, National Medicines Agency (ANM)
- Russia: Association of International Pharmaceutical Manufacturers, Ministry of Health and Social Development
- Saudi Arabia: Ministry of Health
- Senegal: Ministry of Health and Prevention

- Serbia: Medicines and Medical Devices Agency (ALIMS), Ministry of Health
- Singapore: Health Sciences Authority (HSA), Ministry of Health
- Slovak Republic: Ministry of Health, State Institute for Drug Control (SIDC)
- Slovenia: Ministry of Health Agency for Medicinal Products
- Sri Lanka: Ministry of Healthcare and Nutrition
- South Africa: Department of Health, Medicines Control Council (MCC)
- South Korea: Food and Drug Administration
- Spain: Medicines and Health Products Agency (AEMPS), Ministry of Health
- Sweden: Medical Products Agency (MPA)
- Switzerland: Swiss Agency for Therapeutic Products
- Taiwain: Department of Health
- Tanzania: Ministry of Health and Social Welfare
- Thailand: Ministry of Public Health
- Trinidad and Tobago: Ministry of Public Health
- Tunisia: Ministry of Public Health, Office of Pharmacy and Medicine
- Turkey: Ministry of Health
- Uganda: National Council for Science and Technology (UNCST)
- Ukraine: Ministry of Health
- United Arab Emirates: Ministry of Health U
- UK: Department of Health, Medicines and Healthcare Products Regulatory Agency (MHRA)
- Uruguay: Ministry of Public Health
- USA: Centers for Disease Control and Prevention, The Food and Drug Administration (FDA)
- Venezuela: Ministry of Public Health
- Vietnam: Ministry of Health
- Yemen: Ministry of Public Health and Population

Not only is the list of regulatory agencies extensive, the different inspectorate bodies conduct inspections differently and have different remits. What is also noticeable is that in some countries, all functions related to drug regulation come under the jurisdiction of a single agency, which has full authority in the command and control of these functions, as well as bearing the responsibility for their effectiveness. In other countries, drug regulatory functions are assigned to two or more agencies, at either the same or different levels of government[15].

In relation to the remit of different regulatory agencies, whilst legal structures form the foundation of drug regulation, some drug laws traditionally omit or exempt certain areas of pharmaceutical activity from their scope of control, thus resulting in a so-termed 'regulatory gap'. For instance, some countries do not require registration of herbal or homeopathic drugs whilst other countries do. In a second example, with nation states legal mandates are not imposed on the importation of drugs in contrast with other states. Therefore, each regulatory inspection will focus on different aspects of pharmaceutical manufacturing depending upon the type of drug and the distribution chain. It is important to understand the scope of the regulatory agency prior to the inspection taking place and to note such international differences.

In terms of inspections, with EU GMP inspectors for example, the key chapters of the EU GMP guide will be followed[16]; whereas with the FDA the systems based approach to inspections will be followed. The systems based approach utilises a risk-based approach to conducting inspections which identifies six key systems and three critical elements within each system that are common to establishments that produce biological drug products[17]. Although the systems and elements are specific to the requirements of the FDA, by preparing inspections around these points of concern most important aspects of pharmaceutical manufacturing will be covered. The six systems are[18]:

- Quality System

- Production System

- Facilities and Equipment System

- Materials System

- Packaging and Labelling System

- Laboratory Control System

Each system relates to the overall quality system and the aim of producing a quality product (one both efficacious and contamination free)[19]. This relationship is illustrated in the diagram below:

Quality Product

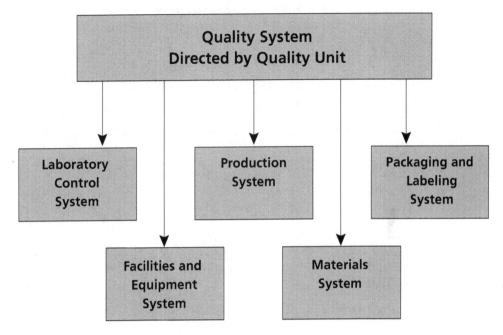

Figure 1: an illustration of the systems approach to inspections.

In addition to the six systems the three key critical elements are:

- Standard Operating Procedures (SOPs)
- Training
- Records

These systems are covered in detail elsewhere within this book. The point here is to identify the differences in approach and it is important for the company receiving the inspection to appreciate and to understand these differences.

Whilst the FDA and EMA provide considerable guidance concerning what to expect during an inspection it is notable that not all drug regulatory authorities provide documented standard procedures for registration, and even fewer provide documented guidelines and checklists for inspection (thus much of the information presented in the chapters within this book will prove of great use for inspection preparation).

8.3   Pre-inspection preparation activities

When notification of the inspection is received, or is as part of a company's preparations in advance of the anticipated date of the inspection, various actions should be undertaken. This section of the chapter identifies these actions.

In terms of pre-inspection planning, the following preparatory steps should be considered[20]:

- Review the notification letter. Ideally, although not in all cases, the letter will spell out the type of inspection the Agency intends to conduct, including the records the inspector expects to review.

- Review previous inspection records. This can be company internal audit reports. In addition, with the case of the FDA, previous establishment inspection reports (EIRs) and FDA Form 483 observations should be available for review.

- For FDA inspections, review the FDA quality system inspection technique (QSIT) manual. For other regulatory agencies, some inspection guides are available to review[21].

- Review the relevant regulatory inspection manuals.

- Review relevant harmonization guidelines. Here documents issued by the International Conference on Harmonization (ICH) and the Global Harmonization Task Force (GHTF) are particularly useful.

- Prioritise likely areas of scrutiny. This may include: the quality system management review SOP, quality system management review summaries and action plans, training SOPs, training records, including effectiveness assessments, for management, organisational charts, management job descriptions, supplier evaluation and selection SOP, documented risk evaluation of various suppliers and supporting documentation showing management involvement, documented decisions and rationales on which suppliers to use and the controls to be put in place, including supporting documentation showing management involvement, quality or technical agreements with critical suppliers, and any supporting documentation showing management involvement, and documentation showing management review of supplier deviations and investigations.

- Hold an inspection expectation overview meeting.

Some companies have an inspection SOP which can be followed in order to help with this planning process. This is a very useful document to have in place. The general content of such SOPs may include:

- Operating procedures
- Hosting regulatory inspections
- Purpose and responsibilities
- Organisational structure
- Composition and training of staff
- Critical documents required for review
- Recruiting materials
- Review process
- Safety reporting
- Informed consent
- Document management
- Quality management

It is important that any such document is a working document and made available to all personnel who are likely to be involved in the inspections.

8.3.1   Inspection preparation team

One of the most important steps prior to the inspection is to establish an interactive inspection preparation team. For this, at least one representative of the following departments should be included:

- Quality Assurance
- Product Development
- Quality Control
- Microbiology
- Engineering
- Production
- Warehousing or Supply Chain Management
- Validation
- Information Technology

The team should be co-ordinated by a senior manager, ideally a representative of the Quality Assurance (QA) department. The team should receive periodic training on[22]:

- What the facility can reasonably expect during an inspection
- The scope of inspectors' authority
- The topics typically addressed during an inspection

It is very important to organise this team according to project management principles. This means having a project leader defined and having a governance structure with frequent meetings with senior management in place.

As a general point, the use of timelines can be useful in weighting tasks and for pinpointing when certain key steps need to be completed by. An example, in relation to engineering and facility inspections, is shown below:

Timeline

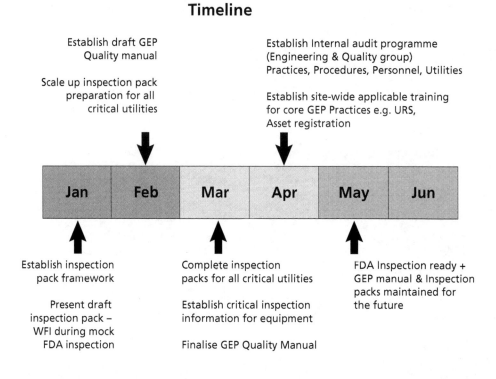

Figure 2: An example of an engineering systems timeline for inspection preparation

Timescales can also be expressed using different formats. A second example is shown below using a tabular approach. This is also in relation to an engineering system, although the principles are applicable to any area of the facility.

Activity		
Phase 1	Schedule	Estimate of days
Equipment of core engineering objectives Consultation with Site Engineering, QA, Production debt Authoring of core engineering policies Feedback engineering review workshops Roll out presentations and GEP framework FDA Awareness briefings to engineers and support groups Ongoing support and Q&A to internal queries	18th August through to 30th September	20 Days
Detailed GEP check point and agreement of objectives and progression into next phase	30th September	1 Day
Phase 2		
In depth assessment of existing Eng. SOPs Rationalisation of existing Eng. SOPs Adjustments to Eng. SOPs against Policy objectives Establishment of site applicable engineering standards Execution of equipment pilot risk assessments FDA Inspection readiness training for Engineering dept Establishment of Engineering training strategy/records	1st October through to 28th November	30 Days
Detailed GEP check point and agreement of objectives and progression into next phase	28th November	1 Day
Phase 3		
Establishment of GEP Key performance indicators Mock FDA audit of GEP department practices, training records and equipment documents	1st December through to 18th December	10 Days

Table 1: Example of a pre-inspection timescale chart

8.3.2 The role of Quality Assurance

As soon as a facility is informed of a forthcoming regulatory inspection the Quality Assurance (QA) group has an important role in ensuring appropriate pre-inspection activities occur. This might involve identifying who will need to be involved in the inspection, providing coaching on what to expect during the inspection and how they should respond when asked questions.

The role of QA is equally important during the inspection. Generally QA host the inspection and act as the facilitators, providing assistance and guidance to site personnel. Working with local operational and quality management, QA are responsible for scheduling inspection activities and ensuring the availability of required individuals. During inspections QA should be present during the inspection interviews and record all key questions and any issues that arise. QA should also ensure that all questions are directed to individuals sufficiently qualified and experienced to provide accurate answers.

In addition, QA should be responsible for ensuring the effective, efficient running of the 'back room', encouraging the maintenance of a calm and focused environment where the retrieval, review and copying of documents can be performed. In this role QA should also prepare interviewees prior to going into the inspection room and also debrief them immediately after they have been interviewed by the inspector to make sure the interviewees are satisfied with all of the responses they provided. This debrief additionally allows for an early indication of potential issues.

8.3.3   Baseline audit

One of the first tasks to undertake prior to an inspection is to conduct a 'baseline audit'. A baseline audit is performed in order to identify any possible gaps and issues. This audit should be performed under the lead of an experienced QA person. External experts can be also hired to strengthen the team in certain areas, e.g. to give the FDA view to a site which had never previously been exposed to FDA inspections. In addition to the audit, it may prove useful to run a full-scale mock inspection. Given the familiarity of the staff to the company and each other, it may be prudent to hire a consultant to carry out a practise inspection.

After the mock inspection or baseline audit, companies should seek to enhance all the critical compliance categories that have been identified as deficient. Where issues cannot be resolve din time, position papers should be written to show that the non-compliance has been thought through and risk assessed.

It is useful to base the audit around key inspection issues, for example, in relation to quality systems[23]:

- Assess overall compliance with cGMPs[24]
- Examine internal procedures, and specifications, including:
    - o   Release of components and in-process materials
    - o   Change control

- o   Reprocessing
- o   Batch release
- o   Annual record review
- o   Validation protocols and reports
- o   Product defect evaluations
- o   Complaint handling
- o   Evaluation of returned and salvaged products

When undertaking a base-line audit, it is useful to note and to record the following activities[25]:

- Date of self-inspection,
- Inspection reference (e.g. inspection number),
- Organisation/department involved (this will extend to section and activities),
- Scope of the self-inspection,
- Attendance list (those involved in the inspection). These individuals should be referred to in the final inspection report,
- Inspectors (their role, name and signature). With respect to a self-inspection report it should only be signed by the lead inspector on behalf of the team,
- List of clauses/areas examined including notes on details observed during the inspection. This can also include the names of staff present during the self-inspection and/or those with whom there has been an interaction,
- List of non-compliances and/or deviations observed. These non-compliances should give a precise indication as to the topic and the measures for correction.

For conducting the audit, the quality risk management approach described in ICH Q9 is useful. This is detailed in the following diagram:

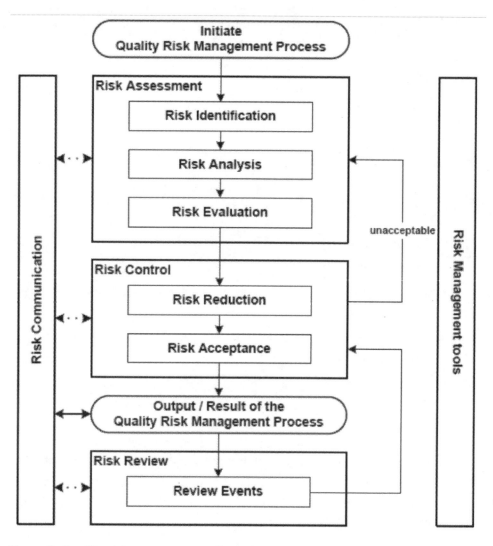

Figure 3: Quality risk management flow chart

Risk management is an important planning tool which can be used to focus and prioritise the audit programme and individual action plans. It is usually worth performing a formal, documented risk assessment for identifying possible risks (firstly on a high level, then focusing towards the details) in the business, and quantifying them by assessing the likelihood of a failure, and the probable impact to the business if the failure took place. The advantage of assessing risks is that they can be ranked, with the highest risks taking the available QA resource for audits.

It is also important to undertake research into 'current' GMP (cGMP) as part of the pre-inspection preparations, particularly those pertinent to the international regulatory agency. This should always be an on-going activity. The review should include material relevant to the body carrying out the inspection and should include papers, articles and guidelines. This is doubly important if the plant has not been inspected by the particular regulatory authority before or if the authority is unfamiliar.

In terms of cGMP a thorough understanding of industry expectations and industry standards is essential. Many industry expectations will not yet have been codified into law; nonetheless, investigators may inspect against them and against that particular trend. For example, in relation to the FDA there is an entire body of documentation that any company preparing for an FDA inspection should study beforehand in addition to cGMPs. Examples of material review includes:

- Professional publications,

- Attending conferences,

- Pertinent CBER [Center for Biologics Evaluation and Research] and CDER [Center for Drug Evaluation and Research] guidances and compliance programs,

- Investigator inspection manuals,

- Obtain any recent 483s issued in situations specific and similar to the company's product types (such as aseptic manufacturing or non-sterile products).

In addition, the FDA publishes warning letters on the internet under U.S. freedom of information clauses. As part of the inspection, a review of FDA warning letters can provide invaluable information relating to common errors made by companies and inspectorate 'hot topics'. This is important even if the FDA are not the regulatory agency coming to inspect the site as, in general, the FDA 'hot topics' are frequently global points of regulatory concern. Some examples of warning letters are:

a) In relation to facilities and equipment:

   "Your disinfectant effectiveness study # FR000-01 dated 01/01/2008, is incomplete.

   The study did not evaluate the effectiveness of the disinfectants in use on fungi and spore forming microorganisms. Spore forming microorganisms have been routinely isolated in your manufacturing facility and accounted for 17% of total isolates identified in 2012 and 2011; 14% in 2010 and 7% in 2009."

b)   In relation to quality systems:

> *"Failures were not fully investigated and documented, nor were they extended to other batches as appropriate. For example:*
>
> *You failed to quarantine numerous process intermediates associated with the use of [redacted] filter membranes that were identified to cause foaming during filtration. This foaming was found to be associated with leaching of [redacted] into process intermediates. These process intermediates were used to further manufacture finished vaccine product lots."*

c)   In relation to the materials system:

> *"Written procedures are not followed for the storage and handling of drug product containers.*
>
> *Specifically, SAP inventory for Glass syringe lot 999999 did not match the physical inventory in the warehouse reject cage on 7/3/12. One box from this sprayer lot was found damaged on 6/18/07, and was placed in the reject cage. However, this transaction was not entered into SAP as required by written procedures."*

d)   In relation to the production system:

> *"Your firm failed to establish an adequate system for monitoring environmental conditions of aseptic processing areas [21 CFR 211.42(c)(10)(iv)]. For example:*
>
> *a)   There is no documentation that monitoring covers all production shifts and is performed during active operations.*
>
> *b)   There is no assurance that monitoring is at the locations where critical operations are performed."*

e)   In relation to the packaging and labelling system:

> *"Examination of labelled product for suitability and correctness is not documented in the batch record. Specifically, during labelling & packaging operations, final product syringes can be rejected by the automated equipment at several stations. The syringes rejected by the automated equipment are manually inspected by production operators and reintroduced to the packaging operations if deemed*

acceptable. However, this visual inspection and reintroduction into the line is not documented in the batch record. "

f)   In relation to laboratory controls:

"Failure to establish the accuracy, sensitivity, specificity, and reproducibility of test methods, in that analytical methods have not been validated [21 CFR 211.165(e)]. For example:

a)   *Sterility test method STR-MTM-0006 has not been validated for sterility testing of [redacted] liquid bulk.*

b)   *bioburden test method S004 has not been validated for bioburden testing of [redacted] pre-filtration bulk.*

Laboratory raw data is recorded onto data sheets. These sheets may be printed at will by analysts from a computerised document management system without tracking as to the number of data sheets printed or used.

Data was not available to support expiration dates assigned to in-house prepared reagent solutions used in the testing of final bulk product."

Other useful information which can be gleaned from public information are the Establishment Inspection Reports (EIR), which provide more detailed information relating to FDA inspectorate views of pharmaceutical manufacturing. A recent review by the authors found that the top ten EIR issues were:

*   21 CFR 211.22(d)The responsibilities and procedures applicable to the quality control unit are not [in writing] [fully followed]. Specifically, ***

*   21 CFR 211.100(b)Written production and process control procedures are not [followed in the execution of production and process control functions] [documented at the time of performance]. Specifically, ***

*   21 CFR 211.110(a)Control procedures are not established which [monitor the output] [validate the performance] of those manufacturing processes that may be responsible for causing variability in the characteristics of in-process material and the drug product. Specifically, ***

*   21 CFR 211.160(b)Laboratory controls do not include the establishment of scientifically sound and appropriate [specifications] [standards] [sampling plans] [test procedures] designed to assure that [components] [drug product containers] [closures] [in-process materials] [labeling] [drug products]

conform to appropriate standards of identity, strength, quality and purity. Specifically, ***

- 21 CFR 211.100(a) There are no written procedures for production and process controls designed to assure that the drug products have the identity, strength, quality, and purity they purport or are represented to possess. Specifically, ***

- 21 CFR 211.192 There is a failure to thoroughly review [any unexplained discrepancy] [the failure of a batch or any of its components to meet any of its specifications] whether or not the batch has been already distributed. Specifically, ***

- 21 CFR 211.165(a) Testing and release of drug product for distribution do not include appropriate laboratory determination of satisfactory conformance to the [final specifications] [identity and strength of each active ingredient] prior to release. Specifically, ***

- 21 CFR 211.25(a) Employees are not given training in [the particular operations they perform as part of their function] [current good manufacturing practices] [written procedures required by current good manufacturing practice regulations]. Specifically, ***

- 21 CFR 211.188 Batch production and control records [are not prepared for each batch of drug product produced] [do not include complete information relating to the production and control of each batch]. Specifically, ***

- 21 CFR 211.67(b) Written procedures are not [established] [followed] for the cleaning and maintenance of equipment, including utensils, used in the manufacture, processing, packing or holding of a drug product. Specifically, ***

The above are examples drawn from FDA warning letters in relation to each of the main systems reviewed as part of the systems based inspection. Whilst the detail is FDA specific, such information can be very useful for reviewing when preparing for an inspection from any of the main international inspectorate bodies and should be incorporated into the baseline audit.

The baseline audit needs to be performed very thoroughly in order to identify any potential issues. It is always better to know the potential findings upfront since this avoids a lot of hectic activity and stress during inspections. Any critical issues identified during the audit must be remediated in an action plan. The plan needs to define clear roles and responsibilities, and timelines. The execution of the remediation plan should be monitored by a governance structure, e.g. senior management meeting, meeting minutes, remediation plan and so on.

8.3.4   Facility upgrade

Any necessary facility upgrades have to be evaluated and the related investments should be reflected in the budget. In case the necessary equipment or upgrade cannot be finalised prior to the inspection, an approved investment and a detailed plan with responsibilities and timelines should be available. This should include an overview of the works to be completed and realistic costing. This can be then be shown to the inspector as necessary and such a plan may help to reassure the inspector and lessen any criticisms levelled in relation to facility investment.

Other areas to check, in relation to facilities, include:

• Verification of the appropriateness and maintenance of buildings and facilities

• Equipment qualification, calibration, maintenance and cleaning (validation and routine)

• Facility utilities (HVAC, water, steam, and compressed air) qualification, routine monitoring and maintenance

8.3.5   Process related issues

As part of the preparation, a number of areas applicable to processing and product formulation should be examined. These include such topics as[26]:

• Validation of computerised inventory control processes

• Product storage

• Distribution controls

• Records for detection of counterfeiting

• Control of facilities used for storage (warehouse, cold rooms, freezers, etc.)

• Batch formulation

• Dosage form production

• Sterile filtration

• Aseptic processing

• In-process testing

• Lot release

• Process validation

- Procedures and documentation of label control to prevent mix ups

- Facilities, equipment, and support systems to maintain proper environmental and processing controls during operations

With the above list it is important to review and update Standard Operating Procedures (SOPs). Reviewing SOPs will not only help the inspection team understand how things operate but also will provide staff with a refresher course on protocol and procedure. This will enable them to better discuss and explain systems when asked.

8.3.6 Document preparation and management

Prior to the inspection the following documents should be reviewed and updated as necessary. If any of the documents have not been written, these should be prepared in advance of the meeting[27].

- Manufactured products during the last two years,

- Validation Master plan,

- Development reports

- Master formulae,

- Individual batch records. It is particularly important to review problematic or rejected batches prior to the inspection,

- Validation protocols and reports,

- Equipment qualification protocols and reports,

- Testing instructions and protocols,

- Manufacturing instructions and reports,

- Annual product reviews (APRs),

- List of all deviations and investigations,

- List of all out of specification results,

- List of all changes over the last two years,

- List of all recalls and Field Alerts over the last two years,

- List of all complaints (medical and technical) over the last two years,

- Stability reports,

- Job descriptions and training records of personnel,

- List of all SOPs with due dates,

- Equipment lists,

- Supplier agreements,

- Contracts for third parties,

- Self-inspection plan and reports,

- Regulatory files.

In addition, specifically for sterile manufacturing:

- Microbiological environmental monitoring data trend reports,

- Media fill reports over the last two years,

- Aseptic operator qualification data,

- Validation data for micro methods,

- Sterility failures over the last two years.

Preparation should also be extended to the Quality Control laboratories. Here, some of the most important topics are:

- SOPs for control of microbiological contamination and environmental monitoring,

- Records for source materials,

- In-process and finished process testing,

- Methods for sampling and testing,

- Validation of test methods.

8.3.7   Recommended presentations

It is also recommended to prepare some overview presentations which can be presented during the inspection. Not all investigators or inspectors will accept presentations. However, if the inspector is open to having data presented a short presentation can prove to be very valuable as means of getting a complex process across in easily explainable terms.

The following list contains some recommended presentations:

- Site overview with maps, product list, organization charts,

- Site Quality meetings, e.g. validation review board, deviation review board, change review board,

- Validation approach, process and cleaning validation, transport validation approach,

- Equipment qualification approach,

- Investigation trending charts, deviations > 30 days,

- Training concept,

- Product release,

- Product distribution (incl. transport companies),

- Critical deviations, complaints.

It is also important to have a copy of the organisation chart to hand to the inspectors. Another useful document is a facility diagram that is coded by environmental control area (such as cleanroom classes). This can help the inspector to get an idea of where they are when doing plant tours, what the controls in those areas are and how they relate to product manufacturing builds. Facility diagrams can also be annotated with process flow route maps.

8.3.8   Facility tour

In most cases a facility tour is a central part of the inspection and therefore it is important to prepare this tour carefully. Normally the tour should mirror the material flow (start in the incoming goods area and then proceed to the warehouse, sampling, production, Quality Control laboratories, engineering area with water, steam and HVAC system etc.). In addition the separated area for rejected goods should be also part of the tour.

The walkthrough should be carefully prepared especially with respect to housekeeping. The first impression that inspectors get when they see the processing areas for the first time is of great importance and therefore the facility should be presented in the best way. Auditors typically request that the tour is performed in operation: that is, they expect to see the machines running and also so that they can check that the operators behave in a competent way. One week before the inspection internal walkthroughs should be performed to train the people and to identify and fix potential smaller gaps.

For aseptic manufacturers, a sterile operations the tour is almost always requested and this is normally extended so that the inspector can review all activities within the aseptic filling area (EU GMP Grades A and B/ISO 14644 class 5 and 7). The request to observe the set-up of the filling machine is a common request, in addition to an overview of an actual batch fill. The training of the aseptic operators plays here an important role and the way that staff stand and behave will be observed.

8.3.9   Training

Trained people are very important in pharmaceutical manufacturing, therefore the check of training documents is an integral part of each inspection. The training records of key personnel, especially the operators involved with product processing, should be reviewed carefully in advance of the inspection.

It may be necessary to organise additional GMP training for staff prior to the inspection. Furthermore, the entire site should be trained prior to the inspection on inspection management. The "does" and "don'ts" should be discussed, especially in reminding staff to only answer the questions asked by an inspector and then only answering if they are 100% sure of the accuracy of the answer. Correcting a wrong answer is very difficult later; it also takes a lot of time and can lead to a reduced level of trust. It is absolutely not a problem if an answer cannot be given immediately.

It is also useful to run through interview techniques with staff. A number of inspection failures have less to do with actual facility management and more to do with simple misunderstandings. Many companies have very complex systems which need to be communicated to an inspector. Staff must be able to explain everything in basic terms; for this task, the use of flow charts and summaries to demonstrate systems is useful.

An example of inspection interview best practice is:

- It is recommended that the interviewee requests appropriate management to be present throughout the interview.

- The interviewee assumes a friendly, cooperative, confident and professional attitude.

- The interviewee does not volunteer information.

- The interviewee does not guess, lie, deny the obvious, make misleading statements or engage in unconstructive arguments.

- The interviewee responds in a concise, factual and accurate manner when the inspector asks a relevant question.

- If the interviewee decides that the question is outside their area of expertise or authority or outside the scope of the Inspector's authority they should consult with their management representative.

- If an interviewee does not understand the question and/or the context they should ask the inspector for clarification.

- If the interviewee realises they have provided erroneous information they should take immediate corrective action when appropriate and have such an action noted by the inspector.

- The interviewee does not solicit opinions from the Inspector.

- The interviewee does not attempt to answer "what if" questions and other hypothetical questions.

- The interviewee does not contradict something said by a colleague. If necessary, the interviewee leaves the room, confirms the correct answer, and then corrects the response with the Inspector.

8.3.10   Data control

Data control is perennially an important topic in inspections. The company needs to assure the inspector that key laboratory instruments and production equipment is password secured and that audit trail is given. Thus electronic data management requirements must be in place[28]. Here, guidance from Good Automated Manufacturing Practice (GAMP) can prove invaluable. GAMP forms a framework of guidance for any GMP computerised system, and includes information on how to ensure such systems are controlled and fit for purpose.

An important set of records which will normally be subject to audit are batch records. These documents are typically used and completed by the manufacturing department. Batch records provide step-by-step instructions for production related tasks and activities as well as include areas on the batch record itself for documenting such tasks.

With batch records, these should be[29]:

- Legible and clear,
- Dated,
- Readily identifiable and retrievable,
- Carry authorisation status,
- Retained for a designated period,
- Protected from damage and deterioration while storage.

Within laboratories, an important data capture system like LIMS (Laboratory Information Management System) will be subject to considerable scrutiny given the criticality of the data for batch release.

Further in terms of control, it is important that both manufacturing and laboratory areas have restricted access in place.

8.3.11   Reconciliation and Traceability

Reconciliation and traceability needs to be assured on production steps. All components used for a specific batch have to be traceable back to their origin, for pieces of product and for labels a proper reconciliation process has to be in place across the entire manufacturing process. If equipment is dedicated to manufacturing one intermediate or API, then individual equipment records are not necessary if batches of the intermediate or API follow in traceable sequence. In cases where dedicated equipment is employed, the records of cleaning, maintenance, and use can be part of the batch record or maintained separately. For this, the inspector will ask for records and certificates pertaining to material control. It is prudent to check these records in advance.

Furthermore, it is important to check that all equipment calibrations should be performed using standards traceable to certified standards (where applicable).

8.3.12   Action plan

All of the above pre-inspection issues should be rolled into an action plan. It is best practice is to have a pre-defined inspection preparation action plan at the ready, rather than assembling and executing one in an ad hoc manner under a tight timeline in an environment suddenly fraught with the tension and anxiety inevitably accompanying the announcement of an impending inspection.

An action plan should have targets, time-points and identifiable responsibilities for each department and team members.

8.4   The inspection

8.4.1   Management Issues

The compliance manager or the QA manager should lead the inspection. This person is the primary contact person and should be in place for the duration of the inspection. To support the QA lead there should be a team of writers (or scribes), who are tasked with taking notes during the inspection. There should be additionally some personnel who act as runners. The role of these staff is to organise each of the requests and to ensure that the information reaches the inspection room in a timely fashion. General rules for runners include:

- The Runner notes the document requested on a log, obscures any confidential information e.g. financial information and makes duplicate copies of all documents requested.

- The runner provides a copy to the Department Manager or delegate for management review.

- The Runner stamps "confidential" on each sheet of the Inspector's photocopy.

- A duplicate set of all documents given to the Inspectors should be maintained.

- At the conclusion of the inspection the documentation may be retained by the inspectors.

The venue or inspection room needs to be carefully selected and prepared. In addition an office space with phone and e-mail account should be offered to the inspectors.

If more than one inspector is coming it is possible that they will elect to split into sub-groups and therefore the company being inspected needs to prepare two independent teams. More than two groups are difficult to manage and therefore this case should be avoided.

8.4.2 Information gathering

During the inspection it is important for the QA escorts to gather as much information as possible so that inspectorate requests can be met and so that there is a clear understanding of the issues during the inspection. There should be note takers following the inspector at all times, and copious notes should be recorded. In addition, a list of all information requests from the inspector should be recorded including each document asked for by the inspector.

8.4.3 Wrap up meetings

A daily wrap up should be agreed with the inspectors to discuss potential findings, outstanding issues and to prepare for the next day. It is also recommended to make at the beginning of the inspection a proposal for a draft agenda.

For international inspections a translation service should be implemented on site and the key SOPs should be already translated, such as deviation handling, OOS, complaints, recall, batch record review, validation and qualification. The use of an interpreter is important for sometimes there are misunderstandings caused because a proper English word is not used or because the question is not clearly understood. It can also be useful to have an English translation of critical documents and standard operating procedures (SOPs) and to have key terms translated into the applicable language, such as "change control", "deviation investigations", and "validation reports".

In addition to the wrap up meeting with the inspectors an internal wrap up meeting on site has to be established with the different functions on site to discuss the necessary preparation work and to define a mitigation strategy for potential findings.

8.4.5   Inspection management

There are a number of social and psychological aspects to an inspection which need to be considered. These include:

- The inspection is key priority for the site; all key people should be available at any time. The requested documents should be presented on time.

- The inspectors have to get the feeling that they are important.

- Each inspector is different therefore the internal inspection management team has to be flexible. Proposals and offers can be made but the inspector is defining the way forward.

To fulfil all these requirements it is useful to establish a "war room", where one QA person is in place together with the subject matter experts to prepare and review the requested documents prior to going into the inspection room.

The documents presented are normally presented as a controlled copy with a stamp. All presented documents are listed in a log book for traceability reasons. The presenters will be coached by the "war room team" before they are going into the inspection room. The location of the war room should be close to the inspection. A professional environment should be generated by a constant document flow and competent presenters.

There are some other important things to consider for the inspection. These include:

- Book meeting rooms in advance for opening and close out meeting, and try to keep one room available for the duration of the inspection for the Inspector's interviews and document review sessions.

- The document review room should preferably be one that can be locked during the course of the day if the inspectors are out; this also saves moving confidential information when the room is unoccupied. It is also useful to have a bin in this room.

- Order lunch and adequate refreshments during the day for the Inspectors if possible. This enables them to 'work through' facilitating their review of the documentation provided for the inspection. It will be important to check if any of the inspectors have any special dietary requirements.

- Arrange briefing and debriefing of teams at the end of each day as required.

- It is important not to underestimate the resource implications during pre-inspection, inspection, and post-inspection periods.

8.4.6   Opening presentation

The site management team is present at the opening meeting and the site head is normally giving a short introduction presentation with the following content:

- Location,
- Organisation
- Products,
- Inspection history,
- Facility overview,
- Technologies,
- Quality system overview,
- Performed or planned projects,
- Changes to the last inspection.

At the opening meeting there should be

- Flexibility on both sides,
- Open dialogue from the beginning,

- On-going verbal feedback throughout the inspection,

- Opportunity to demonstrate how your systems meet the legislation,

- Review of action plans already in place to address known areas of non-compliance.

These types of behaviours allows for a productive and conducive inspection.

8.4.7   Inspection behaviours

It is important that the people involved in the inspection give a professional and competent impression. If an investigator has a finding or an issue it is important that the site people do not argue or appear to defend this issue too much. The strategy to discuss again on a topic should be defined in the daily internal wrap up meeting.

Each inspector has to be managed in the best way, during the inspection the internal team will find out which persons fit best. This depends from the individual character of the inspectors.

8.4.8   End of the inspection

Normally each inspection starts smooth and gets more stressful at the end. It has to be assured that all requests are fulfilled during the inspection. Time control has to be established if this not done by the inspectors (sometimes the inspector will dictate the agenda and timescale; sometimes the inspector will let the company organise the activities).

8.4.9   Close out meeting

The close out meeting is very important. Normally there should be no surprises if the daily wrap up meeting with the inspectors were performed successfully. The inspectors will present their findings and give their impression. The management team should be at the final meeting, the group of people attending the meeting should not be too big.

At the close out meeting there should be no major discussions on the individual findings. It is also important to discuss the way forward, e.g. the timelines and the address where the official response to the findings has to be sent. Depending upon the type of inspection a letter may be presented to the company at the end of the meeting or it is sent on later.

8.4.10   Common inspectorate findings

The types of issue raised during international regulatory inspection vary. Listed below are some common findings taken from a review of several regulatory inspections. In genera the examples relate to quality systems. Other chapters in this book cover areas pertaining to sterile and non-sterile production activities and to laboratory testing.

a)   Organisation oversight of clinical trials

- Over delegation of responsibilities to CI/PI without ensuring appropriate training/expertise in delegated areas
- Lack of systems to appropriate identify trials that fall under the legislation
- Poorly documented/incorrect sponsorship/Legal Representative arrangements
- Lack of R&D approval
- Failure of R&D systems to ensure awareness of all clinical trials
- Approval & oversight of subcontractors
- Failure to obtain MHRA and REC approvals at all, and also for substantial amendments, including one trial conducted that had received a grounds for non-acceptance letter from the regulatory authority.
- Failure to address remarks/conditions on the CTA issued by MHRA
- Failure to issue End of Trial notifications.
- Little or no oversight of pharmacovigilance requirements when delegated to the principal investigator
- Ensuring staff were trained in appropriate legislation
- Failure to report serious breaches of trial protocol to the regulator.
- Lack of oversight and control when undertaking role of co-sponsor
- No document control
- No system for informing researchers of updates to policies, training, systems and legislation
- Not including non-commercial hosted trials in organisation oversight procedures
- Failure to comply with protocol/GCP compliance
- Failure to report serious breaches

b) Pharmacovigilance

- Inadequate pharmacovigilance systems and/or inadequate use of systems in place
- Lack of involvement of Principal or Chief Investigator
- Lack of awareness of, and compliance with, legislative requirements (7 and 15 day reports)
- Failure to distinguish AEs and ADRs
- Failure to identify 'Serious events'
- Failure to consider event expectedness, and hence to identify events which require IMMEDIATE reporting
- Failure to monitor pregnancy to outcome
- Failure to monitor increased severity or frequency through trend analysis
- Failure to notify R&D/Sponsor key issues
- Failure to comply with legislative requirements
- Lack of/ineffective systems to comply with part 5 of the legislation
- Incorrect/outdated reference document for expectedness assessment
- Failure to submit Annual Safety Reports
- Robust documentation, data basing and follow up of SAEs
- Inadequate safety reporting
- Failed response from emergency / out of office telephone numbers when tested

c) Investigational Medicinal Product

- Missing or unsigned documentation (e.g. shipping records, accountability, dosing records)
- Inadequate provisions for storage of IMPs i.e. not kept separate to usual clinical supplies
- Emergency codes not supplied concurrent with supplies or prior to study start
- Insufficient records for the chain of custody (from purchase to destruction) for marketed products used in clinical trials
- Evidence that formal procedure/systems were not in place or were weak:

- Regulatory Green Light, in particular for multicentre trials, but own trials also affected.

- Lack of/inadequate certification.

- Importation and Manufacturing IMP without the appropriate Licence from Competent Authority

- Use of expired/recalled/non GMP manufactured IMP

- Poor/ineffective blinding system

- Uncontrolled site to site transfer

d) Contract management

- Omissions, errors and discrepancies in contracts

- Responsibilities of collaborating parties not clearly defined

- Unclear ownership of documents and data

- Lack of consistency between protocol and contract

e) Quality systems

- Lack of essential SOPs

- Uncontrolled documents used in place of SOPs

- Insufficient review of SOPs / Protocol to ensure adequate reflection of current practice or current legislation

- Insufficient time between issuing and implementing SOPs, leading to training issues

- Meetings and decisions not documented

- In-process checks not documented

- Internal audit programmes inadequate

- Version control logs

- Failure to adequately document trial activities

- Insufficient and poor data recording

- Failure to retain essential documents

- Informed consent

f) Records retention and management

- Issues relating to record management outside the control of the Central Records Department.

- Facilities and offices used to temporarily store Medical Records of trial subjects, and trial-related documents, e.g. consent forms and CRFs, not sufficiently secure

- Tracing system may be inadequate for all records required to reconstruct clinical trial historically

- Inadequate retention period in radiology

- Inadequate retention of evidence of validation for alternative media used to store records

- Inadequate retention of QA and QC data in laboratories

- Inadequate retention of raw source data with implementation of electronic archiving

8.4.11 Information flow

Before and after the inspection all employees should be informed around the inspection. The content of the message before the inspection should include the following:

- Who is coming and why

- Correct behaviour is requested

- Housekeeping is important

- Contact persons for the inspection

An example of a two stage approach to staff communication prior to the inspection is illustrated below:

STAGE 1 – raising awareness (from January, before inspection dates are known)	
Medium	**Activity**
Staff Newsletter Briefing	• January – start the awareness by publishing an item anticipating the FDA inspection sometime in 2013. • Monthly – updates on FDA/compliance related activities, e.g. departmental preparation updates, ERR, rejects, GMP training, successful inspection essential for company success etc.
Company Intranet	• Set-up an FDA mini site to include background information. • Update the FDA related 'frequently asked questions' on the mini site and advertise the process of raising queries. • Include scan of establishment certificate granted by FDA. • Post regular news items referring to the FDA to maintain level of anticipation/awareness.
Posters	• Deploy FDA inspection readiness posters

STAGE 2 – once inspection dates are known (6 weeks or less before the inspection)	
Medium	**Activity**
Staff Newsletter	• Post inspection announcement with dates, how individuals can prepare, 'hot topics' etc.
Company Intranet	• Post inspection announcement with dates etc. • Update FDA mini site with specific inspection information, how individuals can prepare, relevant warning letters etc.
Posters	• Change from FDA inspection prep posters if necessary to product/patient posters
Briefings	• Face-to-face inspection briefings for relevant staff
Email	• Targeted emails, e.g. to system owners/administrators, managers re preparing staff and areas etc.

Table 2: Table displaying internal inspection communication strategy

The content of the messages should include:

- Outcome of the inspection
- Main findings
- Way forward

8.4.12   Potential language issues

The normal inspection language is English; therefore the presenters should talk in English. If a technical expert is not conversant in the foreign language a translator or the QA lead should do the translation. It is dangerous if a presenter talks in English and is not so familiar with the foreign language. This can create a lot of misunderstandings.

For other international inspections a translator should be on site who understands the language of the inspectors.

8.5   Post-inspection activities

8.5.1   Response

It is important that a team is formed directly after the inspection to prepare the response. Each finding should be addressed and the resolution of the issue should be described. Committed timelines should be aggressive but manageable.

In case of requested major inspection findings a team should be set up to work out proposals. The role of QA is pivotal in co-ordinating inspection responses. QA should be involved in ensuring that any information requested during the inspection but that could not be provided at the time is sent to the inspectors as soon as possible. Once an inspection report is received, QA experience is often employed to review and interpret the inspection report. QA are typically involved in the coordination and review of the responses to ensure they are provided in a timely fashion and that they fully address the concerns of the inspectors. Furthermore, QA can often utilise the inspection experience where lessons learnt during the inspection can be shared with, not only those directly involved in the actual inspection, but also with those other parts of the business that may not have been involved but who are also liable to regulatory inspection.

8.6   Conclusion

Good manufacturing practice (GMP) is a production and testing practice that helps to ensure a quality product. Many countries have legislated that pharmaceutical and medical device companies must follow GMP procedures, and have created their own GMP guidelines that correspond with their legislation. For m this global raft of GMP initiatives there are some similarities and differences; this has been complicated by some regulatory authorities working together on some areas and not on others. This has led to a range of different inspection focus points and legislation and guidelines which the pharmaceutical manufacturer needs to grapple with, especially when facing an international inspection.

Although there are differences concerning what is inspected, there are some common areas which this chapter has drawn out. These include:

- Manufacturing processes are clearly defined and controlled. All critical processes are validated to ensure consistency and compliance with specifications.

- Manufacturing processes are controlled, and any changes to the process are evaluated. Changes that have an impact on the quality of the drug are validated as necessary.

- Instructions and procedures are written in clear and unambiguous language. (Good Documentation Practices)

- Operators are trained to carry out and document procedures.

- Records are made, manually or by instruments, during manufacture that demonstrate that all the steps required by the defined procedures and instructions were in fact taken and that the quantity and quality of the drug was as expected. Deviations are investigated and documented.

- Records of manufacture (including distribution) that enable the complete history of a batch to be traced are retained in a comprehensible and accessible form.

- The distribution of the drugs minimises any risk to their quality.

- A system is available for recalling any batch of drug from sale or supply.

- Complaints about marketed drugs are examined, the causes of quality defects are investigated, and appropriate measures are taken with respect to the defective drugs and to prevent recurrence.

In discussing the above areas of commonality and in capturing some of the key differences, this chapter has presented an overview of the key points relating to an international inspection. The intention of the chapter was not to present a detailed point-by-point guide on how to operate an inspection, for this is covered elsewhere in this book. What the chapter has sought to do is to outline the key issues relating to an inspection of a company by an international inspector and to present some best practice guidance for dealing with the inspection. With this, many of the ideas and potential pitfalls identified in this chapter will be of interest to those facing international inspections.

8.7   References

1   Sharp, J. (2000): Quality in the Manufacture of Medicines and Other Healthcare, Pharmaceutical Press

2   Dukes G. (1985). The effects of drug regulation: a survey based on the European studies of drug regulation. Lancaster, MTP Press Ltd.

3   International Conference on Harmonisation of Technical Requirements for Registration of Pharmaceuticals for Human Use, ICH Harmonised Tripartite Guideline, Q9: Quality Risk Management, Version 4, 2005

4   Friedman, R. L. and Mahoney, S.C. (2003): 'Pharmaceutical cGMPs for the 21st Century: A Risk-Based Approach' in Friedman, R. L. and Mahoney, S.C (Eds.): Risk Factors in Aseptic Processing, Food and Drug Administration, Centre for Drug Evaluation and Research, at: http://www.americanpharmaceuticalreview.com/past_articles/1_APR_Spring_2003/Friedman_article.htm)

5   Neckers K. Harmonization of drug and medical device development in the US and Japan: movement towards international cooperation in the postgenomic Era. New England International and Comparative Law Annual 2006;12:65-96

6   International Conference on Harmonisation of Technical Requirements for Registration of Pharmaceuticals for Human Use, ICH Harmonised Tripartite Guideline, Q10: Pharmaceutical Quality System, 2008

7   FDA Inspection Guides, U.S. FDA, www.fda.gov/ICECI/Inspections/ Inspection Guides/default.htm, accessed on 28th May 2012

8   Abraham, J. and Lewis, G. (2000) Regulating Medicines in Europe: Competition, Expertise and Public Health, London: Routledge, UK.

9   Gardner, J.S. (1996) 'The European Agency for the Evaluation of Medicines and European regulation of pharmaceuticals', European Law Journal 2(1): 48–82

10   Ratanawijitrasin S, Wondemagegnehu E. (2010). Effective drug regulation. A multi-country study. Geneva: World Health Organization

11   World Health Organization (1999). Effective drug regulation: what can countries do? (discussion paper). Geneva, WHO Essential Drugs and Medicines Programme, (Document WHO/HTP/EDM/MAC(11)/99.6).

12   McAuslane N, Anderson C, Walker S. (2004). The Changing Regulatory Environment: Reality and Perception. Epsom, UK: CMR International; R+D Briefing 42.

13   WHO (1992). Good manufacturing practices for pharmaceutical products. In: WHO Expert Committee on Specifications for Pharmaceutical Preparations. Thirty-second report. Geneva, World Health Organization, 1992 (WHO Technical Report Series, No. 823, Annex 1).

14   Friedman MA et al. (1999). The safety of newly approved medicines: do recent market removals mean there is a problem? Journal of the American Medical Association, 218(18):1728-34.

15   Cornips C, Rago L, Azatyan S and Laing R. (2010). Medicines regulatory authority websites: review of progress made since 2001. International Journal of risk safety and medicine, (22). 77-78

16   Euradlex. The Rules Governing Medicinal Products in the European Community, Annex 1, published by the European Commission

17   US Food and Drug Administration Pharmaceutical GMPs for the 21st Century – A Risk Based Approach. Final Report, Rockville, MD. September 2004

18   Code of Federal Regulations, Title 21, Food and Drugs, Part 820, Quality Management Regulations, 820.75

19   Hargreaves, P. (2007): 'Good Manufacturing Practice in the Control of Contamination' in Deyner, S.P. and Baird, R.M. (eds.) Guide to Microbiological Control in Pharmaceuticals and Medical Devices, 2nd edition, CRC Press, Boca Raton, pp121-142.

20   Hynes, M. (Ed.) (1999). Preparing for FDA Preapproval Inspections, Marcel Dekker, Inc., USA.

21   FDA (1999). FDA Guide to Inspections of Quality Systems, August 1999, U.S. FDA, http://www.fda.gov/downloads/ICECI/ Inspections/UCM142981.pdf,accessed on 28th May 2011

22   Marwah R, Van de Voorde K, Parchman J (2010) Good clinical practice regulatory inspections: Lessons for Indian investigator sites. Perspect Clin Res 1:151-155.

23   Claycamp, H.G. (2012). Probability Concepts in Quality Risk Management, PDA Journal of Pharmaceutical Science and Technology, January/February 2012 vol. 66 no. 1 78-89

24   WHO good manufacturing practices: main principles for pharmaceutical products. In: Quality assurance of pharmaceuticals. A compendium of guidelines and related materials. Volume 2, Second updated edition. Good manufacturing practices and inspection. Geneva, World Health Organization, 2007

25   Seidl C, O'Connell M, Delaney F, McMillan Douglas A, Gorham M, van Krimpen P, Letowska M, Sobaga L, de Wit J, Erhard Seifried E. (2007). European best practice in blood transfusion: Improvement of quality related processes in blood establishments. ISBT Science Series, Vox Sanguinis, Volume 2 (1); 143-9.

26   Deeks, T. (2001): 'Microbiological Validation Master Plan' in Prince, R. (ed.) Microbiology in Pharmaceutical Manufacturing, PDA, USA, pp267-304

27   Grazel, J.G. and Lee, J.Y. (2008). "Product Annual/Quality Review," Pharmaceutical Technology, March 2008 (p88-104)

28   Bolton, F.J. and Howe, S.A. (1999): Documentation in the Laboratory' in Snell, J.J.S., Brown, D.F.J. and Roberts, C. Quality Assurance: Principles and Practice in the Microbiology Laboratory, pp19-28.

29   Sandle, T. 'Best practices for microbiological documentation' in Hodges, N and Hanlon, G. (Eds.) (2009 rev.): Industrial Pharmaceutical Microbiology Standards and Controls, Euromed Communications, England, Supplement 9, S9.1 – S9.24

a   Formerly the European Agency for the Evaluation of Medicinal Products (EMEA).

b   Formerly the Committee for Proprietary Medicinal Products (CPMP).

CHAPTER 9

Handling and responding to post inspectional observations

TIM SANDLE, BIO PRODUCTS LABORATORY, UK
MADHU RAJU SAGHEE, MICRO LABS, INDIA
DAVID BARR, ALKERMES INC, USA

9.1   Introduction

Regulatory agencies undertake inspections of pharmaceutical manufacturers under the requirement of Good Manufacturing Practice (GMP). Several chapters in this book have outlined the procedures involved during the inspection and have provided some useful techniques to deploy during the inspection and have highlighted many of the areas likely to be inspected. This chapter focuses on the aftermath: what happens when the inspection is over and how the observations or warnings from the regulatory agency should be handled.

Inspections are undertaken by regulatory agencies to ensure that pharmaceutical drug products are manufactured in ways which render them safe and efficacious. This applies to any substance or product for human or veterinary use is intended to modify or explore physiological states for the benefit of the recipient.

At the end of an inspection, at the closing meeting, the inspector will report verbally on the findings. Depending on the inspectorate body, a written response will be provided or sent through later. Here, the US Food and Drug Administration (FDA) communicates inspection findings to company officials by presenting the documented Form 483 at the conclusion of the inspection; whereas, EU regulatory agencies, which are co-ordinated by the European Medicines Inspectorate (EMA), like the UK Medicines and Healthcare products Regulatory Agency (MHRA) provide the company with a verbal summary of the inspection findings and later describe the arrangements for the formal notification of the deficiencies to the company. The verbal summary does, however, provide an indication of how the inspection went and of the key issues which need to be responded to.

There has been an international movement towards a global inspectorate, where FDA and EMA would work more closely together through the Pharmaceutical Inspection Convention Scheme (PIC/S), although this remains at the early stages at the time of writing this chapter[1].

Irrespective of the approach, all regulatory inspections will conclude notifying the top management of the company the objectionable findings if any are observed during the course of the inspection. The company is expected to respond to these observations promptly with a detailed corrective action plan.

The post-inspection letter is sent to the company by the regulatory agency listing any objectionable conditions, recommendations and deficiencies noted during the inspection.

This chapter provides guidance on how to respond to the inspection letter and some tips for crafting a thoughtful and effective response to the post inspectional observations. For this, timeliness is the key, as in the case of FDA, the company is required to respond promptly within 15 days after a letter has been issued (in this case, the letter 483), if they wish their comments to be taken into consideration when the FDA is deciding whether or not to issue a Warning Letter. The chapter provides a pragmatic framework for responding to regulatory inspectional observations, mainly to form 483s and Warning Letters sent to firms by FDA. The advice is, however, of equal value for how to respond to the post-inspectional letters of EU regulatory agencies like MHRA. Similar guidance applies and the guidance is relevant to any other regulatory agency in analysing the observations and formulating an effective response.

9.2   FDA inspections

The Food and Drug Administration (FDA) is a regulatory, law enforcement agency whose primary mission is to assure the protection of consumers by regulating the products which by law are under the agency's jurisdiction. A primary function of FDA regulation activities includes conducting facility inspections to assure compliance with applicable Good Manufacturing Practices (GMP) regulations[2]. The FDA is authorised to perform inspections under the Federal Food, Drug, and Cosmetic Act, SEC. 704 (21 USC §374) "Factory Inspection"[3]. Outside of special inspections (such as 'for cause' inspections), the FDA routinely inspects a manufacturer's facilities for cGMP compliance every two years. This applies to domestic and foreign facilities which manufacture drugs for sale within the United States[4].

FDA inspections take place for any company operating with the USA and for companies which supply drug products within US territories.[5] The scope of the FDA inspection reviews "current" GMP (cGMP). cGMPs require manufacturers to have adequately equipped manufacturing facilities, adequately trained personnel, precisely controlled manufacturing processes, appropriate laboratory controls, complete and accurate records and reports, appropriate finished product examination, etc.[6]. The approach is described in the cGMP initiative

"Pharmaceutical cGMPs for the 21st Century: A Risk – Based Approach"[7], which is discussed elsewhere in this book.

The stated objective of FDA inspections is "to minimise consumers exposure to adulterated products"[8].

9.2.1 Handling FDA inspectional observations

FDA inspections usually result in some direct communication between the regulators and the company that is being inspected. This is particularly so when a company must respond to the inspectional findings presented on the FDA 483. It is important to remember that the inspection and any resulting FDA 483 is not the end of a company's dealing with the FDA. How these communications are handled can have an impact on the results. It can mean the difference between the agency taking an administrative or regulatory action or not. Additionally, the context, frequency and substance of communication between the FDA and the manufacturing sponsor can greatly affect the granting a post-approval supplement and/or a product approval.

9.2.2 FDA observations during the inspection

If during the course of an inspection significant problems are found, the investigator will issue a list of observations, the FDA 483. Form FDA 483, titled "Inspectional Observations," is a form used by the FDA to document and communicate concerns discovered during these inspections. Form FDA 483 was created in 1953 by addition of Section 704(b) to the Food Drugs and Cosmetics Act, and is therefore a long-established inspection mechanism.

The FDA 483 will be presented at the conclusion of the inspection. The FDA 483 is a tool by which the FDA investigator provides to the inspected company's management a list of the investigator's observations at the conclusion of the inspection. The 483 observations, in the investigator's judgment, are items that require some corrective action. They are not necessarily violations of law or regulations, but in all likelihood they are deviations (or violations) of GMPs. The investigator's 483 observations receive no prior review by other FDA officials. They are the investigator's own observations of deviations from the requirements in the investigator's best judgment. The Agency's outstanding instructions to its investigators are to only list items that are significant.

During the course of the inspection, investigators will ask questions, review documents and observe the company's operations. It is very important to be proactive in dealing with the investigator's queries. It is critical that the investigators are given correct and complete information so that they do not come to erroneous conclusions resulting in incorrect observations on the FDA

483. It is far easier to correct any misunderstandings or an error during the inspection itself than once the FDA 483 has been written and presented.

FDA Investigators should be encouraged to discuss observations during daily sessions (end of day wrap-up meetings), so there is an opportunity to respond to a finding during the course of the inspection. In the event an investigator discloses a legitimate observation during the inspection, the company should initiate a corrective action as appropriate and ideally during the course of the inspection. If a corrective action is initiated, the FDA investigator should be provided with evidence of the corrections. While this may not result in the elimination of the observation on the FDA 483, it does provide strong evidence of the company's willingness to make compliance improvements and learn from its mistakes.

9.2.3   Reportable and non-reportable observations

During an FDA inspection, some of the matters observed by the FDA are noted and reported on the 483, whereas others are comments by the investigator. There is thus a division between reportable and non-reportable observations.

Reportable observations

Reportable observations are items noted during the inspection which are mentioned to firm's management during the inspection and are raised again during the exit interview. These listed observations are normally significant.

If the investigator deems the observations to warrant attention or correction, they are included on the FDA 483. This will include a comment as to whether the the conditions are objectionable in view of their relation to other conditions or controls at the given time and place.

Non-reportable observations

Non-reportable observations are comments from the investigator which do not end up in a written report.

The FDA provides advice for investigators for what falls within the category of non-reportable observations. This includes:

*   Label and labelling content
*   Promotional materials
*   The classification of a cosmetic or device as a drug

- The classification of a drug as a new drug

- Patient names, donor names, etc. If such identification is necessary, use initials, code numbers, record numbers, etc.

- The use of an unsafe food additive or colour additive in a food product

In addition to providing a form FDA 483, FDA investigators prepare an establishment inspection report (EIR), which is sent to FDA headquarters, which then evaluates the report and determines the corrective action. The FDA then classifies the inspection as "no action indicated," "voluntary action indicated," or "official action indicated." The EIR contains much greater detail than contained in the 483 and is not provided to the manufacturer until after the inspection is deemed closed. The EIR is not sent automatically to the company; instead the company needs to request the document. It is recommended that the company obtains this document as it will prove invaluable in preparing for the next inspection.

9.2.4 The FDA 483 presentation meeting

When an FDA 483 is received, this is the organisation's first opportunity to respond to the FDA in a formal way. The investigator will provide the FDA 483 during a formal meeting at the closure of the inspection. At this time the company's representatives should strive to discuss each observation to assure a mutual and clear understanding of the observations. It is important to be certain that there is a clear understanding of what each observation means[9].

Incorrect items may appear on the FDA 483 because the investigator misunderstood or was provided incorrect or misleading information. If this occurs, the company should provide as much evidence and documentation as possible to correct the investigator's misunderstanding. If it can be demonstrated that an item is incorrect and that there is no significant issue, the investigator should note this on the FDA 483. Thus, most FDA 483 items should not be a surprise at the time the list is issued.

Additionally, during the discussion of the observations the company should point out, for the record, corrective action(s) that have already been taken. If it is possible to commit to corrective actions not yet taken, it is prudent to do so. However, it is important to make sure that the commitments of action and time frames are reasonable to accomplish as the FDA will follow-up on such commitments. All of this discussion will be included in the investigator's Establishment Inspection Report (EIR) which becomes part of the agency's official records.

Given that all 483s are releasable under the US Freedom of Information Act (and as set out in the Code of Federal Regulations 21 CFR 20.101(a)), it is important that the company is satisfied with the information and contents for, after the meeting, the contents enter the public domain.

9.2.5   Anatomy of a 483

The content of a 483 may be handwritten, typed, completed in a PDF file and printed. The 483 has the following standard layout[10]:

- **Header information**: The header identifies the FDA district office that performed the inspection, the date(s) of inspection, name and address of the facility that was inspected, the name and title of the individual to whom the 483 is issued (usually the most responsible individual physically present in the facility), a brief description of the type of facility, and the facility's FDA Establishment Identification number.

- **Observations**: This section starts with a "disclaimer" that the form contains the observations of the investigator and does not necessarily "represent a final Agency determination regarding your compliance." Observations placed on a 483 are the opinion of the FDA investigator and may be subject to review by other FDA personnel.

    The full text is as follows:

    "This document lists observations made by the FDA representative(s) during the inspection of your facility. They are inspectional observations, and do not represent a final Agency determination regarding your compliance. If you have an objection regarding an observation, or have implemented, or plan to implement, corrective action in response to an observation, you may discuss the objection or action with the FDA representative(s) during the inspection or submit this information to FDA at the address above. If you have any questions, please contact FDA at the phone number and address above.

    "The observations noted in this Form FDA 483 [sic] are not an exhaustive listing of objectionable conditions. Under the law, your firm is responsible for conducting internal self-audits to identify and correct any and all violations of the quality system requirements."

The 483 then has a large space for recording the observations, which may be continued on several pages. The observations should be ranked in order of significance. If an observation made during a prior inspection has not been corrected or is a recurring observation, that may be noted on the 483.

The FDA will typically include only significant observations that can be directly linked to a violation of regulations — not suggestions, guidance, or other comments. ("Significant" is somewhat arbitrary and may be subject to the bias of a particular investigator). Observations of questionable significance are not usually recorded on the 483, but are discussed with the company so that the inspected company understands how uncorrected problems could become a violation. The 483 will not normally include actual regulatory references.

- **Annotation**: For medical device inspections, the investigators annotate the 483 with one or more of the following:

  - Reported corrected, not verified

  - Corrected and verified

  - Promised to correct (may be appended with "by xxx date" or "within xxxx days or months")

  - Under consideration

  The actual annotation of the 483 occurs during the final discussion with the firm's management; if the firm prefers no annotation, then annotation will not be performed. The annotations may be after each observation, at the end of each page, or at the bottom of the last page prior to the investigator's signature(s).

- **Signatures**: The investigators' names are printed and signed, and the date of issue is recorded in this section. Titles for the investigators may also be included. If the 483 is multiple pages, the first and last pages have full signatures while the intervening pages are only initialed.

- **Converse side**: The converse side of the form has this text:

  "The observations of objectional conditions and practices listed on the front of this form are reported:

  o *Pursuant to Section 704(b) of the Federal Food, Drug and Cosmetic Act, or*

  o *To assist firms inspected in complying with the Acts and Regulations enforced by the Food and Drug Administration.*

"Section 704(b) of the Federal Food, Drug, and Cosmetic Act (21 USC374(b)) provides: 'Upon completion of any such inspection of a factory, warehouse, consulting laboratory, or other establishment, and prior to leaving the premises, the officer or employee making the inspection shall give to the owner, operator, or agent in charge a report in writing setting forth any conditions or practices observed by him which, in his judgement, indicate that any food, drug, device, cosmetic in such establishment (1) consists in whole or in part of any filthy, putrid, or decomposed substance or (2) has been prepared, packed, or held under insanitary conditions whereby it may have become contaminated with filth, or whereby it may have been rendered injurious to health. A copy of such report shall be sent promptly to the Secretary.' "*

- **Addenda/amendments**: It is possible that an error is discovered by the investigator(s) after issuing the 483. If the 483 was generated via Turbo EIR, then an amendment is created within that system. Otherwise, an addendum is created. If possible, the investigator(s) will personally deliver the addendum/amendment to the firm.[15]

    Addenda/amendments are not normally used for adding observations to a 483 after the inspection has been closed out and the investigator(s) have left the premises

The 483 will refer, where applicable, to CFR clauses which relate to the non-compliances cited. The Code of Federal Regulations (CFR) is a compilation of all federal laws published in the Federal Register by the executive departments and agencies of the federal government. This code is divided into some fifty titles which represent broad areas of federal regulation. Each title is further divided into chapters. Parts 210 and 211 of CFR Title 21 are the laws defining good manufacturing practices for finished pharmaceutical products.

Part 210, "Current Good Manufacturing Practice in Manufacturing, Processing, Packing or Holding of Drugs — General, " provides the framework for the regulations, and Part 211, "Current Good Manufacturing Practice for Finished Pharmaceuticals," states the actual requirements. Part 211 is further divided into 11 subsections, which cover the requirements for:

A.   Scope

B.   Organisation and Personnel

C.   Buildings and Facilities

D.   Equipment

E.   Control of Components and Drug Product Containers and Closures

F. Production and Process Controls

G. Packaging and Labelling Control

H. Holding and Distribution

I. Laboratory Controls

J. Records and Reports

K. Returned and Salvaged Drug Products

These cGMP regulations are written to address the primary potential sources of product variability.

9.2.6 Preparing the response to the FDA 483

While a response to a 483 is not compulsory, an informed and well-written response can usually help a company avoid receiving a Warning Letter from the FDA, withholding of product approval, or even plant shut-down. A recipient of a 483 should respond to the FDA, addressing each item, indicating agreement and either providing a timeline for correction or requesting clarification of what the FDA requires.

In the Federal Register of Tuesday, August 11, 2009[11], the FDA established a revised policy concerning written responses to FDA 483 observations. The register states:

> "FDA is initiating a program to establish a timeframe for the submission of such post-inspection responses to FDA 483 inspectional observations for FDA's consideration in deciding whether to issue a warning letter. Under the program (described in more detail later in this document), the agency will not ordinarily delay the issuance of a warning letter in order to review a response to an FDA 483 that is received more than 15 business days after the FDA 483 was issued."

The policy thus gives companies 15 days to respond in writing to the FDA after a 483 is issued, if they wish their comments to be taken into consideration when the FDA is deciding whether or not to issue a Warning Letter.

If an FDA 483 is issued, it is necessary to prepare a written communication for the FDA. This response should be sent as soon as possible as this allows for the response to be considered in determining any follow-up activities by the FDA. Any follow-up actions are generally determined by FDA Managers (not the FDA Investigator) based on the investigator's report and any samples collected.

Before writing the response the company is advised to put together a team to co-ordinate the response. The team should include a leader and key management persons who have knowledge of the issues. Personnel from different disciplines are valuable to gain multiple perspectives. Utilise your experts and include persons who work with and are most familiar with the affected systems. If the team is lacking in a particular area it may be prudent to get an expert in FDA response preparation to assist.

The team should carefully review the FDA 483, and then develop a corrective action plan which covers each point from a systemic view. The systemic approach is important as the company should see if the deficiency can impact other areas.

When examining each of the points, the team should consider the following:

- Understand significance of observations relating to product quality. This is a three-step process involving:

  o **Determining the scope of the problem**. Does the observation identify a problem that is isolated or systemic? Is the problem limited to a certain lot? Does the problem extend to other lots of the same product or to other products? Is the problem contained within a particular department or does it extend to other departments? Is the problem contained within a particular facility or does it extend to other facilities?

  o **Conducting a thorough investigation and take appropriate corrective action**. Far too often, firms respond to 483s by simply defending their actions or relating the history of a problem. The company should thoroughly investigate FDA's concerns and describe the investigation, investigation results and corrective actions in the final response.

  o **Submit a well-written and precise response**. Do not submit a response that is impossible to understand or omits pertinent information. Responses should be factual, complete and easy to read. They should include reasonable timelines, address issues item-by-item and demonstrate an understanding of regulatory requirements. A sloppy or defensive response will not be well received.

- Consider and describe the corrections to be made

- Undertake immediate corrections if possible, otherwise set realistic time frames

- Provide assurance when possible that the quality of distributed product (public safety) is not a concern

- Address all deficiencies; provide a plan of action with target dates (always with the expectation of an FDA follow-up)

- Ensure that "systemic" issues have been addressed

The team may need to formulate a CAPA plan as part of the response. Corrective and Preventative Action (CAPA) is necessary for significant deficiencies. CAPA is a regulatory concept that focuses on investigating, understanding, and correcting discrepancies while attempting to prevent their recurrence. The CAPA process involves consideration of remedial corrections of an identified problem; root cause analysis with corrective action to help understand the cause of the deviation and prevent recurrence of a similar problem; and preventative action to prevent recurrence of similar problems.

A CAPA plan should address problems completely and in a timely manner. A good CAPA plan should assess the root cause of deficiencies; identify the problems; evaluate the extent of problems; give a clear timeline, describe the CAPA being taken; and reassess the root cause.

In your written response it is advised that you:

- Identify the observations that you recognise that are clear deviations from GMPs and your SOPs and prepare a corrective action plan for each. This plan should include realistic time frames for completion.

- Review the deviations from the perspective of what risk this deviation can present to the patient/user of the drug and triage accordingly.

- Review each deviation to see if it could signal a system problem or if it is a onetime occurrence or an isolated event that is unlikely to reoccur again. If there is a possibility of a system issue or likeliness of reoccurrence, your corrective action should reflect the company's action of a systemic correction.

- List any observations that are not believed to be deficiencies. Be sure to include any observations that you have already responded to during the FDA 483 presentation. Document alternate controls you may use to address the observations and consider expert opinions to support your position.

- When you have done the above triaging you now know which items should take priority for corrective action. Sometimes all things cannot be corrected at the same time. Carry out your corrective action plan as you committed to FDA. If you find that you have problems meeting some of your commitments, communicate them with the FDA. Likewise, keep them informed of your progress and let them know when you are finished.

The response letter should cover:

o   Determinations of the root cause(s) of each deviation noted

o   Your corrective actions and/or action plans including how the plan mitigates the probability of future similar problems

o   Any impact of the deviation noted regarding product quality and any follow-up actions if the product quality has been compromised

In preparing your response letter, consider the readers. Having a third party review the letter is generally advisable to ensure that the message is clear and complete. The FDA officials reading your response may not have the benefit of all the background the company has surrounding the issue. Be factual and specific and put your responses in a broad perspective. It is very important not to be defensive. You should anticipate follow-up questions from the FDA. Respond as soon as you can, but at least within one week after the conclusion of the inspection. You may also offer to meet with the district management or request a meeting if you feel that would be desirable[12].

In **Table 1**, deficiencies a schematic is provided which outlines a response check-list which can be used as guidance for responding to the FDA.

Table 1: Generic FDA response check-list

Post-inspection response checklist			
This checklist can help manufacturers as they respond to FDA-483 observations. (Courtesy: Cerulean Associates)			
Instructions: Answer each question as you go through your response to FDA quality system inspections. As appropriate, use the comments section to document logic behind skipping sections or replacing items.			
Status	15-Day Task List	Assigned To	Comments
	Review FDA Form 483 or other summary inspection documents received from investigator		
	Identify issues requiring a response (typically "inspectional observations")		
	Create an observation–response matrix		
	Meet with inspection prep team • Review inspection summary documents • Review observation-response matrix • Review relevant internal inspection summaries from note-takers • Verify everyone understands timelines		
	Assign accountabilities for response components • Timelines and progress to date • Item-by-item resolutions • Verification of resolutions (or steps to take to verify if a long-term resolution is required) • Additional steps complimentary to each resolution (to go beyond just the inspectional observation)		
	Assemble draft response components and verifications • Has each observation been addressed? • Have you done just the minimum or taken additional steps to strengthen area(s) of weakness identified by investigator? • What proof should be enclosed (such as revised SOPs, timelines, validation protocols, etc.)? • Is all proof dated and signed as appropriate? • Include a table of contents • Include a list of appendixes or attachments • Depending on length and format, provide search or indexing capabilities		
	Review draft response and potential verification records with legal counsel		
	Finalise response and verification records, then submit to FDA		
	Other:		

To assist with the formulation of the response, **Table 2**, below, provides a matrix for responding to FDA observations.

Table 2: Guidance matrix for responding to FDA inspection observations.

Observation-response matrix
(Designed for use with the post-inspection response checklist) (Courtesy: Cerulean Associates)

Instructions: This observation-response matrix is designed to help you rapidly close FDA Form 483 observations. Remember to apply these four basic rules:

1. Focus on the "big" context of the observation. In other words, an FDA 483 observation citing missing signatures on an inventory control log has less to do with missing signatures than it has to do with the inventory control and review process; thus, "inventory control and review process(es)" would be the observation to note in column one (missing signatures are just an example of process and control insufficiencies or breakdowns).

2. The accountable owner of the item to be closed/fixed/improved needs to be someone in management (preferably, director level or above). FDA typically does not like to see line workers held accountable (assuming a company is not a start-up that employs fewer than 20 people).

3. There may be multiple items within a response to one observation *(Examples: fix inventory control and review SOP; revise training materials; train personnel; conduct internal audit within 30 days; etc.)*. Anything that will stretch beyond 30 to 45 days should have a plan associated with it (the plan – and progress on it – will be listed in column four as part of the "proof" to be included in the FDA response).

4. Every item used to respond to an observation must have some sort of documentation or other record associated with it *(Examples: the updated inventory control and review SOP; the training records for the revised training; a summary of the internal audit results; etc.)*.

Inspectional Observation	Accountable Owner	Response	Proof (Records)

9.2.7   The 483 response

The section above has presented advice over what to put in the 483 response. This section provides some advice in relation to how to respond to the 483 in terms of procedure and style.

In formulating a response, it is important to have a plan of action. A successful completion of a 483 provides an opportunity to communicate to the agency that the company understands its compliance obligations. It is also important to answer each observation fully and to have understood what each observation means.

In completing a 483, the main areas to focus upon are[13]:

- Complete the response on time and in writing. As indicated above, companies have 15 days to respond, so it is important to ensure that final proofing and substantive editing is done at least by day 10.

- In the first paragraph of the response letter, be explicit about the company's understanding of and desire to comply with FDA regulations.

- Respond individually to each item that was addressed in the Warning Letter. Be specific. Do not try to solve all issues in one paragraph or the response may be rejected, prompting further action from the FDA.

- Strike a conciliatory and contrite tone in your communications, but, be sensitive to admissions of non-compliance.

- Respond in order of importance i.e. respond individually to those items most likely to affect product quality.

- Be detailed yet concise in each response. Outline how each deficiency will be corrected, and when, rather than how the deficiency came to be. Provide documentation of a corrective action commitment from the person responsible.

- Use positive statements; avoid language that implies fault. Address each item in the Form 483 as an opportunity to fine-tune the quality and compliance systems and personnel.

- Include reference to how the company will be forwarding evidence to support the correction. For example, "Company X will use system B validated monitoring and alarming system to provide reports on temperature recordings taken at 10 minute intervals on a monthly basis." Product specifications and protocols of new systems can be provided or offered in support of the corrective action plan.

- Offer a complete, thorough and compelling remediation plan that addresses the discrete observation, but also identifies and remediates the root cause of the deficiency

- If the FDA investigator noted something that you feel was an isolated incident, document this fact and note it in your response. Be sure the data provided is both complete and accurate. Should you find some of the observations were in error after receiving the 483, there is a formal dispute resolution process which can be followed.

- Be proactive. Reassess internal compliance programmes – Why were 483 deficiencies not detected internally? Mention this in the response letter, noting the company's commitment to QC/QA audit management. The definitive guide to what FDA inspectors are looking for (at least in theory) is the agency's Investigations Operations Manual, which can be reviewed.

- If you need clarification, seek it – in writing and from the correct party.

Things to avoid are:

- Never complain about how the FDA requirements are unfair. The FDA is charged with protecting the public health and safety. This will only serve to decrease cooperation.

- Never be late with your response, you must get the response in timely. If the response is late, the Agency will perceive your lack of concern from the timeliness of your responses.

- Do not challenge the FDA's interpretation of the law or regulation unless you have very good reason and have obtained expert counsel. In most cases it is not advisable to challenge the FDA's interpretation of law or regulation, but certainly not without expert advice.

An example of a response to an FDA 483 is:

- FDA 483 states:

  "Instruments 12, 16, and 382, which were in use during the manufacture of Lots 5, 6, and 7 of Product X had exceeded due dates for their next scheduled calibrations" (cited as an observation against 21 CFR 211.68(a)).

- The company should respond by noting:

- Instruments were calibrated and found to be within limits (records attached)

- Usage in manufacture of Product X has no effect on quality

- Calibration program to be reviewed to assure no other such instances
- Review of program along with any needed corrections will be completed in 60 days; documentation will be submitted

A second example is:

- FDA 483 states:

  "The OQ for the UF/DF was insufficient in that flow rates and pressures were not explored for the design range."

- The company should respond by stating:

  "An OQ protocol for the UF/DF skid will be written and subsequently reviewed and approved by QA. The approved protocol will be executed by the validation team and incorporated into a validation report. The completed OQ validation report will be reviewed and approved by QA.

  The OQ Protocol will be completed and approved by Dec 21, 2009. The OQ Validation will be completed by Jan 15, 2010."

In turn, the FDA will review the response. This is an important step as an unsatisfactory response may trigger a Warning Letter. In reviewing the company's response, the FDA will ask the following:

- Are the FDA 483 observations actual violations of the regulations?
- Are there additional violations in the exhibits submitted with the EIR?
- Are the observations documented with exhibits or discussion in the EIR?
- Are the observations significant?
- Did the inspected party address the issue in their response?
- Is the response adequate?

On this basis, it may be a good idea to review the response from the perspective of the person reviewing it to see if the above questions have been adequately answered. If a 483 response is well written, factually correct and complete, it will establish that the firm has taken or is in the process of taking appropriate corrective and preventive action. This can allay FDA concerns about the firm's compliance, and reflect positively on the organisation. If the response is poorly written, contains errors or omits critical information, it can trigger additional inspections or a Warning Letter.

If there is no immediate response from the FDA, it is unwise to assume that the FDA is satisfied with the response. The company should continue to undergo a comprehensive compliance assessment in order to ensure that when the FDA returns their expectations will be met.

9.2.8 Disputes

In relation to point 10 from the list in section 9.2.7 above, the FDA does recognise that disputes will sometimes occur and has a dispute resolution procedure in place. The introduction to this procedure reads[14]:

"Disputes related to scientific and technical issues may arise during FDA inspections of pharmaceutical manufacturers to determine compliance with CGMP requirements or during the Agency's assessment of corrective actions undertaken as a result of such inspections. As these disputes may involve complex judgments and issues that are scientifically or technologically important, it is critical to have procedures in place that will encourage open, prompt discussion of disputes and lead to their resolution. This guidance describes procedures for raising such disputes to the Office of Regulatory Affairs (ORA) and center levels and for requesting review by the Dispute Resolution Panel for Scientific and Technical Issues Related to Pharmaceutical CGMP (DR Panel)."

Dealing with disputes is outside the scope of this chapter and it is something which should be only taken for very serious events and with appropriate legal advice. It is advisable, where possible, that discrepancies be cleared up together with the investigator on site during the inspection and before a Form 483 is handed over.

9.2.9 Warning Letters

Despite the formal communication actions that are taken to address the FDA 483, if serious deviations were observed the company may still receive a Warning Letter from FDA. This is serious since the agency considers the Warning Letter to be the first step in the compliance process. This may lead to administrative actions including import alerts, which can prevent your products from entering the USA, non-approval of NDAs (New Drug Applications); ANDAs (Abbreviated New Drug Applications) and Supplements[15].

A Warning Letter is the means by which the FDA notifies manufacturers about violations that the FDA has documented during its inspections or investigations. A Warning Letter will notify a responsible individual at the company that the FDA considers one or more products, practices, processes, or other activities to be in violation of the cGMPs.

Warning Letters are only issued for violations of regulatory significance, that is, those that may actually lead to an enforcement action if the documented violations are not promptly and adequately corrected (such as not responding appropriately or effectively to a 483 form). A Warning Letter is one of the FDA's principal means of achieving prompt voluntary compliance. An example would be where the FDA considers that the contamination of drugs with toxic chemicals, drug residues, microbiological contaminants has occurred and which would put the health of a patient at risk[16].

One difference between Form 483 and a Warning Letter is that the 483 is from the investigation team alone, whereas the Warning Letter is issued from a higher level FDA official or officials. The Warning Letter indicates that upon official review, serious violations may exist, which means that Warning Letters are issued for violations of "regulatory significance".

An example of a Warning Letter is:

"This is regarding an April 6-9, 2012, inspection ... The inspection revealed significant violations from US current good manufacturing practice (CGMP) in the manufacture of APIs. The CGMP violations were listed on an Inspectional Observations (FDA 483) form issued to you at the close of the inspection. These violations cause the APIs manufactured by your firm to be adulterated within the meaning of Section 50 I(a)(2)(B) of the Federal Food, Drug, and Cosmetic Act (the Act) [21 USC §351(a)(2)(B)]. Section 501 (a)(2)(B) of the Act requires that all drugs, as defined in the Act, be manufactured, processed, packed, and held according to CGMP.

We have received your firm's responses of May 14 and August 12, 2012, and note that they lack sufficient corrective actions.

Specific violations observed during the inspection include, but are not limited, to:

* *Your firm does not assure that suitable processing (b)(4) is used for the (b)(4) step of the Hydralazine HCl manufacturing process. This API is intended for use in parenteral drug products. Your firm currently uses (b)(4) and does not test this (b)(4) for endotoxins and total microbial count. [FDA 483 Observation 8]*

Your written response states that you do not intend to conduct endotoxin testing for (b)(4) or sanitise your (b)(4) system. It is essential that non-sterile APIs intended for use in parenteral drug products are manufactured using (b)(4) that is suitable for the process stage and that routine monitoring is performed to ensure ongoing (b)(4) system control. Our inspection found that your firm uses (b)(4) at the (b)(4) and (b)(4) stage, and failed to test for total microbial count and endotoxins.

Please refer to ICH Q7A Guidance for Industry for guidance regarding quality of active pharmaceutical ingredients intended for use in parenteral drug products.

Please provide us with a corrective action plan for how you will address these concerns."

Warning letters detailing cGMP violations typically conclude with the following:

"The article(s), (DRUG NAME), is (are) adulterated within the meaning of Section 501(a)(2)(B) of the Act, 21 U.S.C. 351(a)(2)(B), in that the methods used in, or the facilities or controls used for, its manufacture, processing, packing, or holding fails to conform to, or is not operated or administered in conformity with, cGMP regulations [21 CFR 210, 211]. "

With the above example, an active pharmaceutical ingredient is any substance or mixture of substances intended to be used in the manufacture of a pharmaceutical dosage form. It is therefore a critical area for purity and contamination control.

If you receive a Warning Letter (or an Untitled Letter) the following actions must be taken:

- Do not ignore the Warning Letter. The FDA is serious when a Warning Letter is sent and they are prepared to take regulatory action if prompt corrections are not made.

- Respond within the time frame stated in the Warning Letter.

- If you had not previously responded, do so immediately. This may be your last opportunity to intervene prior to FDA initiating a legal or administrative action.

- Respond to all issues noted in the Warning Letter.

- Make sure your corrective actions are completed or give good reason for any delay. If the company has not decided how to comply at this point they should explain why and promise a full response by a given date. Complete your corrective actions as soon as possible.

- Provide the FDA with proposed milestone dates from any ongoing corrective actions.

- If there are items with which you don't agree, request a meeting with FDA to explain your position. Present factual and scientific evidence to support your position. Remember, however, that a Warning Letter is reviewed by an FDA agency manager. If you receive a Warning Letter, the agency manager has concluded that the deviations described in the letter are in fact a violation of law or regulation. Attempt to make your position known anyway, but this is not the time to argue (unless you are prepared to go to court).

- Be realistic; do not promise changes which cannot be delivered.

- Avoid repeatedly relying on excuses without providing evidence (e.g.: personnel errors; retraining).

If the response to the Warning Letter is considered adequate, the FDA is likely to conduct a follow-up inspection. If they find things satisfactory you can expect approval of pending applications (if that was the case) and no further regulatory action.

The key throughout this process is communication. Do not assume that FDA knows what you are doing: keep them informed.

9.2.10   Enforcement

In extreme circumstances the FDA will by-pass a Warning Letter and instigate enforcement action. This is for serious violations, such as a history of repeated or continual conduct of a similar or substantially similar nature during which time the company has been notified of a similar or substantially similar violations. The enforcement leads to judicial action by US courts.

9.2.11   Untitled letter

In addition to the Warning Letter, the FDA may issue other correspondence. This communication is referred to as an 'untitled letter'. In many ways a Warning Letter is indistinguishable from a notice of violation (the 'untitled letter'). An untitled letter cites violations that do not meet the threshold of regulatory significance for a Warning Letter, but the FDA has a need nevertheless to communicate. Unlike a Warning Letter, an untitled letter does not include a warning statement that failure to take prompt correction may result in enforcement action and does not evoke a mandated FDA repeat inspection or follow-up. Furthermore, the untitled letter requests (rather than requires) a written response (from the manufacturer) within a reasonable amount of time (for example, the statement "Please respond within 45 days" is common).

9.3   EU regulatory inspections

The production of drug products (medicinal products) in the European Union (EU) is controlled under Directive 2001/83/EC of the European parliament and of the Council, which states that the holder of a manufacturing authorisation for medicinal products is obliged to comply with good manufacturing practices as laid down by European Community law. The principles and guidelines of GMP for medicinal products are stated by the Commission directive 2003/94/EC, which provides the legal basis for GMP in the EU. The actual GMP code with detailed written procedures is published in The Rules Governing Medicinal Products in the European Union, volume 4[17].

EU GMP code is presented in two parts of basic requirements and 18 annexes. Part I, "Basic Requirements for Medicinal Products," covers GMP principles for the manufacture of drug products. It consists of nine chapters covering the requirements for quality management and control, personnel, premises, equipment, documentation, production, contract services, complaints, product recall, and self-inspection.

EU regulatory inspections lead to a letter stating the inspection findings. In many ways this is similar to the letter received from the FDA. European inspectorate bodies are based in national states and have autonomy, although they follow the guidance and regulations established by the EMA (European Medicines Agency). The EMA is a decentralised agency of the European Union, located in London. The Agency is responsible for the scientific evaluation of medicines developed by pharmaceutical companies for use in the European Union.

One of the leading European inspectorate groups is the MHRA (Medicines and Healthcare products Regulatory Agency) which is the UK government agency responsible for ensuring that medicines and medical devices work, and are acceptably safe. The MHRA is an executive agency of the Department of Health.

As with FDA inspections it is useful to have an interim meeting at the close of each day of the inspection, between the inspectors, and members of the quality team. This debriefing session allows both parties to assess progress, discuss unresolved questions, provide outstanding requested information and plan the next day's agenda[18].

The key difference is that inspection findings from bodies like the MHRA are prioritised. Deficiencies are grouped into:

- **Critical deficiency**: This is a deficiency which has produced, or leads to a significant risk of producing either a product which is harmful to the human or veterinary patient or a product which could result in a harmful residue in a food-producing animal.

- **Major deficiency**: This is a deficiency that has produced or may produce a product, which does not comply with its marketing authorisation or which indicates a major deviation from EU Good Manufacturing Practice or (within EU) which indicates a major deviation from the terms of the manufacturing authorisation or which indicates a failure to carry out satisfactory procedures for release of batches or (within EU) a failure of the Qualified Person to fulfil his legal duties or a combination of several 'other' deficiencies, none of which on their own may be major, but which may together represent a major deficiency and should be explained and reported as such.

- **Other deficiency**: This is a deficiency which cannot be classified as either critical or major, but which indicates a departure from good manufacturing practice. (A deficiency may be "other" either because it is judged as minor or because there is insufficient information to classify it as major or critical).

With the process, several related major or other deficiencies may be taken together to constitute a critical or major deficiency (respectively) and will be reported as such.

One difference with the FDA inspection is that all critical and major deficiencies found will be reported even if remedial action has been taken before the end of the inspection.

The findings are presented at the closing meeting. At the closing meeting the inspector presents the findings. At this stage there is an opportunity for establishment to provide further information and to ask questions. After this discussion, the company normally accepts the findings and is then notified of the next steps. A typical schedule is:

- The inspectors provide verbal feedback summarising observations and findings made during the inspection.

- The company's representatives ensure that there is a clear understanding of the findings.

- Any erroneous findings are corrected at the time.

- A date when a report can be expected and when the company is expected to respond is confirmed.

9.3.1   Handling EU regulatory inspectional findings

In many ways the types of responses to other regulatory agencies will be similar to that presented for the FDA above. A post-inspection letter is sent to provide written confirmation of the deficiencies noted and reported verbally during the closing meeting.

In terms of additional information, the MHRA provides the below information on their website giving advice to the inspected company in preparing a response to the post-inspectional letter[19]:

"The inspected site is expected to provide a written response to the post-inspection letter within the required timeframe. The response should consider the context of the deficiency within the overall quality system rather than just the specific issue identified. The response should include proposals for dealing with the deficiencies, together with a timetable for their implementation. It is helpful for the response to be structured as follows:

- *Restate the deficiency number*

- *State the proposed corrective action*

- *State the proposed target date for the completion of the corrective action(s)*

- *Include any comment the company considers appropriate*

- *Provide an electronic version via email."*

Some examples of EU GMP inspection findings are:

- **In relation to training:**

   Key staff with no training or evidence of training in including the appropriate legislation:

   o   Activities delegated to laboratory staff without appropriate prior training.

   o   Staff involved in water sampling activities had not received training in aseptic technique.

   o   Minimum standards for staff and contractor training had not been defined by the organisation.

- o It was not possible to reconstruct or verify the training received by staff as there was insufficient documentation relating to training undertaken.

- **In relation to quality systems**:

  - o Key processes not documented or not reflective of current practice/legislation, uncontrolled documents used for key processes such as adverse event collection.

  - o Failure to implement in-process checking procedures (quality control).

- **In relation to products and processing**:

  - o Missing documentation – no/partial records to confirm additional of buffer into the production process.

  - o Lot numbers not supplied concurrent with batch paperwork.

The above provides an example of an MHRA letter. Noticeably documentation and training feature as key items and these are perennial inspectorate trends[20]. By law and by ethical commitment, companies that manufacture pharmaceutical products must make sure that what they produce is safe and effective. Ensuring that pharmaceutical manufacturing personnel possess the competencies necessary to perform their jobs correctly and efficiently is critical to a safe and successful manufacturing process (the training requirements for personnel working in a pharmaceutical manufacturing environment are specified in 21 CFR 211.25). Developing required skills, providing discrete knowledge and instilling an ethical and responsible approach to work are critical to training in an environment centred on good manufacturing practices. There are also references made to training and to good aseptic practices, which also tend to feature frequently as non-compliances[21].

9.3.2 Responding to EU inspectorate findings

Once the letter has been received from the inspectorate body, the company has 28 days to respond to the letter. In the response, the company will normally draw up a plan and within this will respond to the timescales set by the inspectorate body or, where no timescales were set, the company will propose the times for the implementation plan.

On receipt of the letter, the inspectorate body will consider the corrective action to be taken and will review the timescale. From this, further information may be requested and, if this is the case, the company will be required to provide evidence of satisfactory correction of the deficiency.

When the inspectorate body are satisfied with the findings, a close-out letter sent once the inspector assured that the deficiencies have been adequately addressed.

9.3.3   Reviewing inspectorate findings

Outside the FDA, it can be difficult to assess regulatory inspection trends and to review the types of issues raised in review meetings. The FDA makes its inspection results available to the public as required by the Freedom of Information Act. However, very rarely are inspection results of European supervisory authorities accessible in any public domain, except for the periodic publication of regulatory and GMP inspection analysis by the United Kingdom MHRA and the EMA. The Pharmaceutical Inspection Cooperation Scheme (PIC/S) also has never published the inspection findings of its Participating Authorities.

From the information which can be gathered, the top global GMP observations from across the main inspectorate bodies, in terms of the key categories, are[22]:

- Documentation—manufacturing

- Design and maintenance of premises

- Documentation—quality systems (elements / procedures)

- Personnel issues—training

- Design and maintenance of equipment

- Cleaning validation

- Process validation

- Product quality review

- Supplier and contractor audit

- Calibration of measuring and test equipment

- Equipment validation

The term 'validation' features regularly in the list. Validation of an analytical procedure is the process by which it is established, by laboratory studies, that the performance characteristics of the procedure meet the requirements for its intended use[23]. This is in keeping with the overarching philosophy in current good manufacturing practices for the 21st century and in robust modern quality systems that quality should be built into the product, and testing alone cannot be relied on to ensure product quality[24].

In terms of areas within the pharmaceutical facility the departments commonly cited are, in order of the number of observations:

- Production
- Quality system
- Quality control
- Premises and equipment
- Validation
- Personnel issues
- Materials management
- Regulatory issues

In terms of more specific observations, common findings from the FDA (compiled from a review of FDA Turbo EIR reports) are:

- Failure to have written procedures for production and process controls.
- Failure to have testing and release of drug product for distribution for determination of satisfactory conformance to the final specifications/identity and strength of each active ingredient prior to release.
- Batch production and control records were not prepared or are incomplete.
- Control procedures are not established to monitor the output/validate the performance of manufacturing processes that may be responsible for causing variability in the characteristics of the drug product.
- Employees were not given appropriate training.
- Laboratory controls do not include the establishment of scientifically sound and appropriate specifications/standards/sampling plans/test procedures.
- Drug product production and control records are not certified by the quality control unit to assure compliance with all established, approved written procedures before a batch is released or distributed.
- Procedures describing the handling of all written and oral complaints regarding a drug product either not established or not followed.

Whereas the most common findings listed by the MHRA are, in terms of general topic areas[25]:

- Quality management, which includes:
  - o Incomplete or tardy recording and investigation of complaints and incidents
  - o No regular management review of quality indicators
  - o Lack of Quality Improvement / CAPA processes
  - o Insufficient control of change
  - o Ineffective Self Inspection systems
  - o Recall systems incomplete and untested
  - o Non-compliance with previous inspection commitments
- Quality system documentation, which includes:
  - o Lack of control of procedures and specifications
  - o SOPs lacking detail or missing for certain activities
  - o Inadequate recording of training effectiveness
  - o Technical Agreements missing or incomplete
  - o Documentation of maintenance and calibration activities
  - o Product Quality Reviews missing or incomplete
- Batch release and QP duties
- Environmental monitoring, for instance:
  - o Wash bays, cold stores and water systems not monitored microbiologically
  - o No monitoring during or following building work
  - o No viable monitoring close to point of fill for aseptic products
  - o Poor handling and positioning of settle plates in critical zones
  - o Poor or no continuous particle monitoring in critical zones
  - o No assessment of recovery or growth promotion
- Supplier and raw material control
- Potential for microbiological contamination, for example:
  - o Poor cleanroom and aseptic practices
  - o Materials passed into sterile areas with insufficient sterility assurance
  - o Transfer of partially sealed vials to lyophiliser

- o   Handling of vials prior to oversealing
- o   No programme for routine drain sanitisation
- Design and maintenance of premises, for example:
  - o   Use of low quality building fabrics leading to easily damaged walls and doors which become difficult to clean
  - o   Incorrect airflows and pressure differentials to prevent cross-contamination
  - o   Sterile area changing rooms
  - o   Goods receipt areas of insufficient size
- Raw material control, including:
  - o   Insufficient assurance of supplier adequacy
  - o   No evidence that APIs have been manufactured to GMP
  - o   TSE risks inadequately controlled
  - o   No vendor re-certification of secondary/back-up suppliers
  - o   No systems to address problems with suppliers, e.g. audit or increase testing
  - o   Poor sampling facilities
  - o   Insufficient identification testing

The above lists will be additionally useful in reviewing actions prior to a regulatory inspection.

9.4   Conclusion

This chapter has presented practical and pragmatic guidance in relation to post-inspection findings and observations. The objective was to explain the process and to provide the reader with a framework which can be used as part of the strategy for dealing with the post-inspection findings. This has included the important things to include in the response, the style in which the response should be written, a list of common pitfalls to avoid, and an emphasis upon timeliness.

The chapter was designed in such a way that, although there was a focus upon FDA, the guidance is applicable to responses for each of the major regulatory agencies. For this reason, the chapter has included some of the differences in the approach taken by the UK MHRA as an example of a European regulatory agency.

If the guidance in this chapter is followed, then the company which receives an inspection letter will be on a firmer footing in terms of their response. Regulatory inspections can be challenging times. It is therefore important not to compound this challenge further with a weak or ill-conceived response.

9.5 References

1   Braithwaite J and Drahos P. *Global Business Regulation*, Cambridge University Press: Cambridge, 2000.

2   Rudolph P M and Ilisa B G. Bernstein. Counterfeit Drugs. *N Engl J Med*. 2004; **350**: 1384-1386.

3   US FDA. SEC. 704. [21 USC §374] Factory Inspection, Food and Drug Administration, Rockville: MD, USA (at: http://www.fda.gov/RegulatoryInformation/Legislation/FederalFoodDrugandCosmeticActFDCAct/FDCActChapterVIIGeneralAuthority/ucm109377.htm) (accessed 23rd May 2012)

4   Vesper J L. So what are GMPs, anyway? *BioProcess Int.*, 2003; **1**(2), 24-29.

5   US Government Accountability Office. (2007, November). Drug Safety: Preliminary Findings Suggest Weakness in FDA's Program for Inspecting Foreign Drug Manufacturers. (Publication No. GAO-08-224T)

6   Willig S H. Production and process controls, in Swarbrick, J., (Ed) *Good Manufacturing Practices for Pharmaceuticals: A Plan for Total Quality Control from Manufacturer to Consumer*. Marcel Dekker, New York, 2001; pp 99-138.

7   FDA. Pharmaceutical cGMPs for the 21st century: A risk-based approach, Rockville, MD, August 21, 2002.

8   FDA. Compliance Program Guidance Manual for FDA Staff: Drug Manufacturing Inspection Program, 7356.002, Food and Drug Administration, Rockville: MD, USA (available: www.fda.gov).

9   Elsevier Business Intelligence (2011) FDA Continues Aggressive Enforcement as Drug GMP Warning Letters Mount, *The Gold Sheet*, Vol. 45, No.4

10 Goebel P W, Whalen M D and Khin-Maung-Gyi F. What a Form 483 Really Means, *Applied Clinical Trials*, September 2001

11 US Government. Federal Register/Vol. 74, No. 153/Tuesday, August 11, 2009: "Review of Post-Inspection Responses", Food and Drug Administration, Rockville: MD, USA, pp 40211

12 Anon. FDA's Enforcement Crackdown To Increase Inspections, Delays, *Drug GMP Report*, 2010; Issue No. 210.

13 Apel, K. Right response to FDA 483 letters, *Manufacturing Chemist*, May 2010.

14 US FDA. (2006) "Formal Dispute Resolution: Scientific and Technical Issues Related to Pharmaceutical CGMP", Food and Drug Administration, Rockville: MD, USA.

15   US Department of Health and Human Services, Food and Drug Administration, Centre for Drug Evaluation and Research (CDER), Guidance for industry: Changes to an approved NDA or ANDA, available: http://www.fda.gov/cder/guidance/3516fnl.pdf (accessed Apr. 15, 2012).

16   Jimenez L. *Microbial Contamination Control in the Pharmaceutical Industry*, Marcel Dekker, New York, 2004.

17   Anonymous (2011), EU guidelines to good manufacturing practice, in The Rules Governing Medicinal Products in the European Union, Vol. 4, European Commission Enterprise and Industry Directorate - General, available: http://pharmacos.eudra.org/F2/eudralex/vol - 4/home.htm (accessed 24th May 2012).

18   MHRA/Department of Health (2006): "Explanatory Memorandum to The Medicines for Human Use (Clinical Trials) Amendment Regulations 2006 No. 1928", available via www.opsi.gov.uk/si/em2006/ uksiem_20061928_en.pdf (accessed 10th May 2012).

19   MHRA (2012). "The inspection process". MHRA, London at: http://www.mhra.gov.uk/Howweregulate/Medicines/Inspectionandstandards/GoodLaboratoryPractice/TheGLPinspectionprocess/index.htm (accessed 25th May 2012).

20   Hampton P. Reducing Administrative Burdens: effective inspection and enforcement HM Treasury, UK, 2005.

21   Levchuck, J W. Training for GMPs, *J Parenter Sci Technol*, 1991; **45**(6), 270-275.

22   Smallenbroek H and Boon Meow Hoe, B M. Inside PIC/S: Top GMP Deficiencies, *Pharmaceutical Technology*, 2012; **36**(4) 135-137.

23   van Zoonen P, Hoogerbrugge R, Gort S M, Van de Wiel H J and Van 't Klooster H A. Some practical examples of method validation in the analytical laboratory, *Trends Anal. Chem.* 1999; **18**, 584 –593.

24   International Conference on Harmonization (ICH) (2005, Nov.), Harmonised tripartite guideline Q2(R1), Validation of analytical procedures: Text and methodology.

25   MHRA. Inspection Findings – General examples, MHRA presentation delivered at Pharmig Annual Conference, 2011.

CHAPTER 10

Preparing for regulatory inspections of sterile facilities: the focal points

TIM SANDLE
HEAD OF MICROBIOLOGY, BIO PRODUCTS LABORATORY, UK

10.1   Introduction

Pharmaceutical preparations are expected to be safe and efficacious[1]. Safety relates to the effectiveness of the medicine, and to the avoidance of contamination. Contamination relates to both chemical and microbiological. This chapter is concerned with microbiological contamination: that is the presence of microorganisms or microbial by-products which render a product intended to be sterile as non-sterile[2].

Sterile manufacturing is arguably the most difficult and important facet of the preparation of pharmaceutical medicines (whether these are manufactured on a large industrial scale by a multinational company or on a named patient basis within a hospital pharmacy). This is because the medicines, due to their route of administration, are required to be sterile and if they are not sterile then this could lead to patient harm or even death[3].

As with any other type of pharmaceutical manufacturing, sterile manufacturing is subject to inspections by regulatory authorities. Due to the importance of maintaining sterility, sterile manufacturing is often subject to the highest level of regulatory assessment[4]. As section III of the FDA Guidance on Sterile Products notes:

"Nearly all drugs recalled due to nonsterility or lack of sterility assurance in the period spanning 1980-2000 were produced via aseptic processing"[5].

The preparation and operation of regulatory inspections represents an important part of the biannual or annual cycle for pharmaceutical organisations[6]. With microbiological contamination, the main risk concerns are:

*   Viable microorganisms

*   Particulate matter

*   Pyrogens[a]

In the context of these risks, sterility is achieved through protective controls and good practices during processing and by the presentation of the final dosage form[b]. 'Sterile products' encompasses the preparation of terminally sterilised products and the aseptic preparation of products through sterile filling. The distinction here is that some products can be terminally sterilised in their final container while others cannot and need to rely on pre-sterilisation of the components and bulk product before being aseptically filled within a cleanroom. For these processes there are different levels of risk and inspectorate concerns. These are explored in this chapter.

Sterile manufacturing is a continuum that stretches from development to manufacturing, to finished product, to marketing and distribution, and to utilisation of drugs and biologics in hospitals, as well as in patients' homes. Although the terms 'sterile manufacture' or 'aseptic manufacturing' are widespread there is no generic approach to the manufacturing of sterile products. Each plant or process will differ in relation to the technologies, products and process steps. If this chapter attempted to embrace all forms of manufacturing it would fail. What the chapter attempts to do is to present an overview of the general topics relating to the manufacture of sterile products and presents some of the more important aspects which the manufacturer should review and prepare for prior to a regulatory inspection by a body like the Federal Drug Administration (FDA), a European inspector (as required by the European Medicines Agency) or by the World Health Organisation (WHO). These selected topics are the 'focal points' and reflect the deficiencies most commonly cited by regulators.

The key focal point is the one feature which is common to all manufacturers of sterile products. That is to produce a product which is sterile. Although sterility can be defined as 'the absence of all viable microorganisms' it can only be expressed in terms of probability for each item of product produced cannot be tested for sterility without destroying each item. Sterile products must also be free of particulates[7], chemical impurities and pyrogenic substances (including microbial by-products like endotoxins).

The manufacture of sterile products involves the philosophy and application of sterility assurance. On one level sterility assurance can be taken as a quantitative assessment of the sterility assurance level (SAL), a term used to describe the probability of a single unit being non-sterile after a batch has been subjected to the sterilisation process[c]. At another level sterility assurance concerns the wider embracement of the aspects of Good Manufacturing Practice (GMP) which are designed to protect the product from contamination at all stages of manufacturing (from in-coming raw materials through to finished products) and thus it forms an integral part of the quality assurance system[d].

To aid the pharmaceutical manufacturer to prepare for a regulatory inspection the chapter is divided:

- 10.2: An overview of regulatory documentation

- 10.3: Regulatory inspections

- 10.4: Types of sterile manufacturing

- 10.5: Examination of the key focal points for a regulatory inspection of sterile manufacturing

- The appendix to this chapter contains a check-list to aid the pharmaceutical manufacturer in preparing for the inspection.

10.2   Regulatory Overview

Each regulatory inspection differs. This varies according to the agency; the inspector; as standards become revised; through the changing list of current GMP topics; and how the facility performs during the inspection. One commonality is that inspections will be against a standard or guideline. This section of the chapter outlines some of the most important guidelines of relevance to sterile manufacturing.

The key preparation for a regulatory inspection of a sterile manufacturing facility is the review of regulatory documentation and to understand what is required and what an inspector is likely to ask for. This task is made somewhat difficult by the range of different regulatory documents and standards, and due to the fact that these are sometimes contradictory.

The standards and guidelines are outlined below. The list below is not exhaustive:

10.2.1   Federal Drug Administration (FDA)

FDA documentation is divided between US laws, as contained in the Code of Federal Regulations (CFR) and inspectorate guidance documents. The Code of Federal Regulations is the codification of the general and permanent rules and regulations published in the Federal Register by the executive departments and agencies of the Federal Government of the United States. The CFRs applicable to sterile manufacturing are under Title 21 (Food and Drugs). Contained in chapter I of Title 21 are parts 200 and 300, which are regulations pertaining to pharmaceuticals[8]. In particular:

- Good Manufacturing Practice (cGMP) regulations (Code of Federal Regulations sections CFR 210 and 211)[e].

The FDA inspection guides are non-binding documents designed as reference material for investigators and other FDA personnel.

With these there are many documents and this chapter cannot list them all. However, the most important document for sterile manufacturing is:

- US Department of Health and Human Services, Food and Drug Administration, "Guidance for industry: Sterile drug products produced by aseptic processing – current Good Manufacturing Practice, (2004)[9].

Other inspectorate guidelines of relevance are[f]:

- Guidance for Industry Container and Closure System Integrity Testing in Lieu of Sterility Testing as a Component of the Stability Protocol for Sterile Products

- High Purity Water Systems

- Lyophilization of Parenterals

- Microbiological Pharmaceutical Quality Control Labs

- Pharmaceutical Quality Control Labs

- Validation of Cleaning Processes

- Dosage Form Drug Manufacturers cGMPs

- Sterile Drug Substance Manufacturers

There are two FDA drug inspectorates: Center for Biologics Evaluation and Research (CBER) and the other in the Center for Drug Evaluation and Research (CDER). CBER functions to protect and enhance the public's health through the regulation of biologics and related products, including blood and blood products, vaccines, allergenics, and emerging technologies such as human cells, tissues, and cellular and gene therapies. CDER is in place to ensure that all prescription and non-prescription drugs marketed in the United States are safe and effective. CDER evaluates all new drugs before they are sold and monitors drugs on the market to ensure that they continue to meet the standards of purity, potency, and quality.

10.2.2   European Good Manufacturing Practices

European GMP relates to European Commission Directive 2003/94/EC which describes principles and guidelines of Good Manufacturing Practice in respect of medicinal products for human use and investigational medicinal products for human use. European GMP is set-out within:

- Eudralex. The Rules Governing Medicinal Products in the European Community, Annex 1, published by the European Commission[g].

In addition, the following document is of relevance to the pre-sterilisation bioburden:

- European Agency for the Evaluation of Medicinal Products Committee for Proprietary Medicinal Products (CPMP) Note for Guidance on Manufacturing of the Finished Dosage Form, CPMP/QWP/486/95

European GMP is overseen by the European Medicines Agency (EMA) and enforced by national inspection agencies (for example, in the UK this is the Medicines and Healthcare Regulatory Agency).

10.2.3 Pharmaceutical Inspection Convention (PIC) and the Pharmaceutical Inspection Co-operation Scheme (PIC Scheme)

The PIC/S scheme exists to provide an active and constructive co-operation in the field of GMP. The purpose of the PIC/S is to facilitate the networking between participating authorities and for the exchange of information and experience between GMP inspectors. In 2011 the FDA joined the PIC/S scheme which should lead to closer integration of the approach to regulatory inspections and closer alignment of the standards required.

The PIC/S publishes a range of documents; like FDA inspection guides these are aimed at aiding inspectors so an understanding of these by the manufacturer can be beneficial[h]. Of importance to sterile manufacturers are:

- PIC/S GMP Guide PE 009-9

- Aide-Memoire Inspection of Utilities PI 009-3

- Aide Memoire on Inspection of Quality Control Laboratories PI 023-2

- Validation of Aseptic Processes PI 007-5

- Recommendation on Sterility Testing PI 012-3

- Isolators Used for Aseptic Processing and Sterility Testing PI 014-3

- Technical Interpretation of Revised Annex 1 To PIC/S GMP Guide PI 032-1

10.2.4   World Health Organisation

The World Health Organization (WHO) is a specialised agency of the United Nations (UN) that acts as a coordinating authority on international public health. WHO enforces similar requirements to European Union's GMP (EU-GMP).

The primary WHO document is:

*   WHO. "Quality Assurance of Pharmaceuticals: A compendium of guidelines and related materials". Volume 2. Good Manufacturing Practices and Inspection. 2nd edition. WHO Library Cataloguing-in-Publication Data, Geneva.

Within the GMP guideline are a number of annexes. These are updated at different intervals and are published on-line[i]. Those which are relevant to pharmaceutical manufacturing include:

*   WHO good manufacturing practices: main principles for pharmaceutical products. Annex 3, WHO Technical Report Series 961, 2011

*   Active pharmaceutical ingredients (bulk drug substances). Annex 2, WHO Technical Report Series 957, 2010

*   Pharmaceutical excipients. Annex 5, WHO Technical Report Series 885, 1999

*   WHO good manufacturing practices for sterile pharmaceutical products. Annex 6, WHO Technical Report Series 961, 2011

*   Water for pharmaceutical use. Annex 3, WHO Technical Report Series 929, 2005

*   Application of Hazard Analysis and Critical Control Point (HACCP) Methodology in Pharmaceuticals. Annex 7, WHO Technical Report Series 908, 2003

10.2.5   ISO

The International Standard Organisation publishes a number of standards of relevance to pharmaceutical manufacturing. Not all of these standards tie in with GMP. However, one standard in particular is of a great importance. That is the standard for cleanrooms – ISO 14644, for this series of standards is referenced both in EU GMP and the FDA Sterile Drug Products guide.

A second important standard is the one pertaining to biocontamination control.

- EN ISO 14644-1: Cleanrooms and associated controlled environments; Part 1: Classification of air cleanliness (May 1999).

- ISO 14698-1 & 2 (Part 1 – Cleanrooms and associated controlled environments – biocontamination control – General principles and methods and Part 2 – Evaluation and interpretation of biocontamination data

There are a host of other ISO standards pertaining to pharmaceutical manufacturer, ranging from HEPA filter standards to irradiation guidance.

10.2.6   ICH

The International Conference on Harmonization (ICH) publishes quality and GMP documentation. ICH guidance is applicable to those countries and trade groupings that are signatories to ICH (including the EU, Japan and the U.S.A.). The ICH has produced a number of guidelines relating to the quality of medicines[j]. These include:

- Q7 "Good manufacturing practice for active pharmaceutical ingredients"

- Q11 "Development and manufacture of drug substances (chemical entities and biotechnological/biological entities), Step 3"

Some important ICH documents have been 'adopted' by regulatory agencies as part of their formal GMP systems. For example:

- Q8 "Pharmaceutical Development"

- Q9 "Quality Risk Management", which was adopted as part of EU GMP in 2008 and by the FDA in 2010.

- Q10 "Note for Guidance on Pharmaceutical Quality System"

are established as part of EU GMP.

10.2.7   Pharmacopoeias

Of the three main international pharmacopoeias – United States, European and Japanese – it is the U.S. pharmacopoeia which contains the greatest number of chapters of relevance to sterile manufacturing, covering the spectrum from laboratory tests to pharmaceutical manufacturing instructions.

USP includes two distinct types of chapters: standards (chapter numbered below 1000) and informational documents (chapters numbered above 1000). Amongst the chapters, those of direct relevance to sterile manufacturing are[10]:

- <55> Biological Indicators—Resistance Performance Tests: Total Viable Spore Count

- <61> Microbial Examination of Nonsterile Products: Microbial Enumeration Tests

- <62> Microbial Examination of Nonsterile Products: Tests for Specified Microorganisms

- <63> Mycoplasma Tests

- <71> Sterility testing

- <85> Bacterial Endotoxins Tests

- <610> Alternative Microbiological Sampling

- Methods for Nonsterile Inhaled and Nasal Products

- <1072> Disinfectants and Antiseptics

- <1111> Microbiological examination of nonsterile products: Acceptance criteria for Pharmaceutical preparations and substances for Pharmaceutical use

- <1112> Application of Water Activity Determinations to Non-sterile Pharmaceutical Products

- <1113> Microbial Characterisation, Identification, and Strain Typing

- <1116> Microbial Control and Monitoring Environments Used for the Manufacture of Healthcare Products

- <1117>Microbiological Best Laboratory Practices

- <1208> Sterility Testing-Validation of Isolator Systems

- <1211> Sterilisation & Sterility Assurance of Compendial Articles

- <1222> Terminally Sterilised Pharmaceutical Products-Parametric Release

- <1223> Validation of Alternative Microbiological Methods

- <1227> Validation of Microbial Recovery from Pharmacopoeial Articles

There are some equivalent chapters within the European and Japanese pharmacopoeias in relation to sterility testing, testing of non-sterile products and endotoxin testing. In general, these pharmacopoeias do not cover processing or manufacturing in any great detail and instead focus on laboratory test methods.

10.2.8   National standards

Within the European Union, GMP inspections are performed by National Regulatory Agencies. Sometimes these agencies publish additional advice and guidance. Elsewhere, many other countries have similar GMPs to the EU and FDA GMP standards, as with Australia, Canada, Japan and Singapore; whereas many other regions adopt WHO GMP (which is itself strongly influenced by European GMP).

In addition to the documents described above, inspectors anticipate that sterile manufacturers will be aware of and keep up-to-date with "current good manufacturing practices" (cGMP)[11]. This is a term to describe the evolvement of GMPs inbetween the update of regulatory guidelines. Recent examples include the need for manufacturers to show evidence of a risk based approach to pharmaceutical processing together with the use of Process Analytical Technology (PAT), which allows for 'real time process data to be collected.

10.3   Regulatory inspections and risk management

Regulatory inspections of pharmaceutical manufacturers are undertaken by those regulatory organisations with which the manufacturer's licence is registered with. For products licensed in the USA this is the Federal Drug Administration (FDA). In other regions this is by the national body. Different regulators sometimes have different requirements[12] and some of the important differences are presented in this chapter.

The timescales for inspections vary and in part relate to how well the previous inspection went. Most regulatory authorities adopt a risk based approach to inspections and will schedule visits according based on how well they perceive the manufacturer to be operating. The typical frequency is every two years for each agency.

Importantly an inspection is not an audit. Audits are systematic and independent examinations of manufacturing related activities and documents to determine whether the process is being conducted and records made according to protocols and SOPs. Audits are part of the quality system and are designed to be part of continuous improvement. They are undertaken in a way which is generally co-operative[13]. In contrast, an inspection is the act by a regulatory authority of conducting an official review of documents, facilities, records, and any other resources that are deemed by the authority to be related to the manufacture of the product.

There are different types of inspections which fall within GMP: general GMP inspection, routine re-inspection, product related inspections, and 'for cause' or 'directed' inspections. This chapter is based on the requirements for general GMP inspections. With general GMP inspection, inspectors assess whether the manufacturer is in compliance with GMP; that is ensuring that all manufacturing operations are performed in accordance with the relevant marketing authorisation and that the manufacture and control of a product is undertaken according to licences and are carried out in a way which does not put the product at a contamination risk.

The approach by regulatory authorities differs in terms of how an inspection is approached. Generally the FDA adopts an approach which focuses on documentation, along the lines of "have you done what you said you would do?" Whereas other authorities put more emphasis on inspecting facilities. Beyond this, each inspector will have a different approach and different area of expertise and consequently each inspection will run differently. In part this rests on how the pharmaceutical organisation has performed since the previous inspection. For example, if there have been licence variations, product recalls, or if the manufacturer has raised its own critical deviations, then the reasons for these are likely to take up a great deal of inspection time.

The format of each inspection will vary. However, inspections will have:

- An opening meeting. At the opening meeting the inspection plan will be reviewed. This may include inspection schedule, products or operations to be examined and the procedures and records to be reviewed.

- A facility tour.

- An assessment phase which will consist of a review of procedures, operations, records and interviews with personnel to assess compliance with requirements.

- A closing meeting, at which the deficiencies or failures to comply with GMP are presented formally.

At the closing meeting, the deficiencies observed during the inspection will be discussed. Their importance will also be discussed so that deadlines for remedial actions can be fixed.

- A final letter from the inspector where any deficiencies are confirmed to the manufacturer in writing and to which the pharmaceutical organisation needs to respond.

Different inspectorates have different ways of classifying GMP non-compliances. With the European Medicines Agency (EMA), these are classified as "critical", "major" and "other significant deficiencies". A critical GMP failure occurs when a practice could give rise to a product which could or would be harmful to the patient. A combination of major deficiencies, which indicates a serious system failure, may also be classified as a critical deficiency. Deficiencies which are classified as "other" represent deficiencies which cannot be classified as critical or major, possibly because of lack of information, but which nevertheless indicate departures from GMP.

The FDA uses a record to note the inspection findings. The FDA 483, [2] "Inspectional Observations," is a form used by the FDA to document and communicate concerns discovered during the inspection. With the 483, the observations are ranked in order of significance. Unlike inspections by the EMA, the FDA 483 form will typically include only significant observations that can be directly linked to a violation of regulations and not suggestions, guidance, or other comments.

The course of each inspection will differ. However, in general, most routine GMP inspections will consist of the inspector reviewing at least some of the following:

- A review of the Quality Policy and the robustness of the quality system (including quality assurance, the combination of independent checks that systems in place achieve to set standard and quality control, the ongoing checks to make sure outputs are to set standard);

- Organisational charts and responsibilities summaries,

- Description of archiving arrangements,

- A review of major or critical deviations,

- A review of change control procedures,

- A review of customer complaints procedure,

- A review of product recall procedures ,

- A review of product recalls,

- A review of significant changes to manufacturing operations since the previous inspection,

- List of current projects,

- Standard Operation Procedures (SOP) index and copies of specific SOPs

- A tour of the manufacturing facility. Here the inspector will assess the performance of systems to meet specified standards,

- A tour of the quality control laboratories,

- Evaluation of data, as requested by the inspection,

- A review of validation protocols and reports,

- Review of documents, as requested by the inspector (which can range from laboratory SOPs to in-process specifications to training records),

- Many inspectors will check whether the manufacturer has the resources required to manufacturer,

- Training systems,

- Quality control testing,

- Environmental monitoring test results and approaches.

The preparation for inspections is a combination of prior knowledge of previous inspections, information gathered from reading reports, and from reviewing other inspections[k], drawing upon technical guidelines, and an anticipation of what is likely to be asked for. During an investigation, the inspector can review everything from the receipt of in-coming raw materials, through each manufacturing step and any component associated with laboratory testing.

Some areas to examine which can help to prepare for an inspection include:

- Carrying out research on current inspectorate trends by reviewing FDA 483s, reading papers and inspectorate guides, and attending conferences.

- Adopting a global risk management approach to the pharmaceutical organisation (risk management is discussed below).

- Ensuring conformance to international regulations.

- Preparing and reviewing documents including: batch records, media fill records, in-process sampling plans and specifications and manufacturing equipment records. The review of documentation is very important. Good Documentation Systems underpin and constitute an essential part of the quality assurance system. Documents should be easily understandable and clearly written in order to prevent error from spoken communication and permit traceability with regards to all aspects of the preparation of a medicinal product.

- Checking validation data and reports, including master validation plans. Records should be accurate, complete, and legible. It is important to know where all reports are located as these should be presented in a timely fashion. An inspector will be interested in how validations are executed and whether each type of product or item of equipment has been validated or whether a bracketing approach has been used. Bracketing approaches are controversial and problems will arise if the inspector is unconvinced of the similarity of the items bracketed together. A further important point is the frequency of re-validation and whether this has led to any changes.

- Considering if any significant changes have been made in the process or equipment, which includes reviewing change controls.

- Checking laboratory test method validation and reports.

- Checking data records. Data should be consistent with the source documents, or where discrepancies are apparent, a justification written.

- Ensuring SOPs are up-to-date and available,

- Checking equipment log-books, calibration and maintenance records,

- Reviewing training records.

- Re-reading deviation reports.

- Studying laboratory out-of-specification reports.

- Reviewing change controls.

- Examining stability study test protocols and reports.

- Having a list of any recent batch rejects or recalls available, together with associated reports available.

- Ensuring that key staff are available during the inspection period. This will include staff with management responsibility, R&D, pilot plant, engineers, microbiologists, production staff, quality assurance, and regulatory affairs.

As a general point, when reviewing documentation it is important to check that any changes have an explanation of why the change has been made and are initialled, dated and signed.

It is also useful to carry out a mock inspection prior to the actual inspection in order to test systems, practise interviewing and to reinforce staff awareness (particularly the role required during the inspection). It is a good idea to carry out an inspection close enough to the scheduled inspection so that all the data and procedures are fresh but far enough ahead of time that any errors or required adjustments can be made in sufficient time.

10.3.1   The risk based approach

Inspectors frequently adopt a risk based approach to inspections and risk assessment should be firmly built into the pharmaceutical organisation's quality system. The use of risk management for sterile manufacturing is linked to a wider process called 'Quality Risk Management'[14]. Quality risk management is a systematic process for the assessment, control, communication and review of risks to the quality of the medicinal product. It can be applied both proactively and retrospectively. Quality Risk Management has been promoted by European medicines inspectors and by the FDA as key to 21st Century GMPs.

Risk management has always been an intrinsic part of the world of pharmaceutical and healthcare[15], in terms of drug and patient safety, but it has not always been systematic in application or documented. Two important points to remember for any risk assessment are that, first, there is no such thing as 'zero risk' and therefore a decision is required as to what is 'acceptable risk'. Secondly, risk assessment is not an exact science – different people will have a different perspective on the same hazard. A significant change happened when the ICH (International Conference on Harmonisation) published a document called ICH Q9 (which was later 'adopted' by the FDA and as Annex 20 of the EU GMP Guide).

It is important to emphasise ICH Q9 as inspectors will expect risk assessment to be the basis of manufacturing operations and for the various aspects of processing to have been studied using risk assessment methodologies. With risk assessment methodologies, as ICH Q9 defines, there are three key definitions which help to contextualise what is meant by 'risks':

- Risk: "The combination of the probability of occurrence of harm and the severity of that harm"

- Harm: "Damage to health, including the damage that can occur from loss of product quality or availability"

- Hazard: "The potential source of harm" (ICH Q9; ISO/IEC Guide 51)

There are various risk assessment methodologies including Hazard Analysis and Critical Control Points (HACCP) and FMEA (Failure Modes and Effects Analysis). The pharmaceutical organisation should be familiar with these approaches and be employing them for validation and process risk assessment activities.

Risk Management is fundamentally about understanding what is most important for the control of product quality and then focusing resources on managing and controlling these things to ensure that risks are reduced and contained. Before risks can be managed, or controlled, they need to be assessed[16].

When the pharmaceutical organisation presents its risk assessments to an inspector, it is important that they are:

- Based on systematic identifications of possible risk factors,

- Take full account of current scientific knowledge,

- Conducted by people with experience in the risk assessment process and the process being risk assessed,

- Based on factual evidence supported by expert assessment to reach conclusions,

- Free from any unjustified assumptions,

- Comprehensive in that all reasonably expected risks have been identified, outlined simply and clearly, along with a factual assessment and mitigation where required

- Documented to an appropriate level and controlled and approved,

- Linked to the protection of the patient,

- Objective and contain a risk mitigation plan.

To summarise this section of the chapter: preparation is the key for any regulatory inspection irrespective of which agency is conducting the inspection and a risk assessment should be firmly built into the pharmaceutical organisation's quality systems.

10.4   Types of sterile products and manufacturing

There are two broad groupings of pharmaceutical preparations which are required to be sterile: parenteral[i] dosage forms and ophthalmic dosage forms. With a third group, aqueous inhalations, these are required to be sterile by some regulators (like the FDA) but not all. There may also be other dosage forms which are not normally supplied sterile but which may be required to be sterile for particular applications.

Broadly, there are two approaches to manufacture of sterile products, terminal sterilisation and aseptic manufacture. The regulatory bodies favour terminal sterilisation, and in the development of new sterile dosage forms the EU regulations demand that a "decision tree" is followed whereby the new dosage must be proven to be unable to withstand various defined processes of terminal sterilisation before it is allowed to be manufactured aseptically. It is important that the organisation has selected the appropriate method of sterile manufacturing and is aware of why that method is in place. The preparation of sterile products up to the filling and sterilisation of the final product are broadly similar.

Terminal sterilisation

Both the FDA Guidance (2004) and the European Pharmacopoeia (in chapter 5.1.1) state that of the methods of sterile manufacture a process in which the product is sterilised in its final container (terminal sterilisation) is the preferred method. This is not possible for all types of products and for this filtration through a bacteria-retentive filter and aseptic processing is used (see below).

Terminal sterilisation involves filling and sealing product containers under high quality environmental conditions. This means that non-parenteral products that are to be terminally sterilised may be filled in a Grade C/ISO class 8 area. With parenteral products these can be filled under the same conditions if the process or product does not pose a high-risk of microbial contamination. Examples of high-risk situations include slow filling operations; the use of wide-necked containers; or the exposure of filled containers to the environment for more than a few seconds before sealing. In these cases products are filled in an aseptic area with at least a Grade B/ISO class 7 environments or in a Grade A/ISO class 5 zone with at least a Grade C/ISO class 8 background, prior to terminal sterilisation.

Products are filled and sealed in this type of environment to minimise the microbiological content of the in-process product and to help ensure that the subsequent sterilisation process is successful. It is accepted that the product, container and closure will probably have low bioburden but they are not sterile. The product in its final container is then subjected to a sterilisation process such as heat or irradiation. A terminally sterilised drug product undergoes a single sterilisation process in a sealed container. The assumption is that the bioburden within the product can be eliminated by the sterilisation process selected[17].

Product formulation is undertaken at a Grade C/ISO class 8 or a Grade D/ISO class 9 environment. For some higher risk products a pre-filtration through a bacteria retentive filter may be advisable in cases, particularly where there is a high bioburden[m]. It is up to the pharmaceutical organisation to define the level of risk and to justify this to an inspector.

10.4.1   Aseptic filling

Aseptic manufacturing is used in cases, where the drug substance is instable when subjected to heat (thus sterilisation in the final container closure system is not possible) or where heat would cause packaging degradation. Aseptic filling is arguably the most difficult of the sterile operations. This is because the end product cannot be terminally sterilised and therefore there are far greater contamination risks during formulation and filling. With aseptic processing there is always a degree of uncertainty, particularly because of the risk posed by personnel to the environment.

In aseptic manufacture, the dosage form and the individual components of the containments system are sterilised separately and then the whole presentation is brought together by methods which ensure that the existing sterility is not compromised. Sterility is normally achieved through sterile filtration of the bulk using a sterilising grade filter (with a pore size of 0.2µm or smaller) in sterile container closure systems and working in a clean area[18]. This is undertaken in a Grade C/ISO class 8 cleanroom environments. The container and closure are also subject to sterilisation methods separately. The sterilised bulk product is filled into the containers, stoppered and sealed under aseptic conditions (under Grade A/ISO class 5 air) within a Grade B/ISO class 7 cleanroom, unless filling is undertaken within a barrier system.

Aseptic processes that exclude human intervention (such as robotics or barrier systems like Rapid Access Barrier Systems (RABS) or isolators) are at a considerably lower risk than operations which consist of filling machines under unidirectional airflow devices where there is a need for periodic human intervention. With isolator systems the background environment for the cleanroom can be at Grade C/ISO class 8, based on an appropriate risk assessment. There are additional risk considerations for isolators in that the decontamination procedures should be validated to ensure full exposure of all isolator surfaces to the chemical agent.

10.4.2   Blow-fill-seal Technology

Blow-Fill-Seal-Technology (BFS) is a type of aseptic filling but one at a lower risk compared with conventional filling. BFS is an automated process where containers are formed, filled and sealed in a continuous operation without human intervention. This is performed in an aseptic enclosed area inside a machine. The technology can be used to aseptically manufacture certain pharmaceutical liquid dosage forms[19].

BFS operations are undertaken under Grade A/ISO class 5 conditions with the background environment at Grade C/ISO class 8. Where BFS equipment is used for the production of products that are terminally sterilised, the operation can be carried out within a Grade D/ISO class 9 background environments if appropriately risk assessed[20].

10.4.3   Risk considerations

When an inspector examines facilities and processes for risks the emphasis will always be about the risk to the finished product. With terminal sterilisation the most important step is the sterilisation process whereas with aseptic processing the most important step is the filling of the product under Grade A/ISO class 5 conditions. It is at these key steps where the risks are greatest and these steps will be subject to the greatest regulatory scrutiny.

When considering any type of sterile manufacturing the essential risk must never be forgotten: that the objective is to avoid the contamination of the product by microorganisms or microbial by-products (such as endotoxins). It is also important to focus on the most common sources of contamination. These are:

Air: air is not a natural environment for microbial growth (it is too dry and absent of nutrients), but microorganisms such as *Bacillus, Clostridium, Staphylococcus, Penicillin,* and *Aspergillus* can survive. To guard against this products and sterile components must be protected with filtered air supplied at sufficient volume.

Facilities: inadequately sanitised facilities pose a contamination risk. Furthermore, poorly maintained buildings also present a risk such as potential fungal contamination from damp or inadequate seals. The design of buildings and the disinfection regime are thus of importance.

Water: the presence of water in cleanrooms should be avoided. Water is both a growth source and a vector for contamination.

Incoming materials: incoming materials, either as raw materials (which will contain a level of bioburden), or packaged materials, present a contamination risk if they are not properly controlled. Paper and cardboard sources in particular present a potential risk.

People: People are the primary source of contamination within cleanrooms[21]. People generate millions of particles every hour from activities of breathing, talking, and body movements, where particles are shed from hair, skin and spittle. Many of these particles will be carrying microorganisms[22]. As such a considerable proportion of this chapter is concerned with the control and training of personnel.

These points are important to bear in mind because these will frame the remit of the inspector. The inspector's main concern will be with seeking assurance that batches of product were not manufactured in a way which introduces microbial contamination or that the sterilisation process can eliminate contamination and that the finished product is not itself open to a contamination risk throughout its shelf life.

10.5 Examination of the key focal points for a regulatory inspection of sterile manufacturing

This section is the substantial part of this chapter and its aim is to review the key steps involved in the manufacture of sterile products and to draw attention to those parts of manufacturing which are likely to be the greatest concern to inspectors. In highlighting these areas the most important thing is preparation. If a protocol has been prepared, a study or risk assessment executed and report written then the material simply needs to be reviewed prior to the inspection. However, if a report is missing, inadequate or a risk assessment or study have not been undertaken then the appropriate parts of this chapter should be considered carefully and a plan put in place. To aid the reader the appendix to this chapter contains a check-list to assist with the task of preparing for regulatory inspections.

There are many aspects relating to sterile manufacturing which are critical to the final product quality and sterility, however no single chapter can describe every part of the sterile manufacturing process in-depth (to do so would require an entire book). What this chapter does is examine the parts of the process at greatest risk and draw upon the most recent inspectorate trends.

What frames this chapter is the concern that microbial contamination of sterile injectable products (parenterals) presents a great risk for such contamination and may result in the death of the patient; whereas with non-sterile products, aromas, off-flavours or discolorations, caused by microorganisms, may have less adverse affects[23]. Therefore, with respect to sterile products the main concern is with any potential microbial contamination.

10.5.1   Facilities

The design and layout of the facilities in which pharmaceutical manufacturing and laboratory testing take place are critical to the manufacture of sterile products[24]. The facility will be an important feature in the inspection. Most of this centres on cleanroom operations.

10.5.1.1 Cleanroom operations

With the premises used for the manufacturing of sterile pharmaceuticals there should be separate and defined areas of operation to prevent contamination, with controlled access and different grades of cleanrooms appropriate to the activities being undertaken. This requirement is set out in EU GMP and in the Code of Federal Regulations (CFR Section 211.42). An inspector will wish to seek assurance that there is proper directional flow of air, controlled material transfer, and people movement. It will also be expected that the logic of cleanroom design and facility layout conforms to something like **Figure 1**.

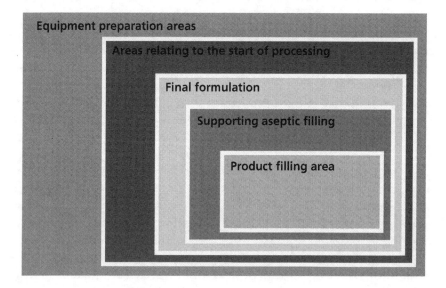

Figure 1: The cascade of cleanroom control, where the inner product filling area is subject to the tightest control

For this the pharmaceutical organisation should ensure that drawings detailing room air classifications; pressure differentials between rooms; and product, material, and personnel flows are accurate and available.

10.5.1.2 Cleanroom classification

Cleanrooms are especially designed rooms designed and built to achieve a cleanliness level defined by the concentration of airborne particles. The class or grade assigned to the cleanroom should reflect the activities in the room and be achieved through physical controls. Evidence as to whether the cleanroom can meet the required class or grade is shown through classification.

One key question that the inspector will consider when examining facilities "Is the cleanroom grade appropriate?" The pharmaceutical manufacturer must be able to demonstrate why a cleanroom has been given a particular grade or class (which will relate to the activities being undertaken within the cleanroom) and provide evidence that the cleanroom has been designed and built in such a way that the classification is met in terms of air cleanliness (particulate levels) and microbial contamination levels[25]. Of particular importance are the critical area in which sterile ingredients are added, aseptic connections are undertaken or where the product is filled and the areas surrounding the critical area. For this there must be in place equipment for adequate control over air pressure, microorganisms, dust, humidity, and temperature together with fit-for-purpose air filtration systems, including prefilters and particulate matter air filters.

A key difference between inspectorate bodies is with cleanroom classification. For the FDA the ISO 14644 standard of cleanroom classification is used[26]. ISO 14644 is a general standard covering all industries which use cleanrooms (from electronics to pharmaceuticals). With Europe, a grading system of A to D is used within the EU GMP guide for standard operation of cleanrooms (although ISO 14644 is used for the validation of cleanrooms). The World Health Organisation uses the same grading system as Europe in its GMP guidelines.

If manufacturers operate in both FDA and European markets then a document should be available which shows equivalence. This is additionally important because the ISO 14644 and EU GMP maximal particle values are not equivalent with the EU GMP levels being of a marginally tighter standard. In order to demonstrate GMP awareness, redundant terms like "Class 100", which related to the withdrawn Federal Standard 209E, should be avoided[27].

For aseptic filling

EU GMP Grade	ISO 14644 Class	Example of operation
A	5^n	Aseptic connection, preparation and filling Including the filling zone, stopper bowls, open vials
B	7	Background environment for the critical zone for aseptic filling
C	8	Preparation of components for sterilisation Final formulation Preparation of solutions to be sterile filtered
D	9	Less critical manufacturing stages, such as equipment washing

For terminal sterilisation

EU GMP Grade	ISO 14644 Class	Example of operation
A	5	Filling of products at an unusual level of risk
C	8	Filling of products
D	9	Preparation of solutions and components prior to filling

10.5.1.3 Cleanroom construction

Cleanrooms should be constructed in a way which makes them easy to clean and disinfect. What an inspector will look for is whether cleanrooms:

- Have a smooth and cleanable finish,

- Have a final coating which is impervious to detergents and disinfectants,

- Have no uncleanable recesses,

- That there are very few projecting ledges,

- There should be very few electrical sockets,

- Pipes and conduits are appropriately boxed in. With this, it has not been unknown for an inspector to 'follow a pipe' particularly for water outlets where there could be concerns about dead-legs (this is discussed further in relation to water systems).

10.5.1.4 Cleanroom and facility operations

To support the above classifications and the cleanroom uses detailed above, data should be available to present during an inspection relating to:

a)   Heating Ventilation and Air Conditioning (HVAC) system operations. Reports for the recent commissioning, as well as the initial qualification, should be available.

With the HVAC system it is important that alarms are tested and shown to be functioning before the inspection. This is especially important for pressure differentials. Any deviations must have been documented and investigated.

b)   The types of HEPA (high efficiency particulate air filters) in place (relating to the appropriate grades of filters along with Certificates of Conformance). The policy for HEPA filter repair and replacement should be available.

c)   The results from the annual or six-monthly re-certification of the cleanroom. This will include data relating to particle classification (which should be to ISO 14644 unless justified otherwise)[28], pressure differentials (P) and room air change rates (air volume replacement and pressure differential data is reciprocal).

With air change rates the FDA Guidance for Industry requires twenty air changes per hour whereas EU GMP no longer species a rate, instead requiring appropriate air change rates. Depending on the inspectorate, a justification may be required although in practise air change rates are normally considerably above twenty and far higher for areas of higher particle generation like changing rooms.

d) In addition, evidence should be available to demonstrate that pressure differentials are monitored continuously throughout processing and that data is recorded. Where any deviations have occurred, a record of the action taken should also be available.

e) In addition, engineering records for HEPA filter leak testing should be available. Depending upon the design of the HEPA filters, leak testing should provide assurance that there are no integrity breaches relating to sealing gaskets, frames or the filter media.

f) For UDAF (unidirectional airflow devices), additional data must be supplied about the air velocity (together with a justification as to where the air velocity is measured) so that the inspector can be satisfied that the air supplied to critical areas is at a velocity sufficient to sweep away particles from the critical zone.

EU GMP requires an air velocity of 0.45 metres per second (within a range of plus or minus 20%) and for readings to be taken at the working height. With the FDA Guidance, the air flow velocity is taken six inches from the filter face and the velocity is required to be of an appropriate speed. As with air changes a separate justification may be required to satisfy different inspectors.

g) For UDAF devices used for aseptic filling, the organisation should have available, or otherwise to prepare, airflow visualisation (or 'smoke') studies of the activities which take place under the UDAF and in relation between the UFAF and the room housing the UDAF. These should be supported by digital recordings together with a concluding report. They key aspect which the inspector will examine is whether there is a unidirectional airflow and if this has a sweeping action over and away from the product with no evidence of eddying.

Although aspects of airflow studies maybe filmed in the static state it is important that studies in the dynamic (in operation state) have been recorded.

10.5.1.5 Facility tour

The layout and process through of operations through the cleanroom is likely to be examined by the inspector through a review of a facility map. It is important to have several maps in order to show different aspects of the process. This could be for equipment flow, for pressures differentials, cleanroom grades and so forth. Such maps must be clear and unambiguous and approved by the QA department.

An inspector will visit at least some of the facility cleanrooms. During the visit the operation and layout of facility will be examined. An inspector is likely to begin with a review of the facility plan and associated documentation and maps. At the time of the inspection there should be no free standing water or sign of leakage of chemicals, otherwise this will raise concerns about contamination control. The organisational should also check and put right any parts of the cleanroom which have become cracked, chipped or damaged, or where a chemical has caused discoloration or oxidation. This is especially important for floors and walls. Exposed concrete would lead to an automatic citation from an inspector (due to the risk from bacterial endospores). There should be no debris on floors at the time of the inspection.

From studying the layout of the facility and from what is seen during the facility tour, the inspector will also note if:

a) If procedures are in place to prevent unauthorised personnel from entering.

b) How clean and tidy the areas appear. There should be no dust or damage. Air exhaust and supply grills must be free from obstruction.

c) There is adequate separation of different parts of the process.

d) The location and design of personnel change areas. Here the final stage of all gowning rooms in the unoccupied or 'at rest' condition should be of the same air classification as the area into which the gowning room leads.

e) That personnel entry is through airlocks

f) That equipment is transferred via airlocks. With airlocks, the pharmaceutical organisation should make a check of door interlocks and of pressure alarms prior to the inspection.

g) If there is suitable airflow from areas of higher cleanliness to adjacent and less clean areas, and that there is a significant positive pressure differential for the higher cleanliness areas relative to the lower cleanliness areas (and, where required, between rooms of the same cleanliness class. In circumstances where more critical activities occur in one area compared with another). Pressure differentials are normally at least 10-15 Pascals. Pressure differentials are normally assessed between areas when doors to cleanrooms are closed. An inspector may wish to note what happens when doors to a higher cleanliness area are opened and whether the outward airflow from the more critical area is sufficient to minimise any ingress of contamination.

h) The layout and positioning of process equipment and that process equipment has been cleaned and sanitised.

i)   Areas with water should be kept to a minimum and ideally equipment wash bays are segregated. There should be no standing water on the floor at the time of the inspection. There should be no water sources in aseptic filling areas.

j)   That there is a separate area for product filling, demarcated from other cleanrooms

k)   Within the context of aseptic filling, that items transported from a steriliser (such as vials) are protected within a Grade A/ ISO class 5 zones prior to entering the critical zone for filling.

10.5.1.6 Specific areas within the facility likely to be inspected

Whilst a definitive list of areas that an inspector will wish to view during a facility tour cannot be constructed it is almost certain that an inspector will examine:

*   Material transfer and flow. Materials and products have to be stored and handled so that the potential for mix-ups of different products or of their ingredients is minimal and that cross contamination is avoided

*   Product filling areas. Although is uncertain if an inspector will wish to enter product filling areas directly as this may require a cleanroom change gowning test to be arrange in advanced, the inspector will as minimum study the process flow and observe some parts of the filling activity through an observation window.

    In particular the inspector will study the filling machines. With filling machines this represents the machinery with the highest potential risk of contaminating the product. The filling machine needs to both provide continuous protection of the critical areas where product is exposed, as well as act as a barrier to minimise contamination ingress. Here the design of the physical barrier is of great importance.

    Where a facility has more than one filling line, an inspector will check whether each filling line is dedicated to a separate area to avoid cross contamination.

    Filling areas should be of the highest standard in terms of personnel controls, design and fabric. Sterile utensils must be used at all times. An inspector will note how items of equipment are transferred from sterilisation devices to filling machines and how the product is connected. Where lyophilisation is a requirement, the transfer must take place under Grade A/ISO class 5 protection.

*   Sterilisation devices such as depyrogenation ovens and autoclaves. Due to the criticality of the sterilisation process an inspector will check validation cycles and records relating to these items. These devices are discussed below.

- Vessels and equipment: all items of equipment should be maintained in a clean and sanitised state (and sterile where appropriate) with appropriate status labels visible. For sterile vessels, these held under positive pressure.

10.5.1.7 Barrier Systems

For aseptic filling, isolators are increasingly used to protect the product and indeed the use of an isolator or a RABS system is now a common cGMP expectation. Isolators are also the most widely used devices for conducting the end product sterility test. Although the barrier concept provides a considerable advantage over a conventional cleanroom the environmental monitoring programme (particles and viable microorganisms) should remain as comprehensive, during operations, as with any cleanroom operation[29].

When examining isolators used for aseptic filling or sterility testing, an inspector will examine the validation package. The key considerations are in demonstrating that the decontamination agent was suitable and that the decontamination process was effective. This is demonstrated through the use of biological indicators (defined preparations of bacterial spores) and chemical indicators. An inspector will check if these have been placed in the appropriate locations and, with biological indicators, that a six-log reduction from a population of 10^{-6} has been achieved[30].

Other aspect of isolator operations which will be checked will include the integrity and seams of gloves (and half-suits, depending upon the isolator design). This is because gloves represent the weakest point of any isolator in terms of air leaks. The risk of air leakage is that particles from the surrounding environment may ingress into the isolator environment.

The inspector will check whether any breach of the isolator integrity, should it occur, leads to a cessation of the operation and commencement of a decontamination cycle. Integrity can be affected by power failures, valve failure, inadequate overpressure, holes in gloves and seams, or other leaks. The organisation should have a comprehensive maintenance programme in place, describing how important parts of the isolator like transfer systems, gaskets, and seals are replaced before they degrade.

10.5.1.8 QC Laboratories

QC laboratories, particularly the microbiology laboratory are likely to be included in the facility tour. An inspector will expect that laboratories are housed in a dedicated area, separate from Production. The microbiology laboratory should be designed appropriately with different activities segregated, such as microbial identification located in a separate area to the processing and reading of environmental monitoring plates.

In the microbiology laboratory, an inspector will probably wish to see locations for and test results relating to:

- Sterility testing.

- Endotoxin testing.

- Incubators for environmental monitoring samples.

- Areas where environmental monitoring samples are read.

- Water testing.

These areas of testing, which are important measures for demonstrating sterility assurance, are examined below.

10.5.2　Cleaning and disinfection regimes

In order to achieve microbial control, in cleanrooms the use of defined cleaning techniques together with the application of detergents and disinfectants, is important. As such this subject is likely to come under regulatory scrutiny.

There are a number of aspects of cleaning and disinfection which an inspector could examine. These are:

a)　The types of detergents and disinfectants used and whether these have been selected on the basis of a sound rationale[31]. Some pertinent advice is found in USP-NF chapter <1072> "Disinfectants and Antiseptics".

With detergents, these should be non-foaming and preferably neutral. A check should be made as to whether any residues from the application of the detergent interfere with the disinfectant.

With disinfectants, the selected disinfectant(s) should have a wide spectrum of activity (that is, able to destroy a range of different types of microorganisms)[32]. The disinfectant must also be rapid in action with an ideal contact time of less than ten minutes. The disinfectant efficacy should not be impeded by the different operating temperatures of the cleanrooms in which it is used[33].

In addition, disinfectants used in higher grade cleanrooms (EU GMP Grade A/ISO 14644 class and EU GMP Grade B/ISO 14644 class 7) should be sterile. This is normally undertaken by using irradiated ready-to-use sprays or by sterile filtering bulk disinfectants into the clean areas (through a through a 0.2μm filter).

b)   For European inspectors, two disinfectants with different modes of action should be used in rotation (EU GMP Annexe 1 paragraph 3714). This requirement is less likely to occur with the FDA although some FDA inspectors have been known to ask about it[34].

c)   Of the disinfectants rotated one will normally be sporicidal. A case can be made for two disinfectants, with different modes of action which are not sporicidal provided that a sporicide is held in reserve[35].

d)   Disinfectants should be validated according to European CEN norms or to U.S. AOAC guidelines. Although the approaches to disinfectant qualification vary they broadly consist of a suspension test, an examination of the disinfectant in contact with different types of surfaces, and a field trial to determine cleaning and disinfectant frequencies[36].

e)   A programme outlining cleaning and disinfection frequencies should be available.

f)   There should be different levels and frequencies of cleaning and disinfection for different situations such as routine cleaning, periodic intensive or high level cleaning (such as of walls and ceilings as well as floors and working surfaces) and cleaning following a production shutdown or periods of maintenance activity.

g)   Cleaning and disinfection frequencies should be validated through environmental monitoring field trials and environmental monitoring surface data should be examined in order to verify that the cleaning procedures and frequencies continue to be satisfactory.

h)   In addition to the assessment of microbial counts, a periodic examination of the microorganism recovered should be undertaken to determine if any resistant strains are emerging. The primary concern will be within high numbers of endospore forming bacteria which are theoretically the most resistant to disinfectants[37].

Other aspects which should be checked and prepared for are:

•   A written procedure for cleaning should be in place

•   A policy for the selection of disinfectant selection should be drawn up

•   Responsibilities for cleaning should be assigned

•   An inspector will check if the agents are approved, how they have been prepared, and how they are labelled, particularly with respect to expiration dating

•   Staff must be trained in cleaning techniques and have a training record

- Detail of cleaning frequencies, methods, equipment and materials must be recorded in written procedures

- Inspection of equipment for cleanliness before use should be part of routine operations

- A cleaning log should be kept. The purpose is to keep a record of the areas cleaned, agents used and the identity of the operator

- The monitoring for microbial contamination in disinfectant and detergent solutions should be periodically undertaken

- The storing of disinfectant and detergent solutions should be for defined [and short] periods

- There should be a technical agreement with the company who supplies the disinfectant. Ideally the disinfectants purchased should be lot tracked.

10.5.3   Utilities

The utilities which input into pharmaceutical manufacturing will be subject to regulatory inspection. These include:

- Water
- Steam
- Compressed air

Air is also classed as utility in relation to HVAC systems. HVAC has been examined above in relation to cleanroom operations.

10.5.3.1 Water

There are different types of water used within the pharmaceutical industry. The two types which relate to sterile manufacturing, in terms of the water found in process areas, are purified water and Water-for-Injections (WFI). Purified water is used for equipment rinsing and for the preparation of some cleaning agents (normally in Grade C/ISO class 8 areas, some distance away from the aseptic core).

For sterile parenteral products WFI should be used for:

- Final rinse for cleaning product contact equipment and tools
- Final rinse for cleaning containers and closures
- For diluting disinfectants
- For buffer preparation

- Product formulation.

A key difference between the FDA and the EMA is that the FDA allows WFI to be generated by reverse osmosis or by distillation, whilst within Europe only distillation is currently permitted.

In terms of water systems, an inspector will be interested in the validation of the water system, reviewing the process and instrumentation diagram, and the routine quality control data.

With validation the objective is to ensure that the water system can produce water of predictable quality when operated at capacity. Validation results should be presented as a package, including[38]:

- Design Qualification (including detailed diagrams).

- Construction Validation (such as material certification).

- Installation Qualification (including verification of instruments, valves, heat exchangers, distillation units and so forth. In addition data relating to cleaning and passivation may be viewed).

- Operational Qualification.

  For this stage test data relating to flow and pressure rates, temperature, sensitisation, alarms, pumps and filter integrity should be available. Test data relating to the Operational Qualification must include the results of two weeks of microbiological and chemical sampling. An inspector will wish to seek assurance that any out of limits results obtained from testing were suitably investigated and corrected prior to the commencement of the Performance Qualification.

- Performance Qualification.

  The Performance Qualification involves monitoring the water systems for microbial and chemical quality for a period of a minimum of four weeks (phase I) and for a period of one year (phase II), based on daily sampling (with chemical testing the sampling can be rotated but with microbiological testing each sample outlet must be sampled each working day). With such testing it is important that the water has been sampled as the water outlets would be used by the production department. A commonly cited failing by inspectors is where QC staff sanitise outlets before samples are taken but production staff do not, leading to the QC sample results not reflecting production use.

A further issue at inspection would be if there were separate 'QC' sampling valves. It is also common for inspectors to wish to see data relating to sampling times which equate to points of high and low usage of the water system.

When examining the water system one of the key concerns of inspectors will be the frequency and method system sanitisation and the presence of any 'dead legs' in relation to the water system piping°. The FDA have defined "dead legs" according to their dimensions in the 1993 Guide to Inspection of High Purity Water Systems but fundamentally the principle is to always minimise the length of branch pipes. A related regulatory concern is that water can also remain stagnant in valves, particularly at user points, especially user points which are not in frequent and regular use. This is counteracted by use of so-called hygienic or "zero dead leg" valves.

Further issues to check prior to an inspection are:

- Ring mains should be sloped (have "drop") from point of origin to the point of return to ensure that systems are completely drainable.

- Avoidance of leakage. Water leaks can cause bridging of water to the external environment through which bacteria may enter the system.

- An inspector may also review data pertaining to the temperature of the WFI recirculation to affirm that it is between 65-80°C and continuously monitored.

The inspector will no doubt examine the water system as presented as a diagram and may wish to inspect parts of the system. Such an inspection will probably include the water generation plant. Water plants can be areas of the facility which do not always meet GMP requirements in terms of order and record keeping. This represents an important area to check prior to the inspection. Prior to the inspection it is important to verify that tanks are fitted with hydrophobic, microbial-retentive filters and these are in good condition. Engineering records should be checked to determine that tank vent filters have been regularly integrity tested. A further inspection concern may also be with water sampling procedures to ensure that sampling is undertaken as per manufacturing practices in relation to flushing, sanitisation and the use of hoses).

In terms of routine monitoring, an inspector is likely to examine water test data collated over different time periods. The routine test data most likely to be required is:

- pH
- Total Organic Carbon

- Conductivity

- Microbial bioburden (total viable aerobic counts)

  o   When used in bulk for manufacturing purposes the pharmacopoeias also apply a microbiological limit to WFI, this is not more than 10 cfu per 100 mL.

- Bacterial Endotoxins

  o   The level across the pharmacopoeias is 0.25 EU/mL.

With microbiological monitoring this could be data collated up to one year in order to demonstrate the affect of any seasonal variation. Data should be prepared as summary tables, as graphs and as lists of individual out of limits results (with references to the appropriate microbiological data deviation investigation). The results of the microbiological identification of water isolates may also be requested.

Such data could lead the inspector to examine QC laboratory tests, test limits, sampling techniques, sample storage conditions, and methodologies (such as membrane filtration test method, volume tested and the use of an appropriate culture medium, which for the European Pharmacopeia is R2A agar). There are differences with the microbiological test requirements between the European and United States Pharmacopoeia in terms of test incubation times and temperatures and many companies adopt a rationale or have validation data available to show equivalence between different standards.

10.5.3.2   Steam

With steam (or 'clean steam') used to supply steam sterilisation devices an inspector is likely to review test records to look at the frequency of testing (which is typically six-monthly or more frequent), the sampling ports (to confirm if they are representative), and test results (relating to bacterial endotoxin, non-condensable gases and dryness)[39].

10.5.3.3   Compressed air

Although compressed air and gasses (such as nitrogen) represent a theoretical low risk of microbial contamination due to the relatively harsh environments, it is important that where such gasses are used in relation to product filling that they are certified as sterile at point of use (which is normally through a 0.2μm sterilising grade filter), and that the point of use filters have been integrity tested. The gas lines should be subject to periodic testing (including tests like purity).

10.5.4   Processing

The processing of the product, prior to terminal sterilisation or aseptic filling, should be undertaken in a manner which minimises product bioburden and which demonstrates endotoxin control. An inspector will assess this through the facility tour and through a review of documentation. This section of the chapter reviews the aspects of processing which most often feature in inspections.

10.5.4.1 Incoming materials

Incoming materials include sterile supplies and starting materials. Incoming materials should be of an appropriate cleanroom grade and the supplier should be identified through an approved supplier list and, in turn, be included in the external audit programme. With starting raw materials these should have an accompanying Certificates of Analysis detailing microbial and chemical testing. Some materials, particularly those of animal or vegetable origin, should be subject to confirmatory testing by the QC laboratories. A document justifying if confirmatory testing undertaken (and if not, why not) should be prepared.

10.5.4.2 Equipment

An inspector will note whether there is adequate cleaning, drying and storage of equipment. Status labels should be placed on equipment and equipment which is clean should be segregated from equipment which is dirty. Equipment cleaning validation studies, using chemical (total organic carbon) and microbial (either swabs or water rinse tests) should be available to present at the inspection. With cleaned equipment, it is also important to have established the time interval between the washing, drying and the sterilisation of components, containers and equipment.

For aseptic processing all equipment entering the critical zone must be sterilised. With such sterilised equipment, the time period between the sterilisation of the equipment and its time of use should be established. For this, validation is normally undertaken by holding equipment and using it in media simulation trials.

For the operation of automated equipment cleaning or sterilisation, an inspector may check certain cycle recipes and parameters.

10.5.4.3 Storage areas

Storage areas should be clean and tidy, with appropriate controlled access. An inspector will probably seek to verify that storage areas for products and raw materials are maintained at an appropriate temperature and that records are kept. Temperature mapping of storage areas and refrigerators should must been completed and be subject to periodic re-validation.

Temperature conditions during transit of products and the duration of the transit time must also be considered under the requirement of Good Distribution Practice.

10.5.4.4 Containers

The sterility of the containers used to fill the final product is of great importance and an inspector will doubtless examine this process step, particularly for aseptic processing. The important steps are the storage of the vials; and then the preparation: cleaning, washing, sterilisation and depyrogenation (of glass containers). These should be validated and reports should be available. The specific requirements of sterilisation and depyrogenation are discussed later in this chapter.

With containers closed by fusion (such as glass or plastic ampoules or Form-Fill-Seal units), these must be subject to 100% integrity testing. With other products, containers must be visually examined. All such records may be requested in an inspection.

10.5.4.5 Container closures

Container-closures (or 'stoppers' or 'bungs') are an important part of the final packaging for pharmaceutical preparations, particularly those which are intended to be sterile. The most commonly used type of stopper is an 'elastomeric' container-closure[40].

Container-closures are either purchased ready-to-use (for which an inspector will check if a reputable company has been used) or require washing and sterilising by the pharmaceutical organisation. If container-closures are prepared in-house the washing and sterilisation validation will probably be examined. The important part of the washing cycle is whether endotoxin can be removed (at least a three-log reduction of 1000 EU/stopper). For this, endotoxin challenge studies should be performed. Issues relating to sterilisation are discussed below.

The drying process for stoppers should take place in a drying cabinet supplied with HEPA filtered air. An assessment of the drying cabinet for particulates should be available for inspection.

10.5.4.6   Container closure validation

The container closure is an important feature of the finished product and so an inspector will be concerned about the risk of leakage[41] or ingress of contamination into the final container and will expect studies to have been completed which demonstrate the integrity for the shelf-life of the product. The design of the study should have been based on a protocol. Given that there are different methods for assessing container-closures (together with different vial sizes and container types) the organisation should have a defined rationale for explaining the method selected. The methods include: microbial (bacterial immersion and aersolisation, with the latter more common for prefilled syringes) and physical (bubble methods, helium mass spectrometry, dye liquid tracer, headspace analysis, vacuum tests, pressure decay, weight distribution and high-voltage leak induction). Of these the helium leak test[p] is the most widely conducted and will thus be the most familiar to inspectors.

The microbial ingress test involves using a microbial challenge. There is no set standard for this test so an inspector may study the approach carefully. One of the key criteria is the selection of the microorganisms[42]. It is more common to use two different microorganisms of different sizes and of different methods of motility. For example, *Brevundimonas diminuta* (a very small bacterium) and *Escherichia coli* (with a relatively powerful motility) are often used in combination[43]. Sometimes the microbial test is not conducted and only physical tests are used. If this is the case, a rationale must be in place to explain why a microbial ingress test has not been included.

Additionally, a justification as to the frequency of container-closure testing, across the product shelf-life, should also be available. This should also demonstrate that that a statistically representative number of samples for containment system challenge testing was selected.

10.5.5   Manufacturing

This chapter has, so far, discussed many of the design features of pharmaceutical facilities. Of equal importance is the way in which facilities are used for processing, filling and sterilising the product. An inspector will seek assurance that the manufacturing processes are undertaken in a controlled manner and are carried out according to detailed instructions. In particular, instructions for handling sterilised materials must be included in the manufacturing instructions. An inspector will wish to confirm that the process is carried out in a way which minimises the risk of product contamination.

10.5.5.1 Processing issues

There are a number of issues relating to processing which could come up during the inspection. These include the inspector assessing:

- Whether time limits are established and justified for each phase of processing period. This is particularly important for the period between the start of bulk product compounding and its sterilisation; for the filtration processes; and the time of product exposure while on the processing line

- Whether contamination or mix-up is likely or unlikely.

- The policy prescribing how many personnel are allowed within any cleanroom.

- The methods and validation of cycles for automated equipment.

- Whether equipment and items which have been sterilised are appropriately covered or held under clean airflows to prevent re-contamination.

10.5.5.2 Bulk solutions

Of the various processing steps, the overriding inspectorate concern is the risk of product contamination. During the preparation of the intermediate product, the product is at risk from microbial contamination. This can arise through open processing, the addition of materials and from improperly cleaned equipment. These risks can be mitigated through chemical additives and process steps like ultrafiltration. The bulk product is at the greatest risk, particularly for aseptically filled products, because there are no subsequent bioburden reduction steps. The bulk product is, therefore liable to be subject to most attention.

An inspector will look at the way in which the material is transferred to the filling area (either by vessels or through transfer lines) and will examine the sterile filtration process (see below). The inspector will also check the bioburden test results in terms of the numbers of microorganism(s) recovered and the species identified against regulatory and licence limits. Such data should be summarised and available.

10.5.5.3 Product filtration

Product filtration is a critical manufacturing step, especially for the final filtration step prior to filling. This is especially important for aseptic filling as it is the only means by which the bulk is rendered sterile. Specific issues relating to the validation of sterile filtration are discussed below. Aside from the validation and the assessment of the bioburden of the bulk solution, an inspector will check whether the maximum time between the start of the preparation of a solution and its filtration through the sterilising grade filter has been established and qualified. The filtration time, together with the pressure across the filter, should match the parameters established during validation and an inspector will probably check the validation parameters against the batch record. Further points which may be looked at are the methods and results from the integrity testing of the filter before and after use.

10.5.5.4 In-process testing

During manufacture a number of in-process tests will be conducted as part of quality control. Records for these tests should be reviewed in advance of the inspection. Where electronic records are used (such as Laboratory Information Management Systems) the validation of the electronic documentation system should be available and an inspector is likely to seek assurance that a full audit trail is in place. For paper records, the records must be reviewed to ensure that they are accurate and legible.

Examples of in-process tests include:

- pH of buffers
- In-process viable counts
- Bacterial endotoxin tests of process rinses
- Bioburden assessment of the pre-filtration bulk
- Specific tests relating to the product, often for biochemical or inorganic chemical quality

Of these, the pre-final filtration bioburden test is the most important as this determines whether the bulk product is likely to present a problematic challenge to the sterilising grade filter. With the FDA, there is no pre-set limit for the bioburden assessment. With Europe, a Committee for Proprietary Medicinal Products (CPMP) guideline[44] sets a limit of not more than 10 cfu/mL. If this requirement cannot met, it is necessary to use a pre-filtration through a bacteria-retaining filter in order to obtain a sufficiently low bioburden for the challenge to the final filter.

10.5.6   Sterilisation

Sterilisation, unsurprisingly, is critical to all types of sterile manufacturing[45]. The most common types of sterilisation are:

- Thermal
  - o   Moist heat (saturated steam under pressure)
  - o   Dry heat

- Chemical
  - o Gaseous
  - o Radiation
  - o Bright Light
- Filtration

Some sterilisation processes will take place at the manufacturing site, others will be contracted out, and other activities will relate to materials purchased as sterile items. However organised it is the responsibility of the manufacturer to ensure the sterility of all materials.

10.5.6.1 Sterilisation devices

Sterilisation devices play an important function in sterile manufacturing. In order to achieve effective sterilisation the entire batch of the material must be subjected to the required treatment and the process should be designed to ensure that this is achieved. For this the operating parameters of sterilisation devices must be clearly established and monitored at appropriate intervals. In addition, where devices have a dwell time (such as gassing devices) the time must be measured.

The parameters of sterilisation cycles must be validated and the established parameters should be periodically reassessed (possibly at annual intervals and always following significant changes to load configurations). In order to prepare for an inspection the organisation should prepare an inspection pack relating to each item of sterilisation equipment[46]. Inspection packs must contain the protocol, report and data relating to the validation, including:

a) Justification of locations and the results of biological indicator[q] studies

b) Justification of locations and the results of temperature probes and sensors

The locations of biological indicators and thermometric probes will be studied by an inspector. Here the points deemed to be the greatest risk should be monitored (such as downstream from filters). The results from the temperature probes should be used to guide the placement of biological indicators (such as placing biological indicators in the coldest spots). There should be sufficient numbers of probes and biological indicators in order to sufficiently map the sterilisation load[47].

The biological indicators used should be appropriate to the type of sterilisation device[48]: Biological indicators and culture collection reference numbers[r] are:

- For moist heat: *Geobacillus stearothermophilus*[s] (ATCC 7953) (or alternatively *Clostridium sporogenes, Bacillus atrophaeus*[t] or *Bacillus coagulans*),

- For Vapour Hydrogen Peroxide: *Geobacillus stearothermophilus* (ATCC 7953) or alternatively *Clostridium sporogenes* or *Bacillus atrophaeus*),

- For dry heat: *Bacillus atrophaeus* (ATCC 9372)

- For gamma irradiation: *Bacillus pumilus* (ATCC 27.142)

- For ethylene oxide: *Bacillus atrophaeus* (ATCC 9372)

In addition, appropriate certification must be available for biological indicators. There is some debate within industry as to whether the manufacturer's certificate of analysis can be taken or whether biological indicators require verification by an independent laboratory. There is no definitive answer for this depends upon the inspector. If a company chooses not to independently confirm a lot of biological indicators, evidence that the transport and storage requirements have not affected the properties of the biological indicator should be available.

The population[u] and resistance (D-value[v]) of the biological indicators should be sufficient to demonstrate that a sterility assurance level of at least 10^{-6} is achieved[w] (cGMP requirements are often for a sterility assurance level of 10^{-12})[49]. These requirements are set out in the pharmacopoeias. It should be noted that there are some differences between European and United States Pharmacopoeial criteria:

European Pharmacopoeia:
Population of more than 1×10^5
D-value 121°C of more than 1.5 minutes

United States Pharmacopoeia:
Population of between 1×10^5 and 5×10^6
D-value 121°C of between 1.5 and 3.0 minutes

In addition to the characterisation of biological indicators, calibration certificates must be available for the thermocouples and other temperature monitoring equipment. For validation runs it is typical to calibrate temperature monitoring probes before and after the validation study.

With validation reports it is important to check, prior to the inspection that these are current and that, where bracketing approaches have been used for material loads, that the worst case loads and cycles have been included. The maintenance programme of sterilisation devices (such as valve and gasket replacement) is also likely to be examined and records should be cross-checked prior to the inspection.

Where terminal sterilisation is undertaken the rationale and basis for parametric release[x] will probably feature in the inspection. Furthermore, product quality test data must also be available for inspection. This may include potency (for vaccines), moisture content, visual aspect, reconstitution time, colour and pH change.

Different types of sterilisation devices and sterilisation methods are discussed below.

10.5.6.2 Moist heat: Steam sterilisation

Moist heat sterilisation is probably the most widespread method of sterilisation in pharmaceutical manufacturing. It is the main method for terminal sterilisation and is also used for the preparation of equipment and components for use in manufacturing or for aseptic filling (where single-use technology is not in place). With moist heat, as with other types of thermal sterilisation, the lethality depends upon[50]:

- Degree of heat,
- Duration of exposure,
- Humidity.

Such devices function through a cycle, which follows[51]:

- Preconditioning (which must remove air),
- Application of heat,
- Sterilisation,
- Drying (the removal of air and release of pressure).

Therefore, these are the steps which an inspector will focus on.

Within pharmaceutical facilities the most common steam sterilisation devices are autoclaves. Autoclaves are used for sterilising equipment such as tanks, filling nozzles, aseptic attachments, process tubing, filters and apparatuses, rubber closures, and for the terminal sterilisation of product.

The critical function of autoclave operations is air removal (for the insulating properties of air interfere with the ability of steam to transfer energy to the item to be sterilised, leading to lower lethality and the possibility of microbial survival). Therefore, the parameters pertaining to steam sterilisation and aspects of the mechanics likely to lead to pressure problems (such as filter installations) and leak tests are each likely to be subject to the greatest scrutiny.

The validation of autoclaves will also feature in an inspection. Important areas to prepare for include ensuring that[52]:

- Installation Qualification results are available, particularly utility connections and instrument specifications;

- With the Operational Qualification records, it is important to have evidence showing that thermocouples were positioned throughout the chamber within hot and cold regions;

- With the Performance Qualification the results of the thermocouples and biological indicator studies will be scrutinised in order to show that the device can be operated consistently and achieve the required lethality levels.

With steam-in-place systems there are additional concerns that an inspector will explore. These are often in relation to the length of piping and whether long pipes, which can be a risk due to air entrapment, have been proven to sterilise through biological indicator studies. A further concern with piping is whether the piping has been constructed in a way to satisfactorily allow for condensate drainage.

10.5.6.3 Freeze-drying

Freeze-drying (or lyophilisation) is applied to certain liquid products in order to prepare them as freeze-dried powders for reconstitution. Freeze-drying is a dehydration process typically used to preserve a perishable material or make the material more convenient for transport. Freeze-drying works by freezing the material and then reducing the surrounding pressure to allow the frozen water in the material to sublime directly from the solid phase to the gas phase.

Freeze-dryer cycles require validation in a similar way to autoclave cycles. The key functions to be tested are the vacuum pump, which is used to reduce the ambient gas pressure, and the condenser, which functions to remove the moisture by condensation on a surface cooled to -40 to -80°C. An inspector will check whether the freeze-dryer has been validated to demonstrate that the correct temperature and biological lethality are attained within the condenser and chamber. As with autoclaves, the location of temperature probes and biological indicators is of great importance[53].

10.5.6.4 Dry heat: Depyrogenation[y]

Dry heat destroys microorganisms and microbial by-products primarily the result of an oxidation process. Dry heat can either be used to sterilise vegetative cells and endospores but not microbial by-products like endotoxin or to depyrogenate, which will destroy endotoxin and other pyrogenic substances. This is an important concept to understand: endotoxin is not destroyed to any significant extent by sterilisation treatments such as steam sterilisation, gamma radiation, ethylene oxide, hydrogen peroxide – only by depyrogenation.

Depyrogenation cannot be used for drug product sterilisation. It is used for the sterilisation of processing items such as glass and stainless steel[54]. Thus depyrogenation ovens are an essential part of the aseptic filling process in relation to the preparation of primary packaging articles: glass vials and bottles. Depyrogenation devices are typically tunnels or ovens. A depyrogenation tunnel uses unidirectional hot air and typically operates at 250°C for 30 minutes[z].

For depyrogenation validation, USP <1211> "Sterilisation and Sterility Assurance of Compendial Articles," is a good starting point. The validation of depyrogenation ovens is an important aspect and documentation must be available for an inspection. A depyrogenation study is a test of the physical capabilities of a device to depyrogenate an article or device. It is demonstrated by physical measurements (including temperature) and biological (using bacterial endotoxin[aa] applied to a suitable surface as an Endotoxin Indicator (EI))[55]. Depyrogenation, as defined by the USP and the FDA Guide to Sterile Drug Products, is demonstrated by a minimum of a three-log reduction of a minimum of a 1000 EU/article challenge (where an article is a glass vial/bottle or rubber closure). Endotoxin Indicators are prepared using a preparation of Control Standard Endotoxin (CSE) from *Escherichia coli* O113:H10[56].

An inspector will probably check if the correct level of endotoxin has been used and verify how the positions for endotoxin challenge vials were selected. The locations for endotoxin indicators are typically determined by a number of thermometric runs to determine cold spots. The number of EIs used is based on the size of the depyrogenation device and this can be a subject of much debate (with the capacity of the testing laboratory often a limiting step). Typically, 5 – 10 EIs are sufficient to assess the depyrogenation capabilities of most devices undergoing testing[57].

There is no clear consensus as to whether the actual device (vial or bottle) needs to be challenged with a high concentration of endotoxin or whether a commercially prepared endotoxin indicator can be used (the argument against the latter is that it is not a direct test of the article to be depyrogenated). The organisation should prepare a rationale for situations where a direct inoculation does not take place.

As with autoclave validation, an inspector will check whether worst-case conditions have been assessed (such as worst-case process cycles, container size and mass characteristics, and specific loading configurations) and whether these reflect actual production use.

10.5.6.7 Radiation sterilisation

Radiation destroys microorganisms by damaging the nucleoproteins of microbial cells. The main types are gamma irradiation, ultraviolet light and accelerated electrons (like beta particles). The items subject to radiation sterilisation include plastic devices, gowns, and some drug products[58].

For sterilisation by radiation few pharmaceutical companies (if any) will have the facility to do this and items requiring sterilisation by a process like gamma irradiation are carried out at specialist facilities. Consequently an inspector will expect the pharmaceutical organisation to have audited the radiation plant and the inspector may, if the plant is unfamiliar or if the inspector has concerns, extended the inspection to include the radiation plant.

Many inspectors will be concerned with the formation of radiolytic by-products (such as OH), which in turn can cause damage to ingredients and surfaces. As with other sterilisation methods, the validation process will be of importance[59]. An inspector will ascertain whether each load has been correctly dose mapped using dosimeters and validated as per ISO 11137-1 ("Sterilisation of health care products – Radiation"), and that a suitable dose has been used. The starting point for many sterilisation cycles is a dose of 25 kGy but lower (and higher) doses may emerge from development studies.

10.5.6.8 Gaseous sterilisation

There are different types of gaseous sterilisation. Sterilising gases include formaldehyde, ethylene oxide, propylene oxide, ozone, peracetic acid, vapour hydrogen peroxide and chlorine dioxide[60]. Most common to sterile manufacturing are ethylene oxide, which is used to sterilise many plastics, and vapour hydrogen peroxide which is used to decontaminate barrier systems. When preparing validation reports key parameters should be established including: temperature, relative humidity, and gas concentration[61].

Ethylene oxide is an alkylating agent which functions as a very potent and highly penetrating gas. Validation of processes is normally undertaken to ISO 11135-1. The main concern of an inspector will be the presence of residuals: ethylene glycol and ethylene chlorhydrin, which can remain as toxic substances. The inspector will probably examine the policy and specification for residue levels, which are typically 1 mcg/mL or g for ethylene oxide and 50 mcg/mL or g for ethylene chlorhydrin, as set out in ISO 10993-7.

With vapour phase hydrogen peroxide, this is a common method for decontaminating isolators, due to the ability of the gas to destroy spores. Although the technology is well established the main concerns of an inspector will relate to the fact that some gas generators do not produce a consistent concentration of hydrogen peroxide. A second concern is in relation to the absorption of the gas into plastics and other materials. Here suitable tests should be conducted to show that levels of absorption are minimal and do not adulterate the product[62].

10.5.6.9 Sterile filtration

Several parts of the manufacturing process have steps which involve sterile filtration. The most critical of these is in relation to aseptic filling where the bulk product is subjected to a final sterilising filtration immediately prior to the point of filling. This involves the use of filters where the cGMP requirement is for filters with a pore size of 0.22μm or smaller.

It is essential that the sterilising grade filter has been validated to reproducibly remove viable microorganisms from the process stream in order to produce a sterile effluent. This involves using a bacterial challenge. Ideally the bacterial challenge should be from a challenge made to the actual product rather than a placebo (although alternative solutions can be used if justified). The regulatory expectation is that each type of product is challenged against each type of sterile filter. The recommended microorganism is *Brevundimonas diminuta* (strain ATCC 19146 or equivalent), which is selected due to its relatively small cellular size (0.3 μm mean diameter) at a concentration of 1×10^7 microbial cells per cm^2 of the filter surface. It is important, when considering the microorganism, that a review has been undertaken of microorganisms isolated from bioburden tests of each product routinely filtered to show that *Brevundimonas diminuta* remains the most effective challenge[63].

Sterilising grade filter validation must be undertaken under worst case conditions. That is: the filtration of the maximum volume of product, for the maximum filter time and at the appropriate pressure flow rate. Supporting data relating to membrane compatibility screening, the effect of the sterilisation method on the filter, absorption and binding studies, and the impact of any extractables when the filter comes into contact with the product, should be reviewed and be available[64]. Additionally in presenting filter validation data, inspectors will note whether there were any effects on the product formulation such as adsorption of preservatives or active drug substances, or extractables.

In addition to filter validation, all filters must be subject to integrity testing for each use. There are different non-destructive methodologies for filter integrity testing, such as bubble point (which tends to be used for filters with low membrane surface areas) and diffusion rate (which is commonly used for filters with high surface areas). The appropriate method should be justified.

Most regulators will be most familiar with the bubble point method. With this method, the filter is tested either by increasing the pressure on a wetted filter until the wetting liquid is displaced (determination of an actual bubble point) or increasing pressure to the level given by the filter supplier as the bubble point, and as long as the wetting liquid is not displaced the filter can be safely assumed to meet the requirement.

Filter use and validation is a specialist area and the organisation should ensure that the appropriate expert is available to present the material.

10.5.7   Product filling

Many aspects of product filling have been discussed in relation to cleanroom design. Nonetheless there are some operational aspects which an inspector will focus upon. The fill is undertaken within environmental conditions appropriate to the product type: terminal sterilised, blow-fill-seal or aseptic filling.

For batch filling, an inspector may ask to view records or visually inspect:

a)   Method of filling machine and clean zone cleaning, sanitisation and (if appropriate) decontamination,

b)   Batch records,

c)   Facility layout,

d)   Cleanliness of cleanrooms,

e)   Review justifications for cleanliness classes,

f)   Cleanroom operational parameters,

g)   Review of interventions performed (with cross-checking to media trials),

h)   Environmental monitoring locations and data (refer to the environmental monitoring section in this chapter),

i)   Number of personnel permitted in the filling area,

j)   Staff training records,

k)   Observations of personnel practise,

l)   Weight checks.

The capping area, which can be integral to the filling machine or housed in a separate environment, will be inspected separately. With bottle and vial capping there is a difference between the EMA requirements, which require capping areas to be supplied with Grade A/ISO class 5 air, and the FDA which regard capping as taking place in an unclassified environment. For the EU GMP requirement of Grade A air, this normally indicates supply at rest rather than verifying Grade A air during operation.

One of the key assessments of filling is the media fill or aseptic simulation trial. This is discussed below.

10.5.7.1 Media fills

For aseptic filling, the media simulation trial is arguably the most important validation activity as it provides a level of assurance as to the risk of the process and the operators to the product. Media simulation will form a main part of the inspection. The media trial should cover all aspects for operations from sterile filtration of the media into a sterile hold tank, through to filling machine set-up and operation, to oversealing and incubation[65].

An inspector will not only review the results of media simulation trials (and should a failure have occurred within the most recent inspection a detailed investigation report must be available), the inspector will examine the study design in detail.

A study design for media trials should be captured in a clear and unambiguous protocol. It will be important to demonstrate to the inspector that the media trials represent 'worst case' and are a sufficient process challenge and closely simulate aseptic filling. The factors required to show this will vary between facilities but would normally include consideration of[66]:

a)   Initial validation should consist of three media fills, followed by semi-annual re-validation

b)   The longest filling run, in order to test operator fatigue. This is either demonstrated by time or by the number of vials filled

c)   Suitable containers and closures

d)   Appropriate line speeds

e)   Representative number of interventions, including those which the manufacturer considers to be of the greatest risk. This may include:

    a.   Off-loading of stoppers from autoclaves.

    b.   Replenishment of stoppers in the hoppers

    c.   Replenishment of containers in the container-feed if simulating a manual operation from a depyrogenating oven – this is not an issue with tunnel depyrogenation

    d.   Filling machine adjustment in response to, for example, weight checks. These in-process machine adjustments must be simulated even though they may not be necessary in the actual media fill.

    e.   Removal of containers that have fallen over or have missing stopper etc and unblocking jams etc

    f.   Collection of samples

f)   Appropriate line stoppages are simulated (the maximum stoppage in a media trial determines the maximum permissible stoppage for product fills),

g)   Aseptic assembly of equipment and filling equipment preparation

h)   Aseptic sample connections and disconnections

i)   Representative numbers of personnel undertaking normal duties

j)   Personnel shift changes and handovers (including gown changes where applicable)

k)   Traceability of vials (for example through laser etching)

l)   Environmental monitoring

m)   A simulation of freeze-drying (if applicable). The lyophilisation process must not be simulated exactly. This is because the freezing of vials and consequent formation of ice crystals kills microorganisms. Freezing should not be simulated.

The incubation parameters for the filled media containers should be suitable in order to detect microbial growth. This is normally seven days at 30-35°C followed by seven days at 20-25°C, with a 100% inspection for turbidity undertaken at the end of the incubation period.

Prior to incubation the filled vials should be inspected and inverted. The procedure for vial rejection should be the same as per product fills and an inspector will no doubt wish to seek assurance that vials which would normally be incubated are in fact incubated.

In addition, an inspector will also examine the culture media used. The culture medium is normally soyabean casein digest medium (commercial name: tryptone soya broth) or an equivalent (such as a vegetable peptone broth, used in cases where there is concern over the use of animal products being used within aseptic filling lines). In special circumstances, such as the use of nitrogen gas as part of the filling process, an alternative anaerobic culture medium (like fluid thioglycollate broth) should be considered or a suitable justification written as to why such a medium has not been used.

The inspector will also seek evidence that the appropriateness of the culture media has been demonstrated by growth promotion testing (before use and at the end of the incubation of the filled containers). As a minimum, the number of microorganisms used to assess growth promotion should match those to release media for sterility testing. In common with cGMP many inspectors would expect representative isolates of the cleanroom facility to be included within the growth promotion test set.

The acceptance criteria for media simulation trials vary according to different standards. In the past it was commonplace to establish a 10^{-3} sterility assurance level for aseptic filing (that is no vials contamination for 5,000 units and one vial was permitted to be contaminated for 10,000 filled units). However, with the advent of modern filling technologies and the use of RABS or isolation technology the cGMP requirement is normally zero growth of the filled containers. The standard formula for assessing growth is:

Percentage contamination = number of vials with microbial growth ÷ (number of vials filled – number of rejected damaged vials) x 100

10.5.8   Personnel

The training of personnel to work in cleanrooms, and their practices and behaviours are of the utmost importance and this is outlined in EU GMP and a number of FDA CFRs (211.22; 211.192; 211.25; 211.28). An inspector will be aware that personnel are the primary source of contamination within cleanrooms (due to the continuous shedding of epidermal cells, many of which contain microorganisms) and will be conscious of any signs of improper gowning, breaches of procedure or behaviours which indicate hygiene concerns.

During the walk-through part of the inspection the inspector will take note of the changing procedure and the effectiveness of gowning. Proper gowning is important as the gown provides a barrier between bodies and critical parts of the production process. Gowns used for sterile manufacturer must be sterilised and worn only once. An inspector will see if the skin and hair are covered by the gown, long-legged boots, face-mask, hoods, and gloves[67].

The highest level of gowning is required for aseptic processing areas and goggles are essential for these areas in order reduce the possibility particles from around the eyes from entering the air stream. For filling suites an inspector will probably ask about the gowning procedure and may ask questions about the training procedure. It is typical for staff working in aseptic areas to undergo gowning training on a six-monthly basis (which will include both observational assessment and associated microbiological monitoring of the changing room, gown and gloved hands). The inspector may also ask for details about the supplier of the garments (for evidence that they shed no particles or fibres) and for the laundering and repair policy.

It is unlikely than an inspector will enter an aseptic filling area unless a request is put in prior to the inspection for the inspector to be trained in the gowning procedure. However, the inspector will examine the filling area through an observational window or through a closed circuit camera (presumably such means to observe filling are in place, should they not be the organisational places itself in a position where the inspector will probably have to enter the aseptic filling area and the consequences of this could be that the area requires re-sanitisation).

Staff working in the aseptic area should be aware that the inspector is observing and they should carry out their duties as trained. An inspector will note if behaviours are appropriate, which will include checking if:

a)   Gloves are sanitised frequently,

b)   That staff move slowly and deliberately (and avoid rapid movements which will create turbulence),

c)   That if an intervention into the Grade A/ISO class 5 area is required, then whether the entire body stays outside the unidirectional airflow.

Just prior to the inspection a check should be made of staff training records against SOPs. With training all staff must have received periodic GMP training and for staff working in aseptic preparation areas some form of microbiological awareness training.

10.5.9   Quality control tests

There are various quality control checks required for pharmaceutical drugs as they are manufactured and on the final containers. With specific regard to sterile products, the tests of concern are the microbiological[68]: environmental monitoring and the end-product sterility test.

10.5.9.1 Environmental monitoring

Environmental monitoring describes the microbiological testing undertaken in order to detect changing trends of microbial count and microflora within cleanroom environments. It is not the same as environmental control and monitoring should ideally be targeted where the control is weakest[69]. The results obtained from environmental monitoring provide information about the performance of the physical design of the room (most notably the HVAC system) and the performance of the people, equipment and cleaning operations within the cleanroom. The sites where microorganisms are recovered also allow for assessment to be made about the potential impact upon critical parts of the process or the potential risk to the product[70].

Regulatory inspections place a great deal of importance on environmental monitoring and it is commonplace for the company to have an environmental monitoring protocol or rationale, justifying how much monitoring is done, how often it is done and why it is done[71]. The difficulty faced for companies is that environmental monitoring is one of the least defined areas within regulatory documents (notwithstanding that those areas which are defined in the USA and Europe are broadly similar)[72]. Therefore many aspects of the environmental monitoring programme are left for the microbiologist within a company to develop.

Environmental monitoring programmes are normally divided into three categories:

a) Batch related monitoring, which applies to the filling of products

b) Background or routine monitoring programmes, which refers to support areas for sterile filling and to lower grade cleanrooms where the product is manufactured

c) Reinstatement monitoring, which describes how areas are assessed following a period of production shutdown or after more significant maintenance activities.

An environmental monitoring programme should include a number of key topics. These should be defined and in place prior to the inspection:

•   Rationales for how environmental monitoring sites are chosen

•   A rationale describing which type of samples is most appropriate

•   Maps indicating where the environmental monitoring sites are located

•   Methods describing how samples are taken and methods describing how samples are handled

•   Sampling frequencies

- Clear responsibilities describing who can take the samples
- A rationale for the type(s) of culture media used
- Methods describing the incubation regime for samples
- A procedure for data and trend analysis
- A procedure for the investigation of out-of-limits results
- A procedure for assigning warning and action levels using historical data, with reference to specifications
- Review of microflora
- Cleaning and disinfection
- Validation of monitoring methods

There are a number of aspects of the monitoring programme which should be prepared prior to a regulatory inspection. These are discussed below.

10.5.9.2 Monitoring techniques

The inspector will wish to check if a range of monitoring techniques are being used and whether these are applied appropriately. The best way in which to present this is through an SOP listing the techniques and through sampling plans on which the locations of monitoring are described.

The range of techniques required consists of:

a) Passive air-sampling: settle plates

b) Active air-sampling: volumetric air-sampler

c) Surface samples: contact (RODAC) plates

d) Surface samples: swabs

e) Finger plates (required for staff involved in filling activities)

f) Plates of sleeves/gowns (required for staff involved in filling activities)

g) Particle counting

A comprehensive environmental monitoring programme will include each of the above methods.

With particle counting, to meet the requirements of the FDA Guide particles of the size 0.5μm must be measured; whereas to meet EU GMP and WHO requirements particles at two sizes: 0.5μm and 5.0μm must be measured.

Although many particle counters count at a rate of cubic feet per minute, to meet regulatory standards data must be assessed as the total number of particles per cubic metre of air sampled.

An inspector may check to see if the limits assigned to the methods are appropriate and match those detailed in the FDA Guide or EU GMP. These regulatory limits can be used to assign action levels. For alert levels, the regulatory expectation is that these are set based on an historical review of past data (for which the PDA have presented several alternative approaches)[73]. Such reviews are normally undertaken annually and documented.

10.5.9.3 Culture media

For environmental monitoring either one or two culture media are typically used[bb]. The concern of an inspector is that the culture media used can detect both bacteria and fungi from the environment. An inspector may ask about the justification for the incubation regime (incubation time and temperature). This may need to be supported with growth promotion studies. The culture media used for surface monitoring and for taking staff finger plates must contain a disinfectant neutraliser[74]. An inspector may wish to see evidence that the neutraliser in the agar is appropriate for the types of disinfectants used to disinfect the machine or to sanitise the hands of the operators.

An additional concern with media used for aseptic filling is that it is normal for the media to have been irradiated and to be provided triple bagged. Where growth promotion studies are presented, these must have been performed on the media after irradiation[75].

Growth promotion needs to have been performed on each batch of media. The inspection will be straightforward if this has been performed at the site. A case can be made for accepting media where the growth promotion has only be performed by the manufacturer of the media, although this will need to be carefully justified and may not satisfy all inspectors. Some inspectors may also argue that in addition to each lot of media requiring testing that each delivery should also be tested and again a justification may need to be prepared. This can be off-site by obtaining a transportation study from the media supplier.

A further aspect of growth promotion which needs to be considered is the type of microorganisms used. This should reflect the types of microorganisms found in the cleanroom environment and would typically consist of a Gram-positive coccus, a Gram-positive rod, a Gram-negative rod, a filamentous fungus and an ear-like fungus. For this requirement some of the strains listed in both the European and

United States pharmacopoeias make for useful inclusions. An inspector may also wish to see some isolates from the environmental monitoring programme included and a decision to include or excluded these will also need to be justified.

10.5.9.4 Batch fill monitoring

The requirements for batch fill monitoring should rigorous and the inspector will be seeking assurance that the programme is suitably comprehensive as to provide assurance that the environment was under control at the time at which the batch fill took place. This is especially important for batch filling.

Batch related monitoring does not only include the period of filling. It consists of the following stages. For each stage a level of monitoring should take place. These stages are:

a) Filling machine set-up,

b) The period between the machine set-up and the start of the fill,

c) The batch fill activity,

d) Post-fill monitoring of the filling machine, which should include critical surfaces such as stopper bowls and filling needles,

e) An assessment of the post-fill machine clean down.

The most critical phases are, arguably, the set-up of the machine and the batch fill. Machine set-up provides the potential opportunity for contamination to be introduced into the Grade A/ISO class 5 cleanzone. The minimum monitoring requirement should consist of settle plates, active air-samplers, an assessment of particle counts and finger plates taken of the operators' hands.

With batch fill monitoring, the important aspects which an inspector may note are:

a) Continuous particle monitoring takes place using discrete optical particle counters. Particle counting, in some senses, is the most important test as it can dynamically indicate the quality of the air continuously over time and hence the effectiveness of other clean room parameters, such as, pressure differentials, HEPA filtration and so on. The locations selected for monitoring should be those sites where there is the greatest potential risk to the exposed product or to sterilised components[76].

   In reviewing particle counters it is important to check that the distance between the particle counter and the probe is not more than three metres and that particles using to monitor unidirectional airflow devices are fitted with isokinetioc probes, with the probes orientated in such a way as to obtain a meaningful sample.

b)   Settle plates are exposed for the duration of the fill. With settle plates, these should be exposed at meaningful locations which have been selected either on the basis of risk assessment (including areas of the highest risk, such as close to the point of fill) or relate to studies of air-flow visualisation. It is important that settle plates are not exposed for a time which exceeds the validated exposure time otherwise the degree of weight loss through dehydration could adversely affect the nutritive properties of the plate media.

c)   Active air-samples, as a minimum, are taken at near the start and end of the batch.

d)   That finger plates of staff are taken during the fill and following each intervention.

e)   That exit suit plates (a contact plate of the gown) are taken at the end of the filling operation.

f)   That critical surfaces of the filling machine and apparatus are taken at the end of the filling operation to assess surface cleanliness.

10.5.9.5   Routine monitoring

The routine monitor programme applies to background areas which support filling and to all of the other classified cleanrooms in which the product is manufactured, equipment is prepared, autoclaves are unloaded and so on. Unlike batch monitoring such monitoring is rarely continuous or tied to a specific process; instead it represents a snap-shot in time which is taken to be representative of normal operating conditions.

In each of these cleanrooms, monitoring should take place using the standard settle plate, active air, particle counters and surface monitoring methods. For each cleanroom there should be a sampling plan on which the locations of monitoring are indicated and there should be a rationale as to why the locations have been selected. There are different approaches for selecting monitoring locations, ranging form taking samples in grids, through to orientating samples towards activities, and through to more sophisticated risk analysis techniques like Hazard Analysis and Critical Control Points (HACCP) methods.

A further justification required will be with the frequencies of monitoring. There is no regulatory guidance in relation to this and the company microbiologist will need to prepare a statement either justifying how often a set of Grade B/ISO class 7 and Grade C/ISO class 8 rooms are monitored or link the frequency of monitoring to a risk assessment based upon product risk (for example, rooms where open processing takes place may be monitored at a higher frequency than rooms where closed processing take place)[77].

10.5.9.6 Reporting

The results of environmental monitoring should be reported onto suitable documents or into computer systems, along with the appropriate units of measurement. Where incomplete samples have been taken some conversions may be required (such as a settle plate exposed for less than four hours). In addition to recording data, microbial identification should also take place. An inspector may enquire whether all microbiological monitoring isolates from Grade A (ISO class 5) and Grade B (ISO class 7) areas or action levels for any Grades or Classes areas are identified to the species level.

The most important part of environmental monitoring are the data and how they are handled. There is little value in carrying out a lot of sampling if the data are not studied in detail. If an inspector will almost always review environmental monitoring trend data. Given that environmental monitoring programmes generate volumous quantities of data, the microbiologist will need to decide how these are presented therefore careful thought needs to be given as to what information will be captured and how this will be prepared for presentation. This will probably be a combination of tables and graphs. The interval used should be capable of showing short term and longer term trends.

Trend analysis not only involves examining the microbial counts obtained. It also requires an examination of the microorganisms recovered. An inspector may wish to seek assurance that unusual species are not being recovered which might indicate a hygiene concern (enteric Gram-negatives in an aseptic filling suite for example) or the emergence of strains theoretically resistant to the disinfectants used (such as endospore forming Gram-positive rods). Likewise a check should be made to see if the most frequently isolated microorganisms in controlled areas are typical. That is bacteria from the human skin (e.g., the Gram-positive cocci like *Staphylococcus epidermidis, Staphylococcus hominis, Staphylococcus simulans, Micrococcus luteus*, and *Micrococcus varians*), skin diphtheroids (e.g., *Corynebacteria spp.*), and airborne bacterial spores, (e.g., *Bacillus sphaericus* or *Bacillus cereus*). Where there are water sources in cleanrooms (such as equipment preparation areas) then occasional instances of fungi and Gram-negative bacteria may be detected.

10.5.10   QC laboratory testing

Some aspects of quality control laboratory testing will be examined in relation to the inspection. This may include:

- Microbial limits and bioburden testing

- Bacterial endotoxin testing

- Antimicrobial effectiveness testing

- Container and closure integrity testing

- Water bioburden testing

- Examination of water samples for specific microorganisms (such as *Eschericia coli* and *Pseudomonas spp.*)

- Environmental monitoring data

- OOS reports

As with the review of production processes, test records should be available and correct and evidence shown that tests are conducted according to SOPs. Staff training records are likely to be requested.

10.5.10.1 Final product testing

There are several final product tests undertaken on finished goods. These include: appearance, particulate matter, chemical purity and microbiological limits. In relation to sterile manufacturing, the two most important tests are the sterility test and an assessment of pyrogencity. For these tests, samples of the final product should be representative of the batch fill and be drawn from the start, middle and end of the batch. The number of samples for the sterility test is set-out in the pharmacopoeias. For pyrogenicity no such guidance is available, although it is typical to use at least three articles sampled from the batch start, middle and end.

10.5.10.2 Pyrogenicity

Pyrogenicity is either assessed by animal testing (such as the rabbit pyrogen test)[cc] or more commonly by the *Limulus* amebocyte lysate (LAL) test[dd]. The inspection of animal test facilities falls under the remit of Good Laboratory Practice and is not specially addressed here. In Europe, regulatory authorises are more likely to accept the LAL test in lieu of the rabbit pyrogen test. With the FDA, both the rabbit pyrogen test and the LAL test are often required (and sometimes another animal test is required called the 'general safety test' or 'abnormal toxicity test' involving the injection of mice and guinea pigs to assess toxicity). It is important to understand the regulatory expectations in relation to pyrogen and toxicity tests.

With the LAL test an inspector may ask to see evidence of the test validation, particularly with reference to inhibition and enhancement testing in order to show that samples can be tested within their Maximum Valid Dilution (MVD)[ee]. LAL tests are only valid in situations where endotoxin can be shown to be detectable with the same efficiency in a test sample as in a control consisting of water known to be endotoxin free. A regulator will probably seek evidence for the absence of interfering substances for each product as part of method validation.

10.5.10.3 Sterility test

The problems with the Test for Sterility are that sterility can never be proven by end-product testing[78]. Therefore, in a regulatory inspection greater emphasis will be placed on environmental control and, to a lesser extent, environmental monitoring, than on sterility testing.

Nevertheless there is likely to be some focus on the sterility test as, despite the statistical limitations of the test, it is a key product release test and one mandatory for aseptically filled product[79]. Sterility test considerations include having evidence that the products have been validated to show absence of inhibition of a range of microorganisms and that the process of decontaminating the product to test within the sterility test zone (such as transfer into an isolator) does not cause product inhibition[80]. With sterility testing the membrane filtration method is the technique of choice (although rapid microbiological methods are now being implemented in some organisations[ff]).

Investigations must be thorough and should conclude the reason for the failure as either relating to the Sterility Test laboratory or to manufacturing. Both the EMA and the FDA have a strict policy of not allowing a sterility test to be repeated unless it can be categorically proven, through genotypic microbial identification that contamination occurred from the test environment. Where a case can be made only one sterility test is permitted. Due to the sensitivity of this the sterility test re-test policy is likely to be requested in the inspection.

10.5.10.4 Particulate matter

Each container of liquid parenteral product is required to be inspected for evidence of visible particles and any containers which are seen to be contaminated must be rejected. In addition containers are also examined for flaws, cracks, misplaced seals etc. These are material issues which could compromise the integrity of the containers and therefore of the sterility of its contents. For staff tasked with the inspection, the inspector will check whether they can detect different particles (or microbial growth) based on their training and may enquire if frequent eye tests (proper eye tests, not of the number plate reading variety) are carried out.

10.5.10.5 Parametric release

Parametric release is a system of release based on information collected during the manufacturing process and based on verifiable compliance with GMP. It means the release of sterile products without recourse to a pharmacopoeial sterility test. The principle is normally applied to all terminally heat sterilised products but cannot be applied to aseptically filled products[99].

For a sterilisation process to be eligible for parametric release:

- It should have been validated through thermal and biological qualifications, and should demonstrably be capable of achieving 10^{-6} Sterility Assurance Levels referencing a Biological Indicator of defined resistance to the sterilisation process.

- The integrity of the containment system for products proposed for parametric release must have been qualified through microbiological challenges

- The pre-sterilisation bioburden must be tested for each batch of product eligible for parametric release. All spore formers isolated must be identified and have their resistance to the process determined. If any such organism is found to be more resistant than the Biological Indicator used in validation of the process, the steriliser load must be rejected.

Some inspectors are wary about parametric release and the company should prepare an appropriate rationale.

10.5.10.6 Out of limits investigations

A major function of laboratory testing and of microbial monitoring is to identify out-of-trend conditions that can indicate a loss of environmental control or a breakdown in aseptic practices. These situations are described as microbial data deviations or out-of-limits events. Importantly with environmental monitoring, excursions are not out-of-specification incidents. This can sometimes lead to confusion on the part of auditors or regulators who are not microbiologists (such as those with a background in chemistry where pass or fail distinctions are clear cut). Here the company will need to rely upon its microbiologist to explain the importance of trend analysis.

It will be evident from this section on environmental monitoring that there is a mixture of things that must be done in order to meet regulatory expectations and some things which may or may not need to be done. This is a reflection of the regulatory guides being somewhat ill-defined as to certain aspects of the monitoring programme. The key aspect of preparation for an inspection is to prepare an environmental monitoring rationale and to pre-empt some of the questions which might arise.

10.5.11 Batch review

In drawing together the facility tour, validation data and laboratory test records, an inspector will examine some batch records. It is important to check that these are available, complete, up-to-date, legible, and that they have been QA reviewed. Most inspectors will pay particular attention to checking records against validation parameters, manufacturing instructions, and product licences.

10.5.12 Current technologies

Across all aspects of sterile manufacturing, from production processes to laboratory testing, inspectors will expect to see that pharmaceutical organisations are developing new technologies and are considering rapid test methods, especially when these enhance sterility assurance or improve data accuracy. In the aseptic filling market, for example, there has been increased emphasis on single-use equipment throughout the filling line instead of recycling stainless steel.

10.5.12.1 Single-use technologies

Single-use technologies are generally sterile, plastic disposable items implemented to replace traditional pharmaceutical processing items which require recycling, cleaning and in-house sterilisation[81]. The items are generally sterilised using gamma irradiation. Such technologies include tubing, capsule filters, single-use ion exchange membrane chromatography devices, single-use mixers, and bioreactors, product holding sterile bags in place of stainless steel vessels (sterile fluid containment bags), connection devices and sampling receptacles.

The advantages of single-use technology are that the technology eliminates the need for cleaning, eliminates the need for the pharmaceutical company to perform in-house sterilisation, reduces the use of chemicals, reduces storage requirements, reduces process downtime and increases process flexibility, and avoids cross contamination. However, single-use technology is still in its infancy and there are a number of validation steps which need to be undertaken before such technology is adopted by a pharmaceutical manufacturer, and this regulatory overview will be high. These include assessing any leachables or extractables which might arise when the product comes into contact with the single-use technology. The presence of extractables could lead to adulterated product or to the inhibition of any microbial contamination[82].

10.5.12.2 Process Analytical Technology

Process Analytical Technology (PAT) describes the mechanisms in place to design, analyze, and control pharmaceutical manufacturing processes through the measurement of Critical Process Parameters (CPP) which affect Critical Quality Attributes (CQA)[83]. Many PAT systems function on-line and provide real-time data. Inspectors have put considerable emphasis on the use of PAT technology in recent years as 'real time' measurement can help prevent product contamination and the application is part of cGMP expectations.

PAT includes features like Building Management Systems for monitoring HVAC and controls over the temperature and distribution of WFI.

10.6   Summary

Sterile manufacturing represents the most complex part of pharmaceutical production. The importance of maintaining sterility throughout all parts of the process rests on the seriousness of the risk that contaminated medicines could have on patient health as well as the loss of expensive products. This is why regulatory inspections of sterile facilities are thorough and undertaken regularly.

This chapter has presented an overview of the inspection of sterile manufacturing facilities, drawing on the author's experience of regulatory inspections. The chapter has outlined many of the differences between different regulatory agencies and the different standards which are used by regulators. The chapter has also provided some guidance on preparing for inspections, considering both the human aspect in terms of behaviours and the practical in terms of a documentation overview. The main part of the chapter has been a walk-through of sterile manufacturing, drawing out many of the issues likely to be raised by inspectors.

In presenting this the chapter has emphasised the need for preparation and practice, and the importance of remaining up-to-date with inspectorate trends and standards. These are each key for ensuring success at the inspection.

Appendix A: Sterile manufacturing: a checklist for inspection preparation

Quality Systems	
Inspection Point	**Questions to consider**
Quality consideration	Does the facility and its departments (organisational units) operate in a state of control as defined by the GMP regulations?
Organisation	Is an organisation chart available detailing management responsibilities? Does a Quality Assurance unit (department) exist as a separate organisational entity?
Documentation	Is the quality manual suitable? What is the procedure for designing, revising, and obtaining approval for production and testing procedures, forms, and records? Are appropriate policies and procedures in place for (the list below is an example and not intended to be exhaustive): • Product recalls • Deviation reports • Internal and external audits • CAPA • Change controls • Manufacturing instructions • Dress code • Packaging systems • Sterilisation, including recall and storage • Cleaning and disinfection policy • Assessment of product shelf-life • Planned maintenance for equipment • Procedure for record signing and corrections • Archiving • Security policy
Batch release	What is the design of batch records? How are batches released? Who is responsible for the release of batches and which procedure is this against?
Contract testing	If any portion of testing is performed by a contractor, has the Quality Assurance unit inspected the contractor's site and verified that the laboratory space, equipment, qualified personnel and procedures are adequate?
Internal auditing	Does a formal auditing function exist in the Quality Assurance department?
Facility Maps	
Inspection Point	**Questions to consider**
Facility maps	Are the drawing complete? Do the drawings show cleanroom classes / grades? Do the drawings so pressure differentials / pressure cascade? Is the process workflow clearly set-out? Are air-locks located at suitable locations?
Changing rooms	Do changing rooms have air-locks? Are changing rooms of the same room class (in the 'at rest' state) as the areas they lead into?
Material transfer	Are the routes for incoming materials designed in a way which will not introduce contamination?

| Cleanroom classifications | Are cleanroom classifications carried out to ISO 14644? Is the frequency of cleanroom classification appropriate to the room class? Is classification undertaken in the 'in operation' state? Are the assigned classifications appropriate to the room use? |
| Process segregation | Are wet areas and wash bays located outside of process areas? Are the routes of equipment transfer such that clean and dirty equipment are segregated? Are antibiotic product segregated from other products, through separate process areas? Are filling lines independent? |

Utilities and HVAC

Inspection Point	Questions to consider
Air handling	Have the air handling systems been commissioned and qualified? Is airflow velocity measured at a defined distance proximal to the work surface or at working height as required? Are total airborne particulate levels and monitoring frequencies in accordance with ISO 14644-1 and other GMP standards as appropriate?
Water	Is the manufacture of WFI undertaken according to regulatory standards? Is all of the equipment in the system from the water feed to points of use and all sampling points detailed on a diagram? Is the water generation plant in good order? Are appropriate filters in place? Is the temperature range in hot recirculating systems (e.g. 65-80°C) continuously monitored? Is sanitisation of water systems that are not self-sanitizing, performed at validated intervals with hot water, steam, oxidizing chemicals, regeneration chemicals, or passivation chemicals following a validated method? With QC sampling, are water samples at points of use collected in a manner consistent with manufacturing practices (e.g. flushing, sanitisation, use of hoses)?
Steam	Are documentation and review of clean steam monitoring results (e.g. endotoxin, WFI chemical) performed that includes trending?
Compressed air	Is gas used for direct product contact sterilised using hydrophobic sterilising grade filters? Is the gas tested at regular intervals for purity, particles and microbial counts?

Facility tour

Inspection Point	Questions to consider
General layout	Are all parts of the facility constructed in a way that makes them suitable for the manufacture, testing, and holding of drug products? Are functional work areas physically separated by walls or partitions? Is this facility maintained in a clean and sanitary condition? Are all parts of the facility maintained in a good state of repair?
Design	Are floors level and constructed of materials that will withstand daily or more frequent cleaning? Are ceilings and wall surfaces constructed of non-shedding materials to limit condensation and dust accumulation?
Operations	Are doors and pass-through windows kept closed when not in use? How many air changes per hour for each cleanroom? Is there sufficient space in the facility for the type of work and typical volume of production? Does the layout and organisation of the facility prevent contamination? Are rooms free of standing water?

Environmental controls / HVAC	Are temperature and humidity levels monitored and recorded daily? Are the temperature and humidity levels within the acceptable ranges? Is control of air pressure, dust, humidity and temperature adequate for the manufacture, processing, storage or testing of drug products? Are appropriate devices used to maintain temperature and humidity levels? What is the frequency of inspection and replacement for HEPA filters? Does the facility have separate air handling systems, if required, to prevent contamination? (this is mandatory if penicillin is present)
Personnel	Are processing areas restricted to authorised personnel only? Are personnel using and removing personnel protective equipment properly? Is the appropriate dress code adhered to for the class of room?
QC Laboratories	Are adequate laboratory space, equipment, and qualified personnel available for required testing? Is the microbiology laboratory house within a separate building?

Materials

Inspection Point	Questions to consider
In-coming materials	Are incoming material and components quarantined until approved for use? Are labels for different products, strengths, dosage forms, etc., stored separately with suitable identification?
Control	Are all materials handled in such a way to prevent contamination?

Equipment

Inspection Point	Questions to consider
Records	Are equipment records and log books complete? Are there any deviations relating to process equipment? Have there been any recent equipment change controls?
Design	Is all equipment used to manufacture, process or hold a drug product of appropriate design and size for its intended use? Are items of equipment of appropriate capacity? Are machine surfaces that contact materials or finished goods non-reactive, non-absorptive, and non-additive so as not to affect the product? Are design and operating precautions taken to ensure that lubricants or coolants or other operating substances do not come into contact with drug components or finished product?
Operation	Are all pieces of equipment clearly identified with easily visible markings? Are written procedures available for each piece of equipment used in the manufacturing, processing or holding of components, in-process material or finished product? Is each idle piece of equipment clearly marked "needs cleaning" or "cleaned; ready for service"? Is equipment cleaned promptly after use? Is equipment cleaned promptly after use? Is idle equipment stored in a designated area? Is clean equipment adequately protected against contamination prior to use? Have performance characteristics been identified for each piece of equipment?
Maintenance	Does each piece of equipment have written instructions for maintenance that includes a schedule for maintenance? Is the maintenance log for each piece of equipment kept on or near the equipment? Does the facility have approved written procedures for checking and calibration of each piece of measurement equipment? (Verify procedure and log for each piece of equipment and note exceptions in notebook with cross reference.)

Cleaning	Has the cleaning procedure been properly validated? Do cleaning instructions include disassembly and drainage procedure, if required, to ensure that no cleaning solution or rinse remains in the equipment? Are written procedures established for the cleaning and maintenance of equipment and utensils? Does the cleaning procedure or startup procedure ensure that the equipment is systematically and thoroughly cleaned?

Manufacturing

Inspection Point	Questions to consider
In-process controls	Are in-process materials tested at appropriate phases for identity, strength, quality, purity and are they approved or rejected by Quality Control? Are written procedures established to monitor output and validate the performance of manufacturing procedures that may cause variability in characteristics of in-process materials and finished drug products?
Time limits	Are time limits established and justified for each phase of processing period between the start of bulk product compounding and its sterilisation, filtration processes, product exposure while on the processing line? Have holding time limits for sterile equipment/closures/containers been qualified and documented?

Aseptic manufacturing

Inspection Point	Questions to consider
Design	Is a procedure in place for aseptic filling which describes the controls to be taken? Have airflow visualisation studies been conducted for the aseptic filling zone?
Operation	Are alarm systems in place to signal and/or record out of range conditions when specified alert/action level ranges are exceeded for continuously monitored Critical Process Parameters? Are these alarms periodically challenged? Is the number of line personnel required to run the line kept to a minimum?
Environmental monitoring	Does the Environmental Monitoring program include: • Total Airborne Particulates? • Microbiological Air Monitoring? • Microbiological testing of equipment and surfaces? • Microbiological testing of personnel? At point of fill, is monitoring continuous throughout setup and processing? Are the sampling location selection criteria documented and justified? Does the SOPs address elements such as (1) frequency of sampling, (2) when and where the samples are taken (i.e., during or at the conclusion of operations), (3) duration of sampling, (4) alert and action levels, and (5) appropriate response to deviations from alert or action levels? Are time limits defining how long the environmental monitoring results can exceed action levels for total airborne particulates, temperature, humidity, or pressure differential established and justified? Is there a documented rationale for how alert and action levels were established (e.g. use of historical data, regulatory mandate, etc.)?
Media fills	Do protocols/batch records include all significant steps such as interventions planned, type of media used, volume, and number of units filled, shifts, list of personnel participated, etc? Are at least three consecutive and separate successful runs performed for initial line qualification and after major line/process changes? Is the number of units filled at least in the range of 5,000 to 10,000? Is the media run acceptance criteria appropriate? Are all contaminated units investigated and documented according to a procedure?

Electronic systems

Inspection Point	Questions to consider
	When computers are used to automate production or quality testing, have the computer and software been validated?

Sterilisation

Inspection Point	Questions to consider
Validation	Have all different cycles and loads been specified and standard loading patterns and worst case loads for each cycle established? Are steriliser cycles validated? How frequently are cycles validated? Are biological indicators used with thermocouples for each validation run? Is the biological indicator for the sterilisation device appropriate?
Loads	Type of steriliser and cycle used? Are critical parameters for specific sterilisation method checked?
Records	Are the sterilisation records appropriate? Have they been completed satisfactorily?
Storage	What is the expiration time for sterilised items? How is this assessed? Is the packaging for sterile items appropriate? How is sterilisatation indicated? Are sterilised items stored separately from non-sterile items?
Filtration	Are all solutions that are passed through a sterilising grade filter situated immediately before filling wherever possible? Is the maximum time between the start of the preparation of a solution and its filtration through the sterilising grade filter established and qualified?

Depyrogenation

Inspection Point	Questions to consider
Depyrogenation	Do depyrogenation challenge tests use endotoxin indicators as part of the validation? Does the validation study data demonstrate that the process reduces the endotoxin content by at least 3 logs? Does validation of dry heat sterilisation and depyrogenation include appropriate heat distribution and penetration studies as well as the use of worst-case process cycles, container characteristics (e.g. mass), and specific loading configurations to represent actual production runs?

Cleaning and disinfection

Inspection Point	Questions to consider
Cleaning	Does this facility have written procedures that describe in sufficient detail the cleaning schedule, methods, equipment and material? Are the cleaning frequencies appropriate? Are cleaning frequencies supported by satisfactory environmental monitoring data? Are areas kept free of packaging materials? Are the cleaning procedures and processes appropriate? (e.g. double or triple bucket method)
Detergents	Is the detergent used compatible with the disinfectants?

Disinfectants	Are two disinfectants used in rotation? Is a third disinfectant held in reserve? Is one disinfectant a sporicide? Do the disinfectants have different modes of action? Are contact times observered? Have the disinfectants been validated? How are disinfectants prepared and diluted? Is the disinfectant used in the aseptic filling area sterile filtered?
Expiry dates	What is the procedure for controlling expiry dates for detergents and disinfectants?

Laboratory testing

Inspection Point	Questions to consider
Sampling	Is the number of representative samples taken from a container or lot based on statistical criteria and experience with each type of material or component? Is the sampling technique written and followed for each type of sample collected? Is aseptic sampling used as appropriate? How is aseptic sampling assessed?
Control	Does each test have a suitable negative or positive control? How is test performance reviewed? Is a re-test procedure in place? Is an out of specification procedure in place?
Microbiological testing	Are the performance criteria for microbiological tests established? Have samples been validated (e.g. for BET, sterility test and water samples)? Is environmental monitoring ad water test data trended? Are microbial isolates identified?

Personnel

Inspection Point	Questions to consider
Training	Are personnel training records up-to-date? Is all training documented in writing that indicates the date of the training, the type of training, and the signature of both the employee and the trainer? Do personnel hold the appropriate educational qualifications for the job? What is the induction procedure? Is GMP training regularly performed? Does each employee receive retraining on an SOP (procedures) if critical changes have been made in the procedure? Are staff who work in cleanrooms medically assessed for their suitability to work in such areas? Are contractors qualified by experience or training to perform tasks that may influence the production, packaging, or holding of drug products?
Behaviour	Are aseptic practices followed? Are hands sprayed at frequent intervals? In relation to aseptic facilities, are movements slow and deliberate?

Appendix B: FDA Compliance Program Guidance Manual 7356.002A (Sterile Drug Process Inspections)

CHAPTER 56 – DRUG QUALITY ASSURANCE

FIELD REPORTING REQUIREMENTS

Forward a copy of each *Establishment Inspection Report* to H#-30O Attention: Division of Drug Quality Evaluation. Copies of *Samples of Collection Reports* and *Analyst Worksheet* for all samples except those which are classified "1" should also be submitted. (This material will be used in evaluating the program).

As soon as the district becomes aware of any significant adverse inspectional, analytical, or other information which could or should affect the agency's new product approval decisions with respect to a firm, the district should immediately notify HFC-120, Medical Products Quality Assurance Staff, via EMS or fax, and they will, in turn, convey the information by fax or equivalent expeditious means to the appropriate Center regulatory units.

NOTE: Districts should assure that each operation performed by direction of this program circular is entered against the correct Product Code and Program/Assignment Code (P/AC).

PARTI-BACKGOUND

*This program is intended to cover the manufacture of all sterile drug products, including sterile bulk drugs, ophthalmic and ophthalmic dosage forms, Small Volume Parenteral (SVP) products, Large Volume Parenteral (LVP) products, and any other drug products required to be sterile.

Biologicals, veterinary drug products, and bioassay drugs are excluded from coverage under this program.*

PART II – IMPLEMENTATION

OBJECTIVES

To provide guidance for conducting inspections of manufacturers of sterile bulk and finished dosage form drug products to determine compliance with the Food, Drug, and Cosmetic Act and the Good Manufacturing Practice Regulations (GMPs), Title 21, CFR Parts 210 and 211.

To initiate appropriate action against those manufacturers found to be out of compliance.

To obtain information on key practices, to identify practices which need correction or improvement, and to evaluate current good manufacturing practices in the #sterile drugs industry.

PROGRAM #AGEMENT INSTRUCTIONS

*Inspections of sterile product manufacturing firms will be performed as either Full Inspections or Abbreviated inspections.

In the Abbreviated Inspection, coverage will be directed to key points in the major systems affecting the production of the sterile drug product. If the information collected indicates that the firm's practices are in compliance with CGMPs, the inspection may be concluded at this point.

Current Change

It should be pointed out that inspectional coverage under this option is not intended to limit the investigator's initiative in any way. If questionable practices are observed in areas outside of the systems delineated under this option, the investigator is urged to expand the inspection to cover these areas to his/her satisfaction.

The Full Inspection Option involves an in depth inspection of key manufacturing systems and processes and their validation in order to maintain surveillance over the firm's activities.

See **Part III – inspectional** and Attachment A for a complete discussion of the coverage requested under these inspection options.

This program is to be carried out when firms are inspected as part of the regular statutory inspection cycle in accordance with the current ORA work plan. If the sterile drug products to be inspected are radioactive drugs, then CP735G.OO2C, "Radioactive Drugs", should be followed as supplementary guidance.*

- Consider using a team approach in conducting these inspections, utilizing investigators familiar with these processes, and chemists, microbiologists, and engineers, as appropriate.

- *Investigators or team members should be well qualified in sterile product production experience and preferably have completed formal training courses in parenteral drug manufacture, sterilisation methods, procedures and equipment. Microbiologists involved should have experience in sterility/pyrogen testing and some experience in sterile product inspections.*

PART III – INSPECTIONAL

*Refer to CP 7356.002, Drug Process inspections, for general information on CGMP inspections. Refer to CP 7356.OO2C, "Radioactive Drugs" for supplemental instructions specific to radiopharmaceutical drug products.

Foreign inspections should be conducted using the guidance in this program, taking into account the time limitation on these inspections.

This program provides two inspectional options: an Abbreviated inspectional Option and a Full Inspectional Option. To determine which option should be used an evaluation of the following is appropriate:

1. **Review and Evaluation**

A full inspection should be conducted for initial inspections and may also be conducted on a surveillance basis at the District's discretion. Although it is not anticipated that full inspections will be conducted every two years, they should be conducted at less frequent intervals, perhaps at every third or fourth inspection. Also, whenever information becomes known which would question the firm's ability to produce quality products, an appropriate in-depth inspection should be performed.

An abbreviated inspection should **not** be conducted for the initial inspection of a facility, nor when the firm has a past history of fluctuating into and out of compliance. The District should utilise all information at their disposal such as past history results of sample analyses, complaints, recalls, etc. to determine if coverage under the abbreviated inspectional option is appropriate for the specific firm.

a. Determine if changes have occurred by comparing current operations against the EIR for the previous full inspection.

The following type of changes are typical of those that would warrant the Full Inspection Option:

Current Change

1. New potential for cross-contamination arising through change in process or product line.

2. Use of new technology requiring new expertise, significantly new equipment or new facilities.*

b. Review the firm's complaint file, DPPRs, annual product reviews, etc. and determine if the pattern of complaints (or other information available to the District) as well as the firm's records of internal rejection or reworking of batches warrant expanding the inspection to the pull Inspection Option to look for weaknesses in the firm's processes, systems or controls.

c. If no significant changes have occurred and no violative conditions are observed, the Abbreviated Inspection Option may be adequate.

d. If significant changes have occurred, or if violative or potentially violative conditions are noted, the inspection should be expanded to the Full Inspection Option to provide appropriate coverage.

e. If an inspection needs to be expanded to the Full Inspection Option, it need be expanded only for the applicable general product or process area in question.

2. **Abbreviated Inspection Option**

This option involves a more limited inspection of the manufacturer to maintain surveillance over the firm's activities. An Abbreviated Inspection as described below is adequate for routine coverage and will satisfy the biennial inspection requirement. The use of this option will save inspectional and clerical resources.

a. Inspections performed under this option should cover those items delineated under the Full Inspection Option with the exception that validation need not be covered for those systems and processes that have previously been covered under the Full Inspection Option.

Perform an inspection of the firm's manufacturing facility including a review of a representative number of Master and Batch Production Records (minimum of 5 batches) on products manufactured by the firm. Products that appear in the firm's inspectional history of previous problems should be included. A brief inspection of the laboratory should include a spot check of a limited number of test records (at least 10) to assure that batches are being subjected to adequate testing for conformance to specifications.*

Special note should be taken of the firm's packaging and labeling controls. Any observation of inadequate controls will indicate that a Full inspectional Option should be performed. If the following type of procedures are encountered, in-depth inspectional coverage should be given to the firm's Labeling systems:

- The use of labels which are similar in size, shape, and color for different products.

- The use of cut labels which are similar in appearance without some type of 100 percent electronic verification system for the finished product.

- If the use of gang printing of cut labels is not minimised as required by current regulations.

- If the firm has had more than one mislabeling recalling the past two years.

- If the firm fills product into unlabeled containers which are later labeled under multiple private labels.

If the abbreviated inspection reveals no significant objectionable conditions, and there are no other factors requiring the use of the Full Inspection Option, use of the Abbreviated Inspection Option 18 adequate.

Refer to the "Guide to Inspection of Bulk Pharmaceutical Chemical Manufacturing" for guidance on the applicability of CGMPs to bulk operations.

Current Change

3. Full Inspection Option

The Full Inspection Option will be implemented when: (1) this is the initial inspection of the drug firm; (2) this is the first inspection performed following a regulatory action against the firm; or (3) the information collected under the Abbreviated Inspection indicates that the firm's practices are or may be deficient in one or more system areas. An in-depth inspection of all manufacturing, support, and documentation systems at the firm in question should be initiated. However, this in-depth inspection may be limited – at the discretion of the investigator – to only that system area that appears to be deficient.

It is not expected that inspections performed under this option will necessarily result in the preparation of regulatory action recommendations.

INSTRUCTIONS

1. The inspection will focus on the major systems that impact on the safety and effectiveness of all sterile products manufactured by the firm:

    - sterilisation procedures applied to the drug product; components; container/closures; product contact equipment and surfaces
        - water systems
        - air handling
        - environmental monitoring
        - handling of incoming components
        - packaging and labeling
        - laboratory
        - lyophilization (where applicable)

2. It is suggested that one drug product be selected and followed throughout; if the firm utilises more than one type of drug **product** sterilisation process, one drug product representing each type of sterilisation process should be selected.

    When selecting a drug product for review, drugs that are the subject of DPPRs or listed in the firm's complaint files should be considered.

    Drug product information to be reported:

    A. Name of Selected Drug
    B. Dosage Size
    C. Strength
    D. itch sizes
    E. Number of batches per year

    Describe what type of sterile drug products are manufactured by this firm:

    F. SVP
    G. LVP
    H. Ophthalmics
    I. Sterile Otics
    J. Sterile Bulk
    K. Other (identify)

    Please indicate whether any of these products are lyophilized.

Current Change

The report should include separate sections for each unique drug product and sterilisation process investigated.

3.   **Attachment A** has been provided as a reference guide for the type of information that should be evaluated in a sterile process inspection.

SAMPLE COLLECTION

Collect *documentary or physical* samples, including 'in-process samples where possible, to document any suspected adulteration and misbranding problems encountered during the inspection.

If microbiological contamination is suspected, document where possible the conditions which could contribute microbiological contamination to the product *both by collecting records and physical samples taken aseptically at points where such contamination might occur, such as from the WFI system. Products found positive on initial sterility testing should also be considered for sampling.

Physical samples should not be collected if the estimated level of microbial contamination is low.*

Collect samples for particulate matter contamination where inspectional observations indicate poor manufacturing practices have possibly contributed to the introduction of particulate matter into these products or where finished product controls are inadequate to assure rejection of such units.

Sample Size

For guidance in determining sample sizes for endotoxin and sterility evaluation, refer to the respective Drug Surveillance Request (DSR). Such sampling may be accomplished to meet District obligations under that program, as appropriate.

Reporting

*The investigator will utilise sections 590, 591, and 592 of the IOM for guidance in reporting inspectional findings.

For inspections made pursuant to specific assignments from HFD33G, all appropriate program areas should be fully reviewed and reported for all firms inspected, regardless of EI classification.

Attachment of standard operating procedures (SOPs),specifications, or other documentation in response to a question and/or to illustrate a deficiency is acceptable provided the response/deficiency is clearly described in all accompanying narrative.*

Notify supervisors immediately if potentially serious health hazards exist.

Current Change

PART IV – ANALYTICAL

*ANALYZING LABORATORIES

1. Routine chemical analyses: all District laboratories except W#C and MLMI.

2. Sterility Testing:

Region Examining	Laboratory
NE, MA	NYK-RL
SE	SE-RL
MW	MLMI
SW, PA	SAN-DO

3. Other microbiological examinations: WEAC, NYK-RL (FOR NYK and BUF), SJN, BLT, SE-RL, CIN (for NWK and CIN), LOS, SAN,SEA, DAL and DEN (for DAL and DEN). Salmonella Serotyping Lab: MLMI.

4. Chemical cross-contamination analyses by mass spectrometry (MS): NYK-RL, DAL, SE-RL, DET, DEN, and LOS. Non-mass spectrometry laboratories should call one of their own regional MS labs and/or Division of Field Science (HFC-142) to determine the most appropriate MS lab for the determinations to be performed.

5. Chemical cross-contamination analyses by Nuclear Magnetic Resonance (NMR) spectroscopy: SE-RL, NYK-RL, PHI and DET.Non-NMR laboratories should call one of their own regional NMR labs and/or Division of Field Science (HFC-142) to determine the most appropriate NMR lab for the determinations to be performed.

6. Antibiotic Analyses:

Examining Laboratory	Drug Product
DEN-DO	Tetracyclines
	Erythromycins
NYK-RL	Penicillins
	Cephalosporins
Division of Drug	All Other Antibiotics Biology, Antimicrobial Drugs Branch, HFD-178*

7. Bioassys: Division of Research and Testing (HFD-470)

8. Particulate Matter in Injectables: MLMI (HFD-470).*

ANALYSIS

1. Samples are to be examined for compliance with applicable specifications. Check analyses will be by the official method, or when no official method exists, by other validated procedures. See CPG 7152.01

2. The presence of cross-contamination must be confirmed by a second method. Spectroscopic methods, such as MS, NMR, UV-visible, or infrared are preferred. However, a second chromatographic method may be employed, provided the chromatographic mechanisms are different (e.g., ion-pairing vs. conventional reverse phase HPLC).

3. Sterility testing methods should be based on USP XXI and the *Sterility Analytical Manual*, 1981. Other microbiological examinations should be based on appropriate sections of USPXXI and BAM, 6th Edition, Chapter VII, *Salmonella* and current supplement

Current Change

PART V – REGULATORY/ADMINISTRATIVE STRATEGY

The therapeutic significance of the drug product and the potential adverse effect of the GMP deviation on the finished product must be considered in determining whether or not a regulatory action and/or administrative action (e.g., withholding NDA/##, G# QAP non-acceptance) is indicated.

When the nature of the deviations is considered in relation to therapeutic significance of the product(s) and it is determined that they pose a minimal risk to the consumer, voluntary correction by management should be sought as the primary action. *However, this is not to imply that regulatory action will **never** be taken in such cases.* The district should require that all communications for achieving voluntary compliance by firm management be submitted in writing and contain a time schedule for completion. *The field should determine if the schedule is a reasonable time frame and should monitor their progress.*

When voluntary action is not accomplished or when the deviations observed pose a threat to the consumer, formal regulatory and/or administrative action should be recommended. When deciding the type of action to recommend, the initial decision should be based on the seriousness of the problem and the most effective way to protect the consumer (i.e., when non-sterile injectables are found, injunction/recall would be the action(s) of choice). Outstanding instructions in the Regulatory Procedures Manual (RPM) should be followed.

NOTE: The lack of a violative physical sample is **not** a bar to pursuing regulatory and/or administrative action providing the GMP deficiencies have been well documented. Physical samples found to be in compliance likewise are **not** a bar to pursuing action under UP charges.

*The following list represents examples of deficient practices which the Center believes could warrant regulatory and/or administrative action:

1. Contamination with filth, objectionable microorganisms,toxic chemicals or other drug chemicals; or a reasonable potential for contamination by same, with demonstrated avenues of contamination such as contact with unclean equipment or through airborne contamination.
2. Failure to assure that each batch conforms to established specifications, such as NDA, USP, customer specifications, and label claims.
3. Distribution of product which does not conform to established, specifications.*

*4. Use of test methodology which is not adequate or validated.

5. Deliberate blending for the purpose of diluting and hiding pyrogenic, microbiological or other noxious contamination, or where blending of a non-standard batch with one meeting specifications results in one blended batch meeting minimum specifications.
6. Failure to assure that each batch is of uniform character and quality (homogeneous).
7. Conducting packaging and labeling operations in such a manner as to introduce a significant risk of mislabeling, for example, the use of cut labels which are similar in appearance without some type of 100 percents electronic verification system for the finished product.
8. Failure to keep adequate records, including:
   - Date(s) of manufacture
   - Quantity manufactured
   - Lot number
   - Test results and dates
   - Labelling records and specimen of label used
   - signature of person(s) responsible for accomplishing significant steps including:
   1) determining yield

Current Change

    2)   examining labeled containers for correctness of label

    3)   testing for conformance to specifications

    4)   blending, if required

    5)   assuring conformance with established manufacturing procedure (6) reviewing production and testing records and authorizing release for distribution

9.   Failure to record distribution by lot number in a manner which would permit prompt recall.

10.   Failure to have any information which would establish stability for the intended period of use.*

PART VI – REFERENCES, ATTACHMENTS, AND PROGRAM CONTACTS REFERENCES OR AIDS

A.   *Inspection Operations Manual*, Chapter 5, Part 542.58 – Sterile Products.

B.   Proposed CGMP's For Large Volume Parenterals published in the *Federal Register*, June 1, 1976.

C.   *United States Pharmacopeia*, latest revision and its supplements.

D.   *"Guideline on Validation of the Limulus Amebocyte Lysate Test as an End-Product Endotoxin Test for Human and Animal Parenteral Drugs, Biological Products, and Medical Devices," December,1987.*

E.   Inspectors Technical Guide, Number 1, 1/9/13, "Sterilising Symbols (D, z,

F.   "Understanding and Utilizing Values", Akers, Attia and Avis,"Pharmaceutical Technology", May 1978, pages 31-35.

G.   *Inspectors Technical Guide, Number 5, 6/9/72, "Ethylene Oxide a Sterilisations, I. Calculation of Initial Gas Concentration".

H.   *Principles and Methods of Sterilisation in Health Sciences*, Charles C. Thomas Co., (1969), p. 508.*

I.   *Inspectors Technical Guide, Number 6, 4/28/72, "Leak-Testing Sealed Ampuls of Parenteral Solutions"*

J.   *"Parenteral Preparations", Avis, Chapter 36, pp. 498-524 in *Remington's Pharmaceutical Sciences*; Edit. Martin; Mack Publishing Co., (1965).*

K.   *Inspectors Technical Guide, Number 24, 7/30/76, "Air Velocity Meters".*

L.   *Inspectors Technical Guide, Number 25, 9/1/76, "Ethylene Oxide Sterilisations, II. Graphical Aid to Determine Gas Concentration".*

M.   *Inspectors Technical Guide, Number 32, 1/12/79, "Pyrogens, Stilla Danger".*

N.   Inspectors Technical Guide, Number 36, 10/21/80, "Reverse Osmosis".

0.   *Inspectors Technical Guide, Number 41, 10/18/85, "Expiration Dating and Stability Testing for Human Drug Products".*

P.   *Inspectors Technical Guide, Number 43, 4/18/86,."Lyophilization of Parenterals".*

Q.   Federal Standard 209, #current revision,*

R.   **Remington's Pharmaceutical Sciences**, *Current Edition*

S.   *"Guideline on Sterile Drug Products Produced by Aseptic Processing, # June 1987.*

T.   *"Guideline on General Principles of Process Validation," May1987.*

U.   *Regulatory Procedures Manual, Part 8.*

V.   "Guideline to Inspection of Bulk Pharmaceutical Chemical Manufacturing, Revised November 1987.

Current Change

ATTACHMENTS

A   Reference points to be covered as appropriate to the type of inspection being performed, and the type of product and/or manufacturing system being evaluated.

B   To be completed for each type of Biological Indicator and/or Product.

CONTACTS

*The area code for commercial calls to all headquarters contacts is 301.

* A. ORA

1.   Jay S. Allen
      Investigations Operations Branch/DFI/ORO (HFC-I33) Telephone: FTS 443-3340

2.   **Methods Inquiry**

      Division of Field Science/ORO (HFC-I40), Telephone: FTS443-*3007*

B. CDER

Manufacturing Surveillance Branch (HFD-336) Division of Drug Quality Evaluation Telephone: 8-295-8107

PART VII – CENTER RESPONSIBILITIES

The Division of Drug Quality Evaluation (HFD-330) will evaluate all reports. Results of these evaluations will be shared with the field, ORA, and interested headquarters units.

GUIDE TO EVALUATION OF STERILE PROCESS INSPECTIONS

The following reference questions are provided for evaluation of specific drug products, manufacturing systems, and quality control procedures.

The points have been numbered for easier reference in the EIR narrative.

COMPONENT STORAGE AND PREPARATION

1.   Does the firm have adequate written procedures describing the receipt, handling, that are represented to be sterile and/or pyrogen free? (per 21 CFR 211.80 – 211.94; 211.184)

2.   Have these procedures been followed for the selected drug product?

3.   Are any colorants used (none are permitted)?

4.   Does the firm have written control procedures that adequately describe the receipt, storage, sampling, issuance, and reconciliation of labeling and packaging materials? (per 21 CFR 211.122 – 130; 211.134 and 211.137).

5.   Does the firm use cut or roll labels?

6.   Are the labels similar in color, shape, size and format for different products or potencies?

7.   Does the firm use any type of electronic label verification system (bar codes, machine vision systems, etc.)? Describe

8.   Is the label verification on receipt, on line, or both?

9.   Is any printing done on line of label text, lot number,expiration date, etc.?

Current Change

10. Does the firm use dedicated packaging lines?

11. Are the samples of labels used for acceptance (proofing) of labels from vendors based on a statistical plan? Describe sampling plan.

12. Are labels printed by the firm or by an outside vendor?

13. Have these procedures been completely and accurately followed for the subject drug **product**?

EVALUATION SYSTEMS

14. Does the firm have an SOP on vendor audits?

15. Has the firm audited the (a) component, (b) container, (c) closure, and (d) label vendors? Report the dates of last audits.

16. Does the firm have written procedures for the production and process control of drug products? (per 21 CFR 211.100, – 211.115;211.186; 211.188; 211.192)

17. Have these production and process control records been approved by the firm's quality control unit and by designated organisational units?

18. Have these process control records been completely and accurately prepared for the subject drug product?

19. Briefly describe the firm's procedures for changing any of the standard operating procedure documents described above.

20. Does the firm have written procedures for the review and approval of all drug production and control records before release of the batch for distribution? (21 CFR 211.192)

21. Were these procedures followed in the review of the selected drug product?

22. What are the firm's procedures for the investigation to be made following any unexplained discrepancy found in batch production records, or the failure of a batch or any of its components to meet specifications? (21 CFR 211.192)

23. Were these procedures followed accurately and thoroughly concerning any batch discrepancies/failures of the selected drug?

24. Does the firm have written laboratory control mechanisms, including change control procedures, which describe conformance to established specifications and standards for the selected drug product? (21 CFR 211.160-167)

25. Were all specified in-process and end product tests performed on the selected drug product?

26. Were all specifications met?

MAJOR SYSTEMS AND PROCESSES

Monitoring of Environment

27. Is the air supplied to critical areas (exposed product/filling areas) filtered through HEPA filters under positive pressure?

28. Is the air flow in critical areas laminar when delivered to the point of use? At what velocity? Is velocity determined at the critical area or at the filter face?

29. How is the air filtered that is supplied to controlled areas (where unsterilised product, in-process materials, and container/closures are prepared)?

30. What are the firm's air quality classifications for:
    a. exposed product areas
    b. filling area

Current Change

c.   surrounding plant areas

31. Is room classification system based upon Federal Standard 209d or other?

32. Are HEPA filters efficiency tested?

33. How often are HEPA filters integrity tested? What test method is used?

34. How often are air flow velocities checked for each HEPA filter?

35. Does the firm have a written monitoring program for classified areas that included a scientifically sound sampling schedule that describes sampling locations, their relation to the working level, and frequency? Describe the basis for the sampling program. (21 CFR 211.160)

36. Are both viable and non-viable particulate samplings performed in all classified areas during production?

37. Report the frequency of viable sampling using "active" sampling methods for:

a.   exposed product areas

b.   filling areas

c.   surrounding areas

38. Report the limits used, length of sampling period, and if sampling is done during production or at rest.

39. Report the type of viable sampling equipment use (STA, Centrifugal sampler, etc.)

40. Does the firm have data on the ability of these samplers to recover organisms without deleterious effect on survivability such as through impact or dessication of organisms or media?

41. Report the actual volume of air sampled per location.

42. Are settling plates used? Describe the length of exposure period; sampling frequency; location (including proximity to critical operations); microbial limits.

43. Are recovered microorganisms routinely identified? To what level (genus, species)?

44. Are the culture media used in the viable monitoring program shown to be capable of detecting molds and yeasts as well as bacteria by means of growth promotion tests? Is anaerobic monitoring performed?

45. What media are used?

46. Are deactivators (e.g., penicillinase) use for antibiotics or other bacteriocidal/bacteriostatic substances? Has the firm shown that these are effective? (Are records available? Are calculations correct?)

47. What incubation periods are used and at what temperature?

48. How often is non-viable particulate sampling performed in classified areas:

a.   exposed product areas

b.   filling areas

c.   surrounding areas

49. What sampling device is used? What volume of air is sampled?

50. How many samples are collected per location? Are results averaged?

51. When was sampling equipment last calibrated?

52. Were environmental sampling results within specifications during the manufacture of the batches of the selected drug product? (Describe any deviations and firm's response.)

53. How often is monitoring performed on filling room personnel?

54. What are the firm's alert and action limits for personnel monitoring?

55. What type of monitoring is done?

Current Change

56. Does the firm have written procedures for the monitoring of product contact surfaces?

57. What type of contact surface monitoring devices are used (RODAC, swabs, etc.)?

58. Any changes in the air handling or environmental monitoring systems since the last EI? Were changes evaluated by management regarding the need for re-validation?

FACILITY CLEANING/DISINFECTION

59. Are there written procedures describing the cleanup, sanitisation/sterilisation of drug production equipment and utensils?

60. Were these written procedures describing the cleanup, sanitisation/sterilisation of drug production equipment and utensils?

MANUFACTURING FACILITIES

Gowning

61. Briefly describe the firm's procedures for initial gowning and re-gowning after breaks.

FREEZE-DRYING (LYOPHILIZATION)

62. If lyophilization is performed by an outside firm, report the firm's name and address. If lyophilization is performed in-house, report the following:

63. Manufacturer of lyophilizer.

64. Percentage of firm's products which are lyophilized:

65. Describe the heating and cooling systems used in the lyophilizer; the vacuum system; what gas is used to break the vacuum and whether it is sterile; and the temperature controlling system.

66. Briefly describe preparation of the sterile product for drying, including procedures for protecting the product from contamination while loading into the lyophilizer.

67. How is stopper seating vials performed?

68. If performed automatically, is it under vacuum, or if not under vacuum, what gas is used and how is it sterilised?

69. If vials are stoppered outside of chamber, describe how lyophilized product is protected from contamination during this procedure.

70. Is the lyophilizer steam sterilizable?

71. Describe chamber clean-up procedures between batches of the same product and between different products (including sterilant/cleaning agent used and exposure cycle),

72. How are inert gas or air supply lines cleaned? Sterilised?

Lyophilization Validation

73. Is the aseptic handling of lyophilized products validated by:

    a. media fill process

    b. other (describe)

74. If media fills are used, are the fills performed uniquely to evaluate the lyophilization process, or as part of a validation of the aseptic filling process?

75. Are the same acceptance criteria (allowable contamination rate)used as for liquid filling? If not, what criteria are used?

Current Change

76. What number of units are filled for lyophilized product?

77. During validation, what level of vacuum is pulled on the lyophilization chamber?

78. Do media fill vials remain in the lyophilization chamber under vacuum as long as production vials?

79. Is the media frozen?

80. Is environmental monitoring performed during loading of the lyophilizer both during – production as well as during validation?

81. Does the firm have data on growth promotion of the media after the above procedures?

82. Is environmental monitoring performed during unloading of the chamber during production as well as during validation?

83. What is used to break vacuum (nitrogen, air, other gas)?

84. Has the firm validated the lyophilization cycle (e.g., time, rate of heat input, temperatures, eutectic melting point) for each product? (Review validation records for selected drug product and at least three other drug products with different physical and chemical characteristics.)

85. Review at least three lyophilization production records for the products referenced above., are the cycle parameters and observed results within the validated cycles?

86. What are the firm's criteria for acceptable vs. unacceptable runs, including general appearance, moisture, etc.?

UTILITY SYSTEMS

WFI

87. What is the source water for the plant?

88. Briefly describe the treatment applied to the source water before it is considered acceptable for use in manufacture.

89. What type of water is used for

    a. bulk product compounding

    b. non-product contact surfaces

    c. washing of container-closures

    d. final rinse of container-closures

    e. final rinse of product contact surfaces

    f. water used for production of sterile product

90. What process is used to produce Water for Injection (WFI)/sterile WFI?

91. If distilled water is prepared, briefly describe the production, delivery and storage system and temperatures.

92. Briefly describe pyrogen/microbial control in the WFI system,

93. Does the firm have written procedures detailing the specifications and monitoring program for all types of water used in the plant?

94. Review process water sampling results for at least two months preceding and one month following the manufacture of batches of the selected drug product. Were specifications met? (If not, describe any deficiencies and the firm's response.)

95. Have there been any changes in the process water system since the last EI? Have these changes been evaluated for the need for revalidation of the water system?

Current Change

DEPYROGENATION

96. What type of depyrogenation procedures are used? If dry heat is used, report the cycle time/temperature.

97. Is WFI Washing used?

98. If a caustic wash is used, what agent is used?

99. Is ultrafiltration used?

100. Describe any other methods used.

101. Which of the above procedures are used on:

    a. Raw materials

    b. Drug product containers

    c. Drug product container closures

    d. Sterile product contact surfaces

    e. Manufacturing equipment

    f. Drug product

102. What method is used for determining endotoxins? Rabbit or LAL?

103. Have all depyrogenation procedures been validated to demonstrate a minimum of a three-log reduction in endotoxin content? Does the firm have data on recovery of the original endotoxin challenge amount?

104. Have any depyrogenation procedures been changed/added/deleted since the last EI? Have such events been evaluated for revalidation?

105. If pyrogen testing is performed by a facility other than the manufacturer, report the name and address of the facility.

CONTAINER & CLOSURE INTEGRITY

Particulates

106. Evaluate the adequacy of the firm's procedures and criteria used to inspect units for particles.

107. What is the duration of duty of operators doing visual examinations?

108. What is the firm's general rejection rate for particulate matter, and type(s) of **particulates which predominate**?

109. Report results of investigations into sources/types of particulates (other than visual examination.)

110. In evaluating the adequacy of the firm's particulate matter quality control procedures, make a visual examination of are presentative number (at least 100) of units that have passed the firm's inspection. Report the number of units examined; passed; failed. (This examination can be performed on warehouse stock. What is the firm's normal level of rejects for particulates (%)?)

111. If the firm uses an automated method, provide the name and sensitivity levels of the equipment.

112. What are the firm's alert, action, and reject levels for particulate contamination?

113. Briefly describe the firm's procedure when each of these levels is exceeded (attach SOP if appropriate).

114. What is the frequency of testing for particulates?

115. If particulate levels are determined by a facility other than the manufacturer, report the name and address of the facility.

Current Change

116. Has there been a change in particulate contamination testing since the last EI (e.g., equipment)? Has the change been evaluated for re-validation?

STERILISATION SYSTEMS

General

117. What types of parenteral drug products are manufactured by this firm?

    a.   solutions

    b.   suspensions

    c.   lyophilized

    d.   powder fills

118. Are form, fill and seal packages used for any of these products?

119. Which of the above products are manufactured aseptically?

120. Are any products which can be produced using terminal sterilisation produced aseptically? (If so, describe the firm'srationale for producing them aseptically.)

121. Do all of the firm's aseptically filled products contain a preservative; would they pass a USP preservative efficacy test?

122. If the information requested is unavailable at the site of the inspection, determine the name and address of the firm where the information can be gathered.

If the firm does not have a record of a particular parameter, or will not reveal it, a notation to that effect should be made.

NOTE: If biological indicators are used during production and/or validation cycles, complete Attachment A for each type of indicator used.

123. **Contract Sterilisers**: If the firm being inspected is a contract steriliser, choose one drug component/container- closure system and follow it through one complete sterilisation cycle. Complete as many items as applicable, especially under GENERAL INFORMATION and the type of sterilisation procedure used on the drug component/container closure system. Refer to Compliance Policy Guide 7150.16 for Agency policy concerning the status and responsibilities of contract sterilisers.

Steam Sterilisation

124. Report the steam steriliser (autoclave) manufacturer.

125. What is the internal volume of the autoclave?

126. What is the sterilant (e.g., steam, air over pressure, superheated water)?

127. If jacketed, what pressure/temperature is maintained in the jacket as opposed to the chamber?

128. What type of vent filters are used and how often are they integrity tested?

129. Are vent filters hydrophobic? Are the vent filter housings heated to prevent condensation?

130. Is cycle control manual or programmed?

131. What type of monitoring and controlling sensors are used (e.g., mercury-in glass thermometer, thermocouple, RTD, pressure gauge)?

132. How are these sensors calibrated? Are the standards NIST traceable (where appropriate)?

133. If the autoclave is equipped with a steam spreader, describe it (more than one steam entry line would be considered in this category).

Current Change

134. If more than one autoclave is used by the firm, what is the system's capacity for steam production in relation to all autoclaves being in operation at the same time?

135. What are the sterilisation cycle parameters? (Compare Master Process Record/SOP specifications against processing records completed for the selected drug product.)

136. What are the firm's specifications and observed parameters for:

    a. Time

    b. Temperature

    c. Pressure (psi, in. Hg)

    d. Pressure Come Down Rate (specify pressure and time)

136. Where is the cycle controller sensor located?

137. How are each of the above parameters monitored (specify when not monitored)?

138. Is the "cold spot" in each load monitored during each autoclave cycle?

139. Report other characteristics (e.g., air quality, water quality, alarms, etc.).

140. Have any changes in the steam sterilisation system occurred since the last EI? Have these changes been evaluated for the need for re-validation?

Steam Sterilisation Validation

141. Does the firm have written procedures for validation that include:

    1) installation qualification of equipment

    2) operational qualification of equipment

    3) performance qualification with product

    4) description of circumstances requiring re-validation of the system and procedures to do so

142. Does validation documentation include:

    1) empty chamber heat distribution studies:

        a. number of runs?

        b. was cold spot determined?

        c. report firm's allowable variation and actual variation found

    2) heat penetration studies performed:

        a. for each type of loading pattern/for each container size utilised?.

        b. number of runs per pattern?

        c. was the "cold spot" determined for each pattern?

    3) what type of temperature measurement system was used? Does it provide a separate printed reading for each thermocouple?

    4) what type of thermocouples were used, and were they calibrated before and after each run?

    5) was an ice-point reference standard used for calibration?

    6) was the high temperature reference standard NBS traceable?

    7) If biological indicators were used during validation runs:

        a. type of indicator

Current Change

      b.   source of indicator

      c.   organism used

          1)   concentration

          2)   D Value.

      d.   were BIs used in an "end point" or "count reductions"mode?

          1)   If any positive BIs were found (when not expected), what was the firm's response? Include records and response in exhibits to EIR

143. In the event a heat distribution or penetration variance was disclosed during the studies, how did the firm correct or allow for it?

144. Has the firm determined lag times for all container sizes,product viscosities, etc. and adjusted their cycles accordingly?

Dry Heat Sterilisation (separate from depyrogenation)

145. If dry heat sterilisation is performed by an outside firm, report the name and address.

146. If dry heat sterilisation is performed in-house, complete the following:

      a.   Steriliser manufacturer

      b.   Size (internal dimensions)

      c.   Location of the heat source

147. Is steriliser equipped with a fan or is heat distributed by convection only?

148. Is the cooling air HEPA filtered?

149. How often are the HEPA filters integrity tested?

150. Is the cycle control manual or automatic?

151. What type of monitoring and controlling sensors are used (e.g., mercury-in-glass thermometer, thermocouple, RTD, pressure gauge)? How often are they calibrated?

Sterilisation Cycle Parameters

152. What are the sterilisation cycle parameters (compare Master Process Record/SOP – specifications against processing records completed for the selected drug product)

153. What are the firm's specifications and observed parameters for time and temperature? 154. Where is the cycle controller sensor located?

155. How are each of the above parameters monitored (specify when not monitored)?

156. Report other characteristics (e.g., air quality, alarms, etc.). If biological indicators are used during regular production cycles, complete Attachment B.

157. Have any changes in the dry heat sterilisation system occurred since the last EI? Have these changes been evaluated for the need for re-validation?

158. Does the firm have written procedures for validation that include:

      a.   installation qualification of equipment

      b.   operational qualification of equipment

      c.   performance qualification with product

      d.   description of circumstances requiring re-validation of the system and . procedures to do so

Current Change

159. Does validation documentation include:

1) empty chamber heat distribution studies:

    a. number of runs

    b. was cold spot determined

    c. report firm's allowable variation and actual variation found

2) heat penetration studies performed:

    a. For each type of loading pattern/container size to be utilised

    b. number of runs per pattern

    c. was the "cold spot" determined for each pattern

3) what type of temperature measurement system was used?

4) were thermocouples calibrated before and after each run?

5) was an ice-point reference standard used for calibration?

6) was the high temperature reference standard NIST traceable?

7) If biological indicators were used during validation runs:

    a. type of indicator

    b. source of indicator

    c. organism used

        1) concentration

        2) D Value

    d. were BIs used in an "end point" or "count reduction" mode?

        1) If any positive BIs were found (when not expected), what was the firm's 1)response?

160. In the event a heat distribution or penetration variance was disclosed during the studies, how did the firm correct or allow for it?

161. Has the firm determined lag times for all container sizes and adjusted their cycles accordingly? Chemical Sterilisation/Disinfection/Sanitisation

162. What product (container/closure; manufacturing equipment) is sterilised by this method?

163. What is the sterilant, sterilant concentration and exposure time?

164. After sterilant exposure, does the item receive a final rinse and/or allowed to air dry?

165. Report the specifications for the final rinse water and/or air quality.

166. Briefly describe how the sterilised item is protected from recontamination before use.

167. Are there written procedures detailing all chemical sterilisation processes?

168. Were these procedures followed with reference to their employment in the manufacture of the selected drug product?

Chemical Sterilisation Validation

169. Describe the firm's validation of all chemical sterilisation processes utilised, including completeness of documentation, the type of microbial challenges employed, and results.

170. Have any chemical sterilisation processes changed since the last EI? Have these changes been evaluated for the need for revalidation?

Current Change

Ethylene Oxide Gas Sterilisation (EtO)

171. If EtO sterilisation is performed by an outside firm, report the name and address.

172. If EtO sterilisation is performed in-house, report the sterilise r(autoclave) manufacturer.

173. What is the internal volume of the EtO sterilisation?

174. What is the ratio of EtO to the carrier (%)(e.g., 12% EtO/88% Freon; 10% EtO/90% CO, etc.)?

175. Is a Certificate of Analysis received with the gas or analyzed by the user (specify which)?

176. Identify equipment/container/closure/other that is EtO sterilised as part of manufacture of selected drug product.

177. What is the sterilisation cycle? (Compare Master Process Record/SOP specifications against processing records completed for the selected drug product.)

178. Is there **preconditioning** (specify external or in chamber, or both, at start of cycle)?

179. What are the firm's Master specifications and observed parameters for time, relative humidity and temperature?

180. Are biological indicators included in prehumidification cycle?

Cycle Parameters

181. What are the Master specifications and observed parameters for the following:

    a. Vacuum (mm Hg, in. H_2O)

    b. Air Venting other than by vacuum (prior to or during gas charging)

    c. Temperature

    d. Operating Pressure

    e. Relative Humidity (%)

        – at start of cycle

        – during cycle

    f. Preheating (heat exchanger) or holding temperature of gas when injected into chamber

    g. Gas concentration in chamber (mg/Liter)

    h. Use of a circulation fan

    i. Exposure to sterilant (hrs.)

    j. Use of Multiple Evacuation cycles (# of cycles)

    k. Come-Down or Evacuation Rate

182. How are each of the above parameters monitored? (specify when not monitored)

183. Specify method used to control addition of EtO to chamber (e.g., chamber pressure, EtO concentration analysis, EtO weight measurement, other).

184. Specify method of moisture addition during cycle.

185. If more than one EtO steriliser is used by the firm, or there are multiple EtO sterilisation points within one EtO sterilisation system, what is the system's capacity for maintaining established EtO levels when all chambers/sterilisation points are operating at the same time?

186. Explain length of supply line from bulk source, inside diameter, number of equipment serviced by supply line.

187. Are EtO concentration levels monitored in aeration and sterilisation work areas? Specify levels.

    Report information requested in Attachment B for biological indicators used.

Current Change

Residue Levels

188. Report procedures used to assure EtO residue removal (specify time, tamp., etc.)(e.g., hold in forced aeration area, hold in warehouse-ambient, etc.):

189. Report firm's specifications and conformance for residue levels, if any.

190. What are the firm's specifications and observed levels of ethylene oxide (ppm), ethylene glycol (ppm) and ethylenechlorohydrin (ppm)?

191. Are the residue levels established according to any standard?

192. Obtain copy of dissipation curves (specify when not available).

193. If residue levels on the products are determined by a facility other than the manufacturer, report the name and address of the facility.

194. Have there been any changes in the EtO sterilisation system since the last EI? Have these changes been evaluated for the need for re-validation?

Ethylene Oxide Validation

195. Does the firm have written procedures for validation that include:

    a.   installation qualification of equipment

    b.   operational qualification of equipment

    c.   performance qualification with items to be sterilised

    d.   description of circumstances requiring re-validation of the system and procedures to do so.

196. Does validation documentation include:

    1)   empty chamber temperature distribution studies

        a.   number of studies performed

        b.   number of probes used and their location

        c.   report firm's allowable variation and actual variation found

    2)   empty chamber EtO concentration distribution studies

        a.   number of studies performed

        b.   number and location of probes utilised

        c.   report firm's allowable variation and actual variation found

    3)   empty chamber Relative Humidity Measurement studies

        a.   number of studies performed

        b.   number and location of probes utilised

        c.   report firm's allowable variation and actual variation found

    4)   Heat/EtO penetration studies performed:

        a.   for each type of loading pattern to be utilised

        b.   number of runs per pattern

        c.   was the "cold spot" determined for each loading configuration

    5)   What type of temperature/EtO/RH measurement systems were used?

    6)   Is this equipment calibrated according to an established schedule, and traceable to an NIST standard where practible?

Current Change

7) Were measurement systems calibrated before and after each study?

8) If biological indicators were used during validation run:

    a. type of indicator used

    b. source of indicator

    c. organism used

        1) concentration

        2) D value

    d. was load placement based upon heat/EtO penetration data?

9) were BIs used in an "end point" or "count reduction" mode?

        1) if any positive BIs were found when not expected, what was the firm's response?

10) if BIs were not used to determine sterilisation effectiveness, what methods were used?

Radiation Sterilisation

197. Identify the equipment/container/closure/other that is radiationsterilised as part of the manufacture of the selected drug product.

198. Report the following:

    a. Steriliser manufacturer

    b. Radiation type (e.g., beta, gamma)

    c. Radiation Source (e.g., cobalt 60)

    d. Dosimeter type and supplier

    e. Placement of dosimeters within the load

199. If radiation sterilisation is performed by an outside firm, report the name and address.

200. Method for Certifying Dosimeter (e.g., traceability to a NIST (National Institute of Standards and Technology) primary reference, etc.).

201. If sterilisation is performed in-house, what are the radiation cycle parameters? (Compare reference to the selected drug product.)

202. What are the firm's specifications and observed levels for the following:

    a. Exposure time

        1) by batch, report time

        2) by continuous process, report convey or speed time

    b. Dose rate (Mrad/hr.)

    c. Uniformity of dose rate (+%)

    d. Total dose (Mrad) Acceptable maximum Acceptable minimum

    e. Temperature

203. How does the firm adjust exposure time for source decay and when does this occur?

204. How often does the firm do a dose mapping of the chamber?

205. How is each of the above parameters monitored (specify when not monitored)?

206. Are dosimeters and/or BIs used in routine sterilisation runs? Are they used in determining whether a sterilisation run may be released? If so, attach release specifications.

207. Have there been any changes in the radiation sterilisation system since the last EI? If so, have these changes been evaluated for the need for re-validation?

Current Change

Radiation Sterilisation Validation

208. **Dose Setting**: has a minimum sterilising dose been established for each material?

What method is used to establish the sterilising dose? (e.g., AAMIBI, 82, etc. Any dose-setting method used must take into account the quantity and resistance of the natural bioburden of the material being sterilised.)

Has a sterility assurance level (SAL) been established for the material?

209. **Product Loading Pattern**: has a loading pattern been established for each material to be sterilised? The specification for each loading pattern should describe the number and position of material units within the irradiation chamber.

210. **Dose Distribution Mapping**: has the dose distribution within the material been determined in the irradiator, using either actual material or a simulated material that approximates the density of the actual material?

    With irradiators that offer a variety of conveyor paths, dose mapping of the material must be performed for each conveyor path to be used.

    Have the zones of minimum and maximum dosages been determined (for each conveyor path utilised)?

211. **Cycle Timer Setting**: has a cycle timer setting been established for **each** material that will yield the minimum required sterilisation dose?

212. If biological indicators were used during the validation runs:

    a.   type of indicator used

    b.   source of indicator

    c.   organism used

        1)   concentration

        2)   D value

213. How was dosimeter placement correlated with BI placement in the validation loading patterns?

Aseptic Sterilisation Systems

Dry Powder Filling

214. Briefly describe the processes used for preparing the sterile drug powder (i.e., sterile filtration, crystallisation, spray drying, EtO gassing, etc.

215. Is the production facility dedicated to the product? If not, determine the potential for cross contamination with other products manufactured by the firm.

216. Are any penicillin products produced in the same facility as non-penicillin products?

217. Briefly describe the environmental monitoring performed by the firm in critical areas during actual production. (e.g., how are Class 100 conditions maintained? Where are the sampling sites?Is non-viable particulate monitoring performed?

218. Review monitoring data for several representative months of production. Were results within specifications? If not, what was the firm's response? How many months were reviewed?

219. Does the firm have written procedures describing the filling of sterile dry powders? Dry Fill Validation

220. How has the firm validated the filling operation for product homogeneity?

Current Change

221. Is the sterile filling procedure validated by media fill procedure, placebo fill procedure, or other (describe)?

222. Briefly describe media fill procedures or placebo fill procedure(including frequency; whether performed as its own batch or piggy-backed onto a production batch; number of vials routinely filled; etc.).

223. Review results of medial fills/placebo fills performed since last EI (or a minimum of 3 runs, whichever is greater). Are results within specifications? How many fills were reviewed?

224. What are the firm's specifications and procedures following an out-of-limit media fill/placebo fill result?

Placebo Fill:

225. What placebo material is used?

226. How is the placebo material sterilised?

227. Does the firm utilise the same acceptance criteria for placebo fills as for media fills? If not, and higher alert and action levels are permitted for the placebo fills, determine rationale.

228. How long and at what temperature(s) are placebo filled units incubated?

Filtration Sterilisation

I.   Sterilising Filters

229. Provide the following information on all sterilising filters used by the firm:

230. Is the filter assembly pressure tested before and after use?

231. How does the pressure used for testing correlate with the pressure used in production?

232. What type of integrity testing is performed?

233. What in the firm's procedure when the filter fails post-filtration testing?

234. How many filter failures has the firm had in past year?

235. What investigation was performed following filter failures?

236. Does the firm use a single sterilising filter or multiple (redundant) filters?

237. Has the firm or the filter supplier performed the bacterial challenge test on each lot of filter media? Summarise testing procedure (or attach documentation).

238. How are filters/filter assemblies sterilised? Is the sterilisation process validated? (attach SOP)

239. Are filters resterilised and reused? Is this procedure validated? (attach SOP)

240. Are filters changed during manufacture of the batch? Has this change frequency been validated? (attach SOP) Filter Validation

241. Has the firm or an outside supplier performed physical and chemical challenge testing of each filter and product combination in the manufacturing process to validate filter-product compatibility? (If these tests were performed by an outside firm, report the name and address). Summerise study results or attach documentation.

242. Did the test conditions duplicate, as nearly as possible, the actual conditions of production?

243. Following validation of a specific filter for a given process and product, does the firm extrapolate the validation findings to related products having similar attributes and processing conditions? If so, is the justification for such extrapolation documented? Is filter performance data correlated with filter integrity testing data as part of the justification?

Current Change

244. Microbial challenge:

    a.   is a "worst case" organism used?

    b.   do challenge tests cover:

        i.   flow conditions, pressures, volumes

        ii.   **fluid characteristics**, including pH, ionic strength,surface tension. Is there a limit of centipoise for solutions to be filtered. Has the firm determined the effect of elevated viscosity over extended time periods, if applicable?

        iii. **time**, Does validation cover "worst case" conditions? For example, the firm is doing a continuous form/fill/seal operation and using the filter over an extended period of time.

II.   Aseptic Filling

245. Briefly describe the aseptic filling processes from preparation of bulk liquid product to filling and sealing of final dosage form, including the environmental monitoring performed in critical areas during actual production (e.g., how are Class 100 conditions maintained., where are the sampling sites; is bioburden testing performed on the bulk product?)

246. Review monitoring data for several representative months of production, including the period during which batches of the selected drug product were prepared. Were results within specifications? If not, what was the firm's resense?

247. Does the firm have written procedures describing aseptic filling of liquid drug, products? Aseptic Filling Validation

248. Is the aseptic filling procedure validated by media fill procedure or other (describe)?

249. If media fills were used, report the procedures followed and the results, including: number of runs performed; how many vials were filled per run; sizes of vials and fill volume; media used, incubat' periods, temperatures and results; allowable contamination rate.., and what firm's response was to results that failed established limits.

250. Does the firm provide for periodic monitoring or revalidation of filling lines using media fill procedures?

251. If media fills are used, briefly describe procedure (including frequency; whether performed as its own batch or piggy-backed onto a production run; number of vials routinely filled.,allowable contamination rate).

252. Review results of media fills performed since last EI (or a minimum of 3 runs, whichever is greater); are results within specifications?

253. What are the firm's specifications and procedures following an out-of-limit media fill result?

254. Are media fills performed on all shifts?

255. Are all personnel included in the media fill program?

256. What system does the firm have for assuring all personnel are included?

257. If end-line filters are used in actual manufacture, are they also used during media fills?

258. What size vial or ampoule is used for media fills?

261. Is more than one medium used?

262. Will the number of samples used in a media fill vary from line to line? (If yes, please explain)

263. Are growth promotion studies performed on each type of medium used?

264. Are growth promotion studies conducted every time a media fill is done?

265. When are the growth promotion studies performed (before/after filling; after incubation; etc.)?

Current Change

266. Is the source of the growth promotion test USP XXI or other (if other, describe, including organisms used)

267. What temperatures and incubation times are used to incubate media fill samples?

268. Are microorganisms from positive vials identified according to genus?

269. Are such microorganisms correlated to those found during environmental monitoring?

Parametric Release: Parametric release of terminally heat sterilised drug products (in lieu of end product sterility testing) is only permitted pursuant to an approved (supplemental) New Drug Application. A copy of the NDA portion containing the approved parametric release specifications should be requested from the Division of Manufacturing and Product Quality, Sterile Products Branch (HFD-322) prior to initiating inspection.

Refer to Compliance Policy Guide 7132a.l3 for information on this sterility release procedure. Questions or problems should be directed to:

> Sterile Products Branch, HFD-322
> Tkrry Munsen, Branch Chief
> 8/295-8095.

LABORATORY

Stability and Expiration Dating

270. What is the expiration dating on the subject product?

271. Do the stability studies performed on the selected product include preservative effectiveness testing?

272. What is the source of Analytic Method? Sterility Testing

273. If sterility testing is performed by an outside laboratory, report the name and address.

274. Has the firm audited the contract laboratory procedures and test results? What is the date of the last audit?

275. If sterility testing is performed in-house, what are the qualifications off personnel responsible for sterility testing?

276. Does the firm have adequate written procedures for the sampling and testing of products for sterility, potency, pyrogens, particulate matter, and other appropriate tests?

277. Review sampling and testing records for three lots of the selected drug product: were all required tests performed appropriately, and were results within specifications?*

278. Review sterility testing results summary data accumulated since the last EIR, or the last six months, whichever is greater. What is the firm's overall failure rate upon:

    a. initial testing

    b. first retest

    c. second retest

279. How much time routinely elapses between sterilisation of a product and when sterility test samples are put on test? What are the holding conditions of lot samples waiting to be tested?

280. What is the average number of lots sterility tested per month?

281. Describe the firm's procedures for evaluating batches that fail the initial sterility test. How are "false positives" determined? If the cause of a sterility failure cannot be determined as arising from the production environment or laboratory error, what decision is made by the firm concerning the release of the lot in question? (attach retest protocol)

Current Change

282. Are the "false positive" rates similar for aseptically filled products and terminally sterilised products? If the rate for aseptically filled products is markedly higher than for terminally sterilised products that are manipulated in a similar manner during sterility testing, then this rate indicates truely contaminated rather than "false positive", and an in-depth review should be made of the sterilisation process. Pyrogen Testing

283. If the firm is using LAL for pyrogen testing, has the procedure been validated for all products on which it is used? Environment

284. What air quality is provided in the laboratory environment?

285. What air quality is specified for sterility testing areas? Is laminar air flow provided?

286. What type of environmental monitoring is performed in the laboratory (e.g., type and location of sampling; sampling equipment; frequency)?

287. Compare the firm's written environmental specifications for the laboratory with sampling data for the previous three months. Are results within specifications? If not, what action was taken by the firm with reference to:

    a.   environmental specifications;

    b.   product undergoing testing at the time of the out-of-spec results?

CALIBRATION

288. Have all testing, measuring, monitoring equipment (thermometer, thermocouple systems, pressure gauges, pH meter, etc.) used in production and in laboratory testing been calibrated?

289. Is equipment periodically checked for accuracy and recalibrated?

290. Are there written procedures covering the calibration and periodic checking and recalibration of production and laboratory equipment, including set intervals and specifications?

Computers

291. Report if the firm is using an automated process control system. Report if source code documentation is available at the firm.

BIOLOGICAL INDICATOR USAGE USE SEPARATE PAGE OR EACH TYPE OF INDICATOR AND/OR PRODUCT

1.   What type of indicator is used (e.g., inoculated carrier, inoculated product, inoculated simulated product, etc.).?

2.   Is the source of the indicator commercial (report brand name and manufacturer) or prepared in-house? (Identify supplier of organism, describe means of propagation and storage, and method of preparation.)

3.   What organism is used (specify Genus, species)?

4.   What is the challenge level of the biological indicator prior to exposure to sterilant?

5.   Does firm verify viable spore count on each lot BIs?

6.   Does the firm or the indicator labeling claim to meet US performance criteria for steam or EtO biological indicators?

7.   Does the firm perform USP testing on each lot of BIs received?

8.   What is the approximate D-value of the biological indicator?

9.   How many indicators are used per steriliser load?

Current Change

10. Describe the firm's procedure used to assay the indicators after exposure (i.e., USP, NASA, etc.). (Specify growth media used, optimal and actual incubation time and temperature.)

11. How are the indicators packaged for sterilisation?

12. Are these biological indicators located in the most difficult.-to sterilise product sites (explain)?

13. Draw a diagram of the distribution of biological indicators in the loading pattern(s) for the selected drug product.

14. What is the elapsed time (hrs.) between removing indicators from the steriliser and testing? Are there time limits established for this period? What happens if they are exceeded?

15. What is the average number of steriliser loads processed per month?

16. How many steriliser loads with positive indicator test results are there per year?

17. What is the disposition of lots with positive bioindicator tests results (release, relabel, resterilise, destroy, etc.)?

18. Describe biological indicator storage conditions:

    a. Type of room, cabinet, etc. (if stored in freezer or refrigerator, state if frost-free)

    b. Temperature

    c. Relative humidity (if known)

    d. Proximity of storage area to steriliser(s)

19. If biological indicators are used to monitor EtO processes, also describe:

    a. The storage history of the particular lot used in processing selected drug product (including dates of lot expiration lot received, lot sampled, etc.)

    b. How long has this particular lot been held by the manufacturer?

    c. If there a potential for the biological indicators tobe exposed to EtO in the environment before use? (If yes, explain)

20. If the biological indicator testing is performed by a facility other than the primary manufacturer, report the name and address of the facility.

21. Does the firm use chemical process monitor(s) to indicate cycle exposure or to measure one or more cycle parameters? (Report type, brand name and how used.)

Current Change

10.8   References

1.   Halls, N.A. (1994). Achieving Sterility in Medical and Pharmaceutical Products New York: Marcel Dekker.

2.   Sharp, J. (1995): "What do we mean by sterility?", PDA Journal of Pharmaceutical Science and Technology, **49**: 90-92.

3.   Tidswell, E. (2011). "Sterility" in Saghee, M.R., Sandle, T. and Tidswell, E.C. (Eds.) Microbiology and Sterility Assurance in Pharmaceuticals and Medical Devices, New Delhi: Business Horizons, 589-602.

4.   Sharp, J. (1991). Good Manufacturing Practice: Philosophy and Applications, Buffalo Grove: Interpharm Press.

5.   Food and Drug Administration. Guideline on Sterile Drug Products Produced by Aseptic Processing, Food and Drug Administration, Rockville, MD: 2004.

6.   Kallings, L.O., Ringertz, O., Silverstolpe, L. & Ernerfeldt, F. (1966). Microbial contamination of medical preparations. Acta Pharm Suec, **3**: 219-228.

7.   Borchert, S.J., Abe, A., Aldrich, D.S., Fox, L.E., Freeman, J.E., & White, R.D. (1986). Particulate matter in parenteral products: a review. Journal of Parenteral Science and Technology **40**(5): 212-241.

8.   21 CFR 211: Current Good Manufacturing Practice for finished pharmaceuticals. Code of Federal Regulations, Food and Drugs, U.S. Government Printing Office, Washington DC/USA.

9.   FDA. "Guidance for Industry. Sterile Drug Products Produced by Aseptic Processing–Current Good Manufacturing Practice," (FDA, Rockville,MD, August 2004).

10.   Sutton, S. and Tirumalai, R. (2011): "Activities of the USP Microbiology and Sterility Assurance Expert Committee During the 2005–2010 Revision Cycle", American Pharmaceutical Review, July/August 2011, 12-30.

11.   U.S. Food and Drug Administration. "Pharmaceutical cGMPs for the 21st Century – A Risk Based Approach", Food and Drug Administration, Rockville, MD; August 2002.

12.   Groves, M.J. (1998). "US versus EU: Some differences between American and European Parenteral Science". Pharm Technol Europe, **10**: 28-32.

13.   Bliesner, D.M. (2006). Establishing a cGMP Laboratory Audit System: A practical Guide. New Jersey, USA: Wiley-Interscience.

14.   Quality assurance of pharmaceuticals. A compendium of guidelines and related materials. Volume 2. Good manufacturing practices and inspection. Geneva, World Health Organization, 1999.

15.   Winckles, H.W. and Dorpem, J.W. (1994). "Risk Assessment and the Basis for the Definition of Sterility", Med Dev Technol, **5**: 38-43.

16. Sandle, T. (2011): 'Risk Management in Pharmaceutical Microbiology' in Saghee, M.R., Sandle, T. and Tidswell, E.C. (Eds.) (2011): Microbiology and Sterility Assurance in Pharmaceuticals and Medical Devices, New Delhi: Business Horizons, 553-588.

17. Agallocco, J. (2011). "Process Selection for Sterile Products", in Saghee, M.R., Sandle, T. and Tidswell, E.C. (Eds.) Microbiology and Sterility Assurance in Pharmaceuticals and Medical Devices, New Delhi: Business Horizons, 603-614.

18. Meltzer T.H. (1987). Filtration in the Pharmaceutical Industry Marcel Dekker, Inc., New York.

19. Wherry, R.J. "Blow-Fill-Seal(BFS) Sterility Assurance Validation(SAV) – Saving Time and Pain on Your SAV", American Pharmaceutical Review, online: http://americanpharmaceuticalreview.com/ViewArticle.aspx?ContentID=367.

20. Bradley, A., Probert, S.C., Sinclair, C.S. & Tallentire. A. (1991) Airborne microbial challenges of blow/fill/seal equipment: a case study. Journal of Parenteral Science and Technology **45**: 187-192.

21. Reinmüller, B. (2001). "People as a contamination source – Clothing systems". In: Dispersion and risk assessment of airborne contaminants in pharmaceutical cleanrooms. Royal Institute of Technology, Building Services Engineering, Bulletin no. 56, Stockholm (August 2001), 54-77.

22. Sharp, J., Bird, A., Brzozowski, S. and O'Hagan, K. (2010): 'Contamination of cleanrooms by people', European Journal of Parenteral and Pharmaceutical Sciences, **15**(3): 73-81.

23. Tetzlaff, R.F. (1984). "Regulatory Aspects of Aseptic Processing", Pharmaceutical Technology, November 1984.

24. Ljungqvist B. and Reinmüller Berit (1997). Clean room design – Minimizing contamination through proper design. Interpharm Press, Buffalo Grove IL/USA.

25. Ramstorp, M. (2011). "Microbial Contamination Control in Pharmaceutical Manufacturing" in Saghee, M.R., Sandle, T. and Tidswell, E.C. (Eds.) Microbiology and Sterility Assurance in Pharmaceuticals and Medical Devices, New Delhi: Business Horizons, 615-700.

26. ISO 14644-1, "Cleanrooms and Associated Controlled Environments – Part 1: Classification of Air Cleanliness," (ISO, Geneva, Switzerland, 1999).

27. Schicht, H.H. (2003). "The ISO contamination control standards – a tool for implementing regulatory requirements", European Journal of Parenteral & Pharmaceutical Sciences, **8**(2): 37-42.

28. Whyte, W. (2001): Cleanroom Technology: Fundamentals of Design, Testing and Operation, Wiley, London.

29. General Chapter <1116>. "Microbiological Evaluation of Cleanrooms and Other Controlled Environments," USP 27–NF 22. (US Pharmacopeial Convention, Rockville, MD, 2004), 2559-2565.

30. Midcalf, B., Phillips, W.M., Neiger, J.S. and Coles, T.J. (2004): 'Pharmaceutical Isolators', Pharmaceutical Press.

31. Block S. 1977; Disinfection, Sterilisation and Preservation, Third Edition, Lea and Febiger, Philadelphia.

32. McDonnell, D and Russell, A.D. (1999). "Antiseptic and disinfectants: activity, action and resistance". Clinical Microbiology Reviews 12: 147-179.

33. Sandle, T.: 'Selection and use of cleaning and disinfection agents in pharmaceutical manufacturing' in Hodges, N and Hanlon, G. (2003): 'Industrial Pharmaceutical Microbiology Standards and Controls', Euromed Communications, England.

34. Sutton, S.V.W. (2005) "Disinfectant rotation – a microbiologist's view". Controlled Environments July 2005, 9-14.

35. Denny, V. & Marsik, F. (2004) "Disinfection practices in Parenteral manufacturing" In: Microbial Contamination Control in Parenteral Manufacturing Ed K.Williams. New York: Marcel Dekker.

36. Vina, P., Rubio, S. and Sandle, T. (2011): 'Selection and Validation of Disinfectants', in Saghee, M.R., Sandle, T. and Tidswell, E.C. (Eds.) (2011): Microbiology and Sterility Assurance in Pharmaceuticals and Medical Devices, New Delhi: Business Horizons, 219-236.

37. Rutala, W.A. and Weber, D.J. (1999). "Infection Control: the role of disinfection and sterilisation", J Hosp Infection, 43: 43-55.

38. Vincent, D. W. (2003). "Qualification of Purified Water Systems", Journal of Validation Technology, 10(1): 50-61.

39. Latham, T. (1995). "Clean Steam Systems", Pharmaceutical Engineering, 15(2)

40. Food and Drug Administration. Guidance for industry: Container closure systems for packaging drugs and biologics, chemistry, manufacturing, and control documentation. Food and Drug Administration, Rockville, MD; 1999 May.

41. Guazzo, D. M. "Current Approaches in Leak Testing Pharmaceutical Products," J. Pharm. Sci. Technol, 1996, 50(6): 378-385.

42. Lee E. Kirsch, Lida Nguyen, Craig S. Moeckly and Ronald Gerth (1997): "Pharmaceutical Container/Closure Integrity II: The Relationship Between Microbial Ingress and Helium Leak Rates in Rubber-Stoppered Glass Vials", PDA Journal of Pharmaceutical Science and Technology, September-October 1997, 51(5): 195-202.

43. Lee E. Kirsch, Lida Nguyen, Craig S. Moeckly and Ronald Gerth (1997): "Pharmaceutical Container/Closure Integrity II: The Relationship Between Microbial Ingress and Helium Leak Rates in Rubber-Stoppered Glass Vials", PDA Journal of Pharmaceutical Science and Technology, September-October 1997 51(5): 195-202.

44. Committee for Proprietary Medicinal Products. (1996). "Note for Guidance on the Manufacture of Finished Dosage Form", CPMP/QWP/486/95, European Agency for the Evaluation of Medicinal Products, London.

45. Harwood, R.J., Portnoff, J.B. and Sunbery, E.W. (1992). "The Processing of Small Volume Parenterals and Related Sterile Products", in Avis, K.E., Lieberman, H.A. and Lachman, L. (Eds.). Pharmaceutical Dosage Forms: Parenteral Medications, **2**(2): Marcel Dekker: New York.

46. Pflug, I.J. and Odlaug, T.E. (1986) Biological indicators in the pharmaceutical and medical device industry. Journal of Parenteral Science and Technology, **40**: 242-248.

47. Agalloco J.P., Akers J.E., Madsen R.E. Moist heat sterilisation–myths and realities. PDA J. Pharm. Sci. Technol. 1998, **52**(6): 346-35.

48. Nyberg, R. (2011). "Biological Indicators for Sterilisation" in Saghee, M.R., Sandle, T. and Tidswell, E.C. (Eds.) Microbiology and Sterility Assurance in Pharmaceuticals and Medical Devices, New Delhi: Business Horizons, 733-750.

49. Woedtke, T. And Kramer, A. (2008). "The limits of sterility assurance", GMS Krankenhaushygiene Interdisziplinär, **3**(3): 110.

50. Russell, A.D. (1999). "Destruction of Bacterial Spores by Thermal Methods". In Russell, A.D., Hugo, W.B. and Aycliffe, G.A.J. (Eds.) Principles and Practice of Disinfection, Preservation and Sterilisation, 3rd edition. Oxford: Blackwell, 675-702.

51. Sadowski, M.J. (2011). "Moist Heat Sterilisation" in Saghee, M.R., Sandle, T. and Tidswell, E.C. (Eds.) Microbiology and Sterility Assurance in Pharmaceuticals and Medical Devices, New Delhi: Business Horizons, 751-805.

52. Amer, G. and Beane, R. G. (2000): "Autoclave Qualification: Some Practical Advice", Journal of Validation Technology, **7**(1): 9094.

53. Mayeresse, Y., (2006). "Freeze-Drying Process Validation," Drug Manufacturing and Supply, December 2006.

54. Saghee, M.R. and Mitchel, G. R. (2011). "Sterilisation and Depyrogenation by Dry Heat" in Saghee, M.R., Sandle, T. and Tidswell, E.C. (Eds.) Microbiology and Sterility Assurance in Pharmaceuticals and Medical Devices, New Delhi: Business Horizons, 807-839.

55. Guy, D. (2003): 'Endotoxins and Depyrogenation' in Hodges, N. and Hanlon, G., Industrial Pharmaceutical microbiology: Standards and Controls, Euromed, 12.1-12.15.

56. Pearson, F.C. (1985) Pyrogens: Endotoxins, LAL testing, and Depyrogenation. New York: Marcel Dekker.

57. Baird, R. (1988): 'Validation of Dry Heat Tunnels and Ovens', Pharmaceutical Engineering, **8**(2): 31-33.

58. Sebold, M.A. and Williams, J.A. (2011). "Radiation Sterilisation" " in Saghee, M.R., Sandle, T. and Tidswell, E.C. (Eds.) Microbiology and Sterility Assurance in Pharmaceuticals and Medical Devices, New Delhi: Business Horizons, 841-872.

59. Woolston, J. (1999). "Current Issues in Radiation Sterilisation", Med Dev Technol, **10**: 20-22.

60. Alfa, M.J. De Gagne, P. and Olson, N, (1996) "Comparison of Ion Plasma Vaporized Hydrogen Peroxide, and 100% Ethylene Oxide Sterilisers to the 12/88 Ethylene Oxide Gas Steriliser", Infect Control Hosp Epidemiol, **17**: 92-100.

61. Bayliss, C.E. and Waites, W.M. (1979). "The Combined Effect of Hydrogen Peroxide and Ultraviolet Radiation on Bacterial Spores", J Appl Bacteriol, **47**: 263-269.

62. Kokubo M, Inoue T, Akers J. Resistance of common environmental spores of the genus Bacillus to vapor hydrogen peroxide. J Pharm Sci Technol 1998; **52**: 228-231.

63. Jornitz, M.W. and Meltzer, T. H. (2011). "Sterilisation by Filtration" in Saghee, M.R., Sandle, T. and Tidswell, E.C. (Eds.) Microbiology and Sterility Assurance in Pharmaceuticals and Medical Devices, New Delhi: Business Horizons, 873-903.

64. Montalvo, M. (2002). "Validation of Sterilising Filters for a Specific Process", Journal of Validation Technology, **8**(4): 316-319.

65. Halls, N.A. (2002). Microbiological Media Fills Explained. West Sussex: Sue Horwood Publishing Ltd.

66. Budini, M. and Boschi, F. (2011). "Aseptic Process Simulations/Media Fills" in Saghee, M.R., Sandle, T. and Tidswell, E.C. (Eds.) Microbiology and Sterility Assurance in Pharmaceuticals and Medical Devices, New Delhi: Business Horizons, 701-732.

67. Vincent, D. W. (2006). "Regulatory and Validation Considerations for Aseptic Processes", Journal of Validation Technology, **12**(2): 107-133.

68. Cundell, A. M. "Microbial Testing in Support of Aseptic Processing", Pharmaceutical Technology, June 2004, 56-66.

69. Sandle, T. (2011): 'Environmental Monitoring' in Saghee, M.R., Sandle, T. and Tidswell, E.C. (Eds.) Microbiology and Sterility Assurance in Pharmaceuticals and Medical Devices, New Delhi: Business Horizons, 293-326.

70. Ljungqvist, B. and Reinmuller, B. (1996): 'Some observations on Environmental Monitoring of Cleanrooms', European Journal of Parenteral Science, 1996, **1**: 9-13.

71. Parenteral Drug Association (PDA) Technical Report #13, "Revised Fundamentals of an Environmental Monitoring Program," J. Pharm Sci Technol, 2001. **55**(5, Suppl.): 33.

72. Moldenhauer, J. (2008). "Environmental Monitoring" in Prince, R. (Ed.). Microbiology in Pharmaceutical Manufacturing, Parenetral Drug Association, Bethesda, MD, USA, 19-92.

73. PDA Technical Report No. 13 (revised): 'Fundamentals of an Environmental Monitoring Programme', Parenteral Drug Association, September/October 2001.

74. Sandle, T (2003): 'Selection and use of cleaning and disinfection agents in pharmaceutical manufacturing'. In: Hodges, N and Hanlon, G. (2003): Industrial Pharmaceutical Microbiology Standards and Controls, Euromed Communications, England.

75. Sandle, T.: 'Microbiological Culture Media: Designing a Testing Scheme', Pharmaceutical Microbiology Interest Group News, No.2, August 2000.

76. Deeks, T. (1999). "Comparison on state-of-the-at technologies for aseptic processing", Med Dev Technol/BioPharm Europe, **11**: 52-57.

77. Reich, R, Miller, M. and Patterson, H.(2003): "Developing a Viable Microbiological Environmental Monitoring Program for Nonsterile Pharmaceutical Operations", Pharm. Technol., March, 92-100.

78. Gilbert, P.A. and Allison, D.G. (1996). "Redefining the 'sterility' of sterile products", Eur J Parent Sci, **1**: 19-23.

79. Cundell, A. M. "Review of the Media Selection and Incubation Conditions for the Compendial Sterility and Microbial Limit Tests," Pharm Forum, 2002, **28**(6): 2034-2041.

80. Sandle, T. (2004). "Practical Approaches to Sterility Testing", Journal of Validation Technology, 2004 , **10**(2): 131-141.

81. Sandle, T. and Saghee, M.R. (2010): Advances in cleanroom technologies, Express Pharma, 16th-30th September, On-line Paper: http://www.expresspharmaonline.com/20100930/expressbiotech13.shtml

82. Williams, D. (1997). "The Sterile Debate: Then effects of radiation sterilisation on polymers", Med Dev Technol., **8**: 6-9

83. Hinz, D.C. (2006). "Process analytical technologies in the pharmaceutical industry: the FDA's PAT initiative". Anal Bioanal Chem, **384**: 1036-1042

a   These are substances which are capable of raising the mammalian core body temperature when administered by injection.

b   For sterile products this is by hermetically sealed containers designed to protect the product from microbial contamination. The actual design depends upon the way in which the preparation is intended to be used.

c   Some argue that the Sterility Assurance concept was developed for sterilisation processes and it should be limited to terminal sterilisation thus it cannot, as a probalistic concept, be applied to aseptic manufacture.

d   It is necessary to emphasise here that this chapter is concerned with GMP inspections of sterile manufacturing facilities. It is recognised that there are other aspects of GxP which are of relevance to sterile manufacturing and which are subject to regulatory inspections, such as Good Clinical Practise, Good Distribution Practise and Good Laboratory Practise. These are not, however, the primary concern here.

e   Of particular importance are parts Section 211.42, relating to need for separate and defined grades of cleanrooms; 211.46 for the need for physical aspects of cleanroom design to be in place to prevent contamination; and 212.42 relating to the material of construction for cleanrooms.

f   For a complete list of inspection guides and contents go to: http://www.fda.gov/ICECI/Inspections/InspectionGuides/default.htm.

g   The current EU GMP guideline can be found here:
    http://ec.europa.eu/health/documents/eudralex/vol-4/index_en.htm.

h   For a full list go to:
    http://www.picscheme.org/documents/List_of_PICS_Publications.pdf.

i   For a full list go to:
    http://www.who.int/medicines/areas/quality_safety/quality_assurance/production/en/.

j   For a full list, go to:
    http://www.ema.europa.eu/ema/index.jsp?curl=pages/regulation/general/general_c
    ontent_000431.jsp&murl=menus/regulations/regulations.jsp&mid=WC0b01ac05800
    29593&jsenabled=true.

k   The FDA publish 483 Letters, other regulatory authorities do not reveal the results
    of inspections.

l   The administration of a pharmaceutical preparation to a patient by injection under
    or through the skin or mucous membranes.

m   There are different ways for defining bioburden. One approach is:

    No = Initial population of surviving microorganisms per defined unit or surface. In
    sterile manufacturing the typical limit: <10 CFU/100 mL.

n   Due to the difference in the maximal level of permitted particle count
    concentration, EU GMP Grade A is actually equivalent to ISO 4644 class 4.8.

o   Water may stagnate in branch pipes branch from a circulating main if the length
    of the branch is too long to allow the turbulence of the flowing main to disturb
    the contents of the branch pipe.

p   The test measures the rate of helium leak from the vial as well as the actual
    percent of helium that is filled within the vial.

q   A population of microorganisms inoculated onto a suitable carrier. The challenge
    microorganism is selected based upon its resistance to the given process.

r   The culture collection numbers listed are those mentioned in the USP. Other
    pharmacopoeias list alternative culture collection references and a check should be
    made that the biological indicators are equivalent. Many inspectors will be familiar
    with the American Type Culture Collection (ATCC).

s   Formerly classified as *Bacillus stearothermophilus*.

t   Formerly classified as *Bacillus subtilus var. niger*.

u   The number of organisms present on or in a biological indicator.

v   The D-value is the survival rate. This is the time (in minutes) required under
    specified conditions to cause a 90%, or one log reduction of a specific
    microorganism. It is affected by many conditions and is related to the Z-value. The
    Z-value is an expression of resistance and is the number of degrees or dosage
    (Mrad) units required for a one log reduction in the D value.

w   A process that provides sufficient lethality such that there is less than a 10^{-6} probability of a non-sterile unit (that is one viable microorganism surviving in one million).

x   Parametric release is referred to in the General Notices of the Eur. Pharm., where it states: "The manufacturer may obtain assurance that a product is of Pharmacopoeia quality from data derived for example, from validation studies of the manufacturing process and from in-process controls. Parametric release deemed appropriate by the competent authority is thus not precluded by the need to comply with the Pharmacopoeia." and in the USP as "Data derived from manufacturing process validation studies and from in-process controls may provide greater assurance that a batch meets a particular monograph requirement than analytical data derived from an examination of finished units drawn from that batch."

y   Depyrogenation can refer to endotoxin inactivation or endotoxin removal.

z   Although such parameters are well established, dry heat depyrogenation is a complex process which is still poorly understood with contradictory research data.

aa   Bacterial endotoxin is a synonym for lipopolysaccharide.

bb   With two culture media, one medium and set of incubation conditions is designed for the detection and enumeration of bacteria and the other for fungi (such as Sabouraud Dextrose Agar, Malt Extract Agar or Rose Bengal Agar). Where one culture medium is used the medium is normally subject to two incubation temperatures (30-35°C and 20-25°C). Single culture media is normally a general purpose, highly nutritious growth medium like tryptone (or tryptic) soya agar.

cc   The European Pharmacopeia has an alternative monocyte activation test, although this is not, at the time of writing, in widespread use.

dd   The principle of the LAL test is a reaction between lipopolysaccharide and a substance ("clottable protein") contained within amoebocyte cells derived from the blood of the Horseshoe Crab (*Limulus polyphemus*). The reaction is specific.

ee   A method used to determine how much any material being proposed for testing by the LAL method can be diluted while still being able to detect the limit endotoxin concentration.

ff   The Federal Register (Vol. 76, No. 119/Tuesday, June 21, 2011) contained an important announcement about the sterility test and marked a step forward towards the adoption of rapid and alternative methods with an update to the Code of Federal Regulations for the Sterility Test.

gg   Ethylene oxide sterilisation similarly is thought to have too many variables to be confident about parametric release; many processes are biologically monitored through both the sterility test and through routine use of Biological Indicators.

CHAPTER 11

Preparing for regulatory inspections of API facilities

SIEGFRIED SCHMITT, PRINCIPAL CONSULTANT, PAREXEL CONSULTING, UK
RICHARD EINIG, PRESIDENT, EINIG & ASSOCIATES, INC., USA

11.1   Introduction

Active Pharmaceutical Ingredients (APIs), also referred to as drug substances, are the components in a drug product that give the product its pharmacological effect. They are differentiated from excipients in that they impart a measurable pharmacological effect. The active components have been called by a variety of names including bulk pharmaceutical chemicals (BPCs), drug substances, active ingredients (AIs), and other less common names. Some of these names are still used in publications and official documents, although API is now the standard term used in industry. In regulatory submissions that follow the ICH[1] Common Technical Document (CTD) format, the term drug substance is mandated. APIs are typically synthesized through a series of chemical reactions or biological transformations, followed by purification steps that result in a highly purified chemical or biological compound. These compounds are manufactured in equipment and facilities designed for the manufacture of fine chemicals rather than for drug products.

In 1937 there was a tragic event[2] in the United States that occurred when sulfanilamide powder was formulated with diethylene glycol to manufacture a liquid sulfanilamide product. Over 100 patients, mostly children, died from liver toxicity caused by use of the liver toxic excipient in the formulated product. One year later a law was passed in the United States of America, the Food, Drug, and Cosmetic Act, that required all active ingredients and drug products to be manufactured with adequate controls. Since then many regulatory authorities and trade associations around the world have published guidances and regulations that describe adequate controls for the manufacture of APIs.

To aid the pharmaceutical manufacturer to prepare for a regulatory inspection of API facilities the chapter is divided into:

* Part A: An overview of regulatory guidance for API facilities

* Part B: Regulatory inspectional guidance for API manufacturers

* Part C: Examination of the key focal points for a regulatory inspection of API facilities

- Part D: Examples of regulatory citations (483s and WLs) pertaining to API facilities

- Conclusion

- Appendix: Checklist to aid the pharmaceutical manufacturer in preparing for the inspection of API facilities

11.2   Part A: An overview of regulatory guidance for API facilities

European Union and United States of America legislation

The two main agencies developing legislation for drug substances are the United States of America Food and Drug Administration (FDA) and the European Medicines Agency (EMA). In the USA, the legislation is codified in the Code of Federal Regulations. The CFRs applicable to the manufacture of APIs are under Title 21 (Food and Drugs), contained in sections CFR 210 and 211 (current Good Manufacturing Practice (CGMP)[3]. "The CGMP regulations are not direct requirements for manufacture of APIs; the regulations should not be referenced as the basis for a GMP deficiency in the manufacture of Active Pharmaceutical Ingredients (APIs), but they are guidance for CGMP in API manufacture"[4].

The FDA publishes inspection guides and Guidance for Industry documents that are non-binding documents designed as reference material for investigators, other FDA personnel and industry. The following guidances pertain to APIs:

- The FDA published the Guide to Inspections of Bulk Pharmaceutical Chemicals (terminology previously used for APIs) in 1984 and subsequently revised it until the final version in 1994[5]

- The FDA draft Guidance for Industry: Manufacturing, Processing, or Holding Active Pharmaceutical Ingredients of March 1998 was never finalised[6]

- FDA Compliance Program Guidance Manual Program 7356.002F 20084

In the European Union, the Good Manufacturing Practice (GMP) regulations are promulgated in European Commission Directive 2001/83/EC which describes principles and guidelines of Good Manufacturing Practice in respect of medicinal products for human use[7]. This document can be found in EudraLex, the body of European Union legislation in the pharmaceutical sector[8]. European GMP is overseen by the European Medicines Agency (EMA) and enforced by national regulatory authorities (the Competent Authorities) within the member states. In EudraLex Volume 4 Guidelines for good manufacturing practices for medicinal products for human and veterinary use, the GMP requirements for APIs were initially published in Annex 18 in July 2001. This document was identical with ICH Q7A - Good Manufacturing Practice for Active Pharmaceutical Ingredients. This has since been moved and is now listed as Part II - Basic Requirements for Active Substances used as Starting Materials[9]. Two changes should be noted:

a) The ICH document is now ICH Q7

b) The EudraLex document differs from ICH Q7 insofar as it describes the role of the Qualified Person, which does not exist outside the European Union

The EMA (previously EMEA) published a guidance document on the inspection of API manufacturers in 2005[10].

Historical background

The regulations for APIs have been developed over more than two decades. The European Federation of Pharmaceutical Industries and Associations (EFPIA) with the European Chemical Industry Council (CEFIC) published Good Manufacturing Practices for Active Ingredient Manufacturers in 1996[11]. The Pharmaceutical Research and Manufacturers of America (PhRMA) in the USA prepared Guidelines for the Production, Packing, Repacking or Holding of Drug Substances which was later published in two parts in 1995[12] and 1996[13]. Unfortunately some of these published documents were unclear, confusing, and even contradictory in part. Both suppliers and customers became increasingly aware of the difficulties these various documents posed for API manufacture and were vocal in the need to harmonise the disparate approaches. Globalisation of API manufacturing in the last quarter of the 20th century exacerbated the situation.

In September 1997 the Therapeutic Goods Administration (TGA) in Australia held a meeting with various other regulatory authorities to discuss a universal approach to regulating API production. This led to ground breaking harmonisation on API production by the Pharmaceutical Inspection Co-operation Scheme (PIC/S)[14]. Ultimately the task of developing a single universal guidance document for the manufacture of API was undertaken by an expert working group assembled by the International Conference on the Harmonization of Technical Requirements for Registration of Pharmaceuticals for Human Use, commonly referred to as ICH[1]. Agreed ICH guidance has to be transposed into national law in the three member regions, the USA, EU and Japan.

The expert working group used information from a variety of sources including the earlier works cited here. The resulting harmonised document titled ICH Q7A, The Good Manufacturing Practice Guide for Active Pharmaceutical Ingredients, was adopted by the European Committee for Proprietary Medicinal Products (CPMP) in 2000[15]. ICH Q7A was subsequently adopted in 2001 by the US FDA and published in the Federal Register[16]. Later in 2001 it was adopted by the MHLW of Japan[17]. The EU published it as Annex 18 as described above. ICH Q7 (recoded without the "A" by ICH in 2005) is now the standard for API manufacturing worldwide.

Future developments

In May 2011 ICH published ICH Q11 Development and Manufacture of Drug Substances[18]. This document is proposed for Active Pharmaceutical Ingredients (APIs) harmonising the scientific and technical principles relating to the description and justification of the development and manufacturing process (CTD sections S 2.2. – S 2.6) of Drug Substances including both chemical entities and biotechnological/biological entities. It therefore addresses API process development and drug substance scientific understanding "to establish a commercial manufacturing process capable of consistently producing drug substance of the intended quality"[18]. It incorporates the principles of pharmaceutical development, quality risk management, and pharmaceutical quality systems described in ICH Q8[19], ICH Q9[20], and ICH Q10[21] respectively. These three later documents have significantly changed the regulatory landscape by introducing an integrated approach to regulatory compliance built on scientific knowledge and manufacturing experience gained throughout the lifecycle of the drug. This most recent document builds on the prior documents to bring an enhanced approach to compliant API manufacturing. Principles espoused in the more recent documents that were proposed to enhance continual improvement of drug products are now being applied to API manufacturing.

The traditional approach to API manufacturing is described in the ICH Q7 document where GMP requires set points and operating ranges for process parameters to be defined usually in a narrow range, and the drug substance control strategy is typically based on demonstration of process reproducibility through validation and release testing to meet established acceptance criteria.

The enhanced approach which is risk based requires a more complete scientific understanding of the API by determining relationships among synthesis variables initially established during process development. Risk management and more extensive scientific knowledge are used to select process parameters and unit operations that impact critical quality attributes (CQAs) for evaluation in further studies to establish a design space described in ICH Q8 and control strategies applicable over the lifecycle of the drug substance described in ICH Q10. As the knowledge base increases with manufacturing experience, it serves as the basis for risk assessment described in ICH Q9 when changes to the process occur through planned or unplanned deviations. Quality systems ensure that all registration requirements are met throughout the process so that the API has the quality and purity it is purported to possess.

The traditional approach is well documented in ICH Q7 and the risk based approach described in ICH Q11 draft draws heavily from ICH Q8, ICH Q9, and ICH Q10.

Another key aspect of regulatory concern in recent years concerns the security of the supply chain and the prevention of falsified medicines. The US authorities mainly tackle this by requiring unique identifiers and by instigating a more rigorous inspection program. The European regulators have similar concerns, but approach this problem by trying to elevate the regulatory requirements for APIs to the level of finished drugs, i.e. drug products. On 20 January 2012 the European Commission published a concept paper for public consultation "Delegated Act on the Principles and Guidelines of Good Manufacturing Practice for active Substances in medicinal Products for human Use"[22]. The European Commission has to put in place this GMP legislation by 2013. The comment period ends 20 April 2011. Of particular interest are the prerequisites listed in the draft document that would establish equivalence to the EU GMPs in a third country. As until now, GMP inspections of API manufacturers are **not** mandated in the EU, it is only a logical step that the onus of verifying compliance of API manufacturers rest with the Marketing Authorisation holders, typically the drug product manufacturers.

Other regulatory guidance

The Pharmaceutical Inspection Co-operation Scheme (PIC/S)14 is a network between participating authorities and fosters the exchange of information and experience between GMP inspectors. In 2011 the FDA joined the PIC/S scheme which should lead to closer integration of the approach to regulatory inspections and closer alignment of the standards required. PIC/S publishes a range of documents, similar to the FDA inspection guides, which are freely available from their website. The respective guide for GMPs for APIs is identical with ICH Q7[23].

The World Health Organization (WHO) enforces similar requirements to European Union's GMP. In March 2010 WHO published a draft guideline for comment, entitled "Guideline for the Production and Control of Specified Starting Materials"[24]. This guidance includes an extensive introduction section which provides both background and context for the information on "specified starting materials". WHO defines these materials as any substance which is primarily or mainly used as a starting material for the production of an API, but which itself could be used directly as an API.

If the "specified starting material" is itself an API then it should conform to the existing monograph in a recognized pharmacopeia. If the API is not supported by a pharmacopeia monograph, then appropriate specifications should be developed and justified by the manufacturer.

If the material is used in the production of an API then the quality attributes and specifications should be determined by the API manufacturer and the material should be fit for its intended use.

Viral safety and TSE data should be carefully considered if the starting material is animal derived. Quality Control for compounds used as specified starting materials should address impurity profiles, isomers, residual solvents and other impurities that may be carried through to the resulting API.

API manufacturers are encouraged to take a risk based approach in setting specifications for these materials, considering the number and type of unit operations between introduction of the material and production of the resulting API.

The key WHO document covering GMPs for APIs is the WHO Technical Report Series 961[25], which is updated almost annually. The current version is the 45th report. Its structure is similar to the one in EudraLex. The requirements listed are usually similar or identical with ICH and EU legislation.

EU Member States have to transpose EU legislation into national law. By doing so they may impose stricter or additional requirements than the governing law. Some agencies (Competent Authorities) also publish guidance documents that offer valuable insight into the agencies' interpretation of the legislation. Unfortunately, often such guidance is only available in the local language. An example of an outstandingly useful document is the one published by the Zentralstelle der Länder für Gesundheitsschutz bei Arzneimitteln und Medizinprodukten in Germany on the subject of inspection of qualification and validation in pharmaceutical manufacture and quality control[26]. It explicitly disallows retrospective validation, despite it still being an option in EU legislation.

The MHRA publishes guidance documents on API inspections on their website[27], titled: Good Manufacturing Practice (GMP) expectations for Active Pharmaceutical Ingredients (APIs).

In October 2005 new legislation was passed in the European Union requiring that Active Pharmaceutical Ingredients (API) used as starting materials in dose form pharmaceutical manufacture must have been manufactured in compliance with GMP. The relevant legislation is:

Amended EU medicines legislation – Directive 2004/27/EC (2001/83/EC as amended)

This was transposed into law in each member state (in UK as Statutory Instrument S.I. 2005/2789, October 2005).

Article 46(f) of 2004/27/EC introduced an obligation (by law via S.I. 2005/2789 in UK) for manufacturing authorisation holders to use as starting materials only active substances, which have been manufactured in accordance with the detailed guidelines on Good Manufacturing Practice for active substances.

In addition, from this date marketing authorisation applications and variations to change the source of the active substances used as starting materials have had to be supported by a declaration of GMP compliance of the active substance manufacturer by a Qualified Person (QP) of the dosage form manufacturer

The legislation applies to all registered APIs. It is a requirement for all registered drug products that an active ingredient is named on the Marketing Authorisation Application.

In some cases materials named as the active ingredient may be commercially manufactured for uses other than in drug products. These are commonly called atypical actives. Difficulties facing dose form manufacturers in assuring GMP of certain 'atypical' APIs are understood by MHRA. It may be that the primary use of the material is not for drug product use and the primary use may not require standards fully equivalent to GMP. The pharmaceutical user of the material may be a very small customer and does not have the necessary influence to demand or guide the manufacturer toward full GMP manufacturing standards. To date MHRA has taken a pragmatic approach to GMP compliance of these materials and as such has considered each case on its own merits.

Here the MHRA also makes available for download two useful documents:

• API focussed MHRA inspections at dosage form manufacturers

• GMP expectations of Non Traditional APIs

As Japan is a member of ICH, they have adopted ICH Q7 as the standard for GMPs for APIs, which can be found in English in the Pharmaceutical Administration and Regulations in Japan document published in March 2011[28].

11.3   Part B: Regulatory inspectional guidance for API manufacturers

Several industry associations have developed and published a variety of guidance documents for the benefit of API manufacturers. Though by no means all-inclusive, the following is an overview of some of the most active associations in this field.

APIC[29] "the Active Pharmaceutical Ingredients Committee" is a Sector Group within CEFIC (the European Chemical Industry Council). APIC's membership consists of companies from different pharmaceutical industry sectors, all involved in the manufacture of APIs. This provides an ideal basis for developing and communicating a balanced, holistic view on API-related regulations and guidelines.

Some of the documents published by APIC were developed in conjunction with the European Federation of Pharmaceutical Industries and Associations (EFPIA)[30]. EFPIA represents the pharmaceutical industry operating in Europe. Through its direct membership of 31 national associations and 35 leading pharmaceutical companies, EFPIA is the voice on the EU scene of 2,000 companies committed to researching, developing and bringing to patients new medicines that will improve health and the quality of life around the world.

The comprehensive list of publications by APIC comprises guidance for auditing and supplier qualification, both of enormous importance within the realm of GMP:

- APIC list of Abbreviations & Acronyms

- APIC Brochure 2008

- Press Release APIC Quality Agreement and Supplier Qualification and Management" Guidelines, 2010

- Press Release: EFCG and APIC welcome adoption of new Falsified Medicines Directive (Feb 2011)

- Industry Best Practice documents:

- Good Manufacturing Practices for Active Ingredients Manufacturers, with EFPIA, 1996

- Quality System for Active Ingredients Manufacturers, integrating GMP into ISO, 1997

- Manufacture of Sterile Active Pharmaceutical Ingredients, guidance, 1999

- The Active Pharmaceutical Ingredients Starting Material (APISM), 1999

- Cleaning validation in Active Ingredient Manufacturing plants - Policy, 1999

- Good manufacturing practices in Active Pharmaceutical Ingredients development, 1999

- Cleaning validation in active pharmaceutical plants - guidance, 2000

- "How to do" - Interpretation of ICH Q7 document & "Review Form (update May 2011)"

- Computer Validation Guide, December 2002

- Parametric release document, December 2002

- Qualification of existing equipment 2004

- Technical Change Control Guideline 2004

- Quality Management System (QMS) for APIs 2005

- APIC Quick Guide for API Sourcing, 2008

- The Audit Programme, Version 3, August 2010:

    A)   Audit Programme: Procedure

        Annex 1 – Contract Auditor-ACI

        Annex 2 – Agreement Customer, Auditee, Auditors & ACI

        Annex 3 – Auditing Guide-Secrecy Agreement

        Annex 4 – Standardised Letters related to Shared 3rd Party Audits:

            – Letter 1

            – Letter 2

            – Letter 3

        Annex 5 – Feedback Form

    B)   Auditing Guide, August 2010

        Annex 1 – Auditing Guide, Questionnaire

        Annex 2 – Auditing Guide "Aide Mémoire"

        Annex 3 – Auditing Guide, Audit Report Template

        Quality Agreement Guideline 2009

            – Appendix A: Generic APIs

            – Appendix B: Exclusive Substances

- Supplier Qualification & Management Guideline, December 2009

    –   Appendix 1: Examples of Critical/ non Critical Raw Materials

    –   Appendix 2: Supplier Selection Check List

- – Appendix 3: Due Diligence Check List
- – Appendix 4: Supplier Questionnaires
- – A: General Company Information and Quality Management Questionnaire
- – B: BSE/BSE Risk Analysis Survey
- – C: GMP - Vegetable Origin
- – D: Allergen
- – E: Extended Quality Questionnaire for Critical Material
- – F: Packaging Material
- – Appendix 5: Check list for Change Control Assessment

Supplier Qualification Guideline (ZIP File)

The usefulness of these documents can be exemplified as follows: APIC published the guideline and two templates for quality agreements in December 2010[31]: a template for the manufacture of so-called generic APIs and a template for the manufacture of APIs within the framework of an exclusive contract with an API manufacturer. These templates include all points that have to be regulated in such an agreement, like e.g. the right to conduct audits, the notification in case of an imminent official inspection, the keeping of retention samples and production documentation, the customer's right to examine the supplier's annual Product Quality Review, the handling of product reprocessing, deviations and out-of-specification results and many others. Moreover, both documents include a table in which responsibilities of customer and supplier are marked by means of crosses.

Even though the two templates have the same chapters, the template for the exclusive API manufacture is adapted to the special customer-supplier relationship defined by the exclusive contract. Chapter 16 "Raw Materials" goes e.g. into great detail and describes the responsibilities of customer and supplier with regard to the procurement, handling, storage and monitoring of starting materials with considerable exactness.

In principle, these two templates cover all possible constellations of customer-supplier relationships. They are very helpful for the creation of quality agreements insofar as they provide a generally applicable basic framework that can be modified to suit the individual case and to take account of particularities in API production and of specific regulations.

ISPE, the International Society for Pharmaceutical Engineering[32], publishes a variety of guides, predominantly from the viewpoint of engineering and operability, not however from the perspective of the active moiety. The best guide in relation to this chapter is the ISPE Baseline® Guide Volume 1, Active Pharmaceutical Ingredients, Second Edition.

Several organisations have in the past and are still pursuing establishing audit repositories for API manufacturers. The idea is to lessen the burden for these companies, i.e. experiencing fewer audits (and maybe even fewer inspections) as the interested party could simply purchase the audit report. The benefits and risks of these repositories is not subject of this book. The facts are only stated for interest's sake. One of the most active associations at present is Rx360[33]. Rx-360 is a consortium being developed by volunteers from the Pharmaceutical and Biotech industry which includes their suppliers. The purpose is to enhance the security of the pharmaceutical supply chain and to assure the quality and authenticity of the products moving through the supply chain. The individuals developing this concept are working in the best interest of patients. They are a non-profit organization with the mission to create and monitor a global quality system that meets the expectations of industry and regulators that assures patient safety by guaranteeing product quality and authenticity throughout the supply chain. They published a document titled: Rx-360 Audit Standards for Active Pharmaceutical Ingredients and API Intermediates_v1.0[34] that provides the template for their audit approach.

11.4 Part C: Examination of the key focal points for a regulatory inspection of API facilities

Regulatory inspections of API manufacturers are undertaken by those regulatory authorities with which the manufacturer's license or the marketing authorisation is registered. For products licensed in the USA this is the FDA. In other regions this is the national body. Different regulators have different requirements and some of the important differences are presented in this chapter.

The timescales for inspections vary and in part relate to how well the previous inspection went. Most regulatory authorities adopt a risk based approach to inspections and will schedule visits based on how well they perceive the manufacturer to be operating. The typical frequency is once every three years or less. Some have never been inspected and are unlikely to be inspected in future. That does not relieve these companies though from their duties to comply with the regulations and adhere to GMP. In Table 1 of ICH Q7, the table contains a listing of when GMPs start during the manufacturing process for APIs from a variety of starting materials, i.e., chemically synthesized, extracted from natural materials, or produced by cell culture/fermentation. From this starting point, increasing requirements for GMPs are applied through the final steps of API release, storage and transfer of the API for further processing into drug product, and stability testing/monitoring of the API. Steps prior to the designated starting material are expected to be performed in a scientifically sound manner but do not require Quality Assurance oversight. The starting material(s) should be defined by the sponsor and discussed with the regulatory reviewer to achieve agreement before the licensing application is submitted. During a regulatory inspection only those steps after the starting material(s) is introduced into the synthetic scheme are typically reviewed.

It is important that the company employ all the pertinent individuals from R & D, Production, Regulatory, and Quality Assurance to determine the appropriate starting material(s) and prepare a justification document with the rationale and risk assessment clearly described. Data collected to support the starting material decision may be reviewed during a regulatory inspection. Some of the considerations for determining the starting material for chemically synthesized APIs that are given in ICH Q11 and ICH Q7 are:

1) A starting material should be a substance of defined chemical properties and structure. Non-isolated intermediates are usually not considered appropriate starting materials.

2) A starting material is incorporated as a significant structural fragment into the structure of the drug substance. "Significant structural fragment" in this context is intended to distinguish starting materials from reagents, solvents, or other raw materials.

3) Commonly available chemicals used to create salts, esters or other simple derivatives should be considered reagents and are not the starting material.

4) If the molecule has a chiral centre, the racemate is typically not the starting material even if the racemate is purchased and then resolved in-house.

5) Each branch of a convergent drug substance manufacturing process begins with one or more starting materials.

6) The starting material should be prior to or coincident with an intermediate molecule containing genotoxic elements that impact the impurity profile of the drug substance.

For an API which initiates from a compound by fermentation or by extraction from botanical material to which other structural constituents are added by chemical synthesis, the starting material may be the source material (microorganism or botanical material). However, if it can be demonstrated that one of the isolated intermediates in the synthetic process complies with the principles outlined above for the selection of starting materials for synthetic drug substances, that isolated intermediate can be proposed as the starting material. The company should specifically evaluate whether it is possible to analytically characterize the proposed starting material, including its impurity profile, and whether the fermentation or botanical material and extraction process impact the impurity profile of the drug substance. Risks from microbial and other contamination should also be addressed.

When APIs are obtained from pure biotech processes or extractions from natural products, the choice of starting material should be established on a case-by-case basis, using the principles described above when applicable. The impurity profile for extracts of natural products is not required.

The API production facility must have equipment and utilities adequate to produce the specific API and be qualified to demonstrate this capability. Older facilities may be suitable for production of some APIs while totally inadequate to produce others such as biotechnology or highly potent APIs. Production processes and equipment cleaning must be validated. Validation can be performed as described in ICH Q7 where the process is repeated a specified number of times, usually three, and acceptability is based on meeting discrete values for critical process parameters in all the repeats. Validation for drug product manufacturing was modified from this traditional approach when ICH Q8, ICH Q9, and ICH Q10 were approved, and a risk based approach to validating a manufacturing process was described using a design space developed during the R & D phase of the product life cycle. This risk based approach to validation is further described for API manufacturing in the draft ICH Q11 document where the number of process repetitions and discrete values for critical process parameters are not the criteria for acceptability.

APIs are manufactured in equipment and facilities typically used by fine chemical manufacturers. The major difference between the fine chemical manufacturer and the API manufacturer is the need to comply with CGMP regulations and/or guidances to ensure the material has the quality, strength, purity, and identity it is purported to have. During a regulatory inspection API manufacturers must be prepared to demonstrate the quality of their products with written records documenting the manufacturing and testing of all the products under appropriate controls. An atypical active described earlier in this chapter as a material named as the active ingredient that may not have medicinal properties and is commercially manufactured for uses other than in drug products presents a difficult regulatory issue. These special atypical APIs are predominately used in other industries as fine chemicals and are not manufactured according to ICH Q7. This creates an issue for the manufacturers of drug products because their supplier of API is not prepared to be inspected against the ICH standards for Active Pharmaceutical Ingredients. As stated earlier the MHRA has taken a pragmatic approach to GMP compliance of these materials and as such has considered each case on its own merits. However continued use of these APIs in dosage forms has the potential to bring the drug product manufacturer into conflict with current regulatory requirements. This same issue applies to commodity chemicals used as APIs or intermediates. Some of these are specified in pharmacopoeial monographs and meet the testing requirements, but are not usually manufactured under ICH Q7 controls, e.g. Ethylene diamine tetraacetic acid disodium salt dihydrate (EDTA) Ph. Eur., BP, USP, FCC.

In the example above, the same API has multiple differing specifications. While ICH has a separate quality Expert Working Group that is harmonizing pharmacopoeial monographs under ICH Q4 to partially address this difficulty, an API manufactured under ICH Q7 controls and supplied to different generic drug product manufacturers may need to comply with different specifications. The supplier must ensure that all the release testing meets the specifications filed for the API in a DMF or published in the regional Pharmacopoeia. If a customer requires more stringent specifications, these are handled in a Quality Agreement between the parties. In general though Investigators from U.S. FDA and other regional regulatory authorities are not amenable to a manufacturer selecting a lot of API for one customer that does not meet specifications of another customer.

11.5   Part D: Examples of regulatory citations (483s and WLs) pertaining to API facilities

The FDA has attempted to inform the regulated pharmaceutical industry about issues that it found to be of interest by Freedom of Information (FOI) requests and/or significant concerning CGMPs. One of the approaches that it has used is the publication on its website, *www.FDA.gov*, all Warning Letters issued since 1996. The majority of these documents reference regulatory observations obtained during inspections of drug product manufacturing facilities or concerns about off-label promotions of drug products (not a CGMP issue). The FDA has also published selected FDA 483s Inspectional Observations that have been issued by Investigators since 2009. These are available at *http://www.fda.gov/AboutFDA/CentersOffices/OfficeofGlobalRegulatoryOperationsandPolicy/ORA/ORAElectronicReadingRoom/default.htm.* The majority of these selected FDA 483s concern food processing facilities and drug product manufacturing facilities rather than API manufacturing sites.

For this chapter WLs and 483s published during 2010 and 2011 were reviewed. There were 19 WLs for API manufacturers issued since 2010. However only two 483s were published for API manufacturers during this review period, Hospira at

Boulder, CO[53] and Scientific Protein Labs (SPL)[54], and the SPL 483 was later the basis for a WL[48].

While only results of FDA inspections are reviewed here, FDA was a participant in the International API Inspection Program with EMA and TGA to compare API facility inspectional approaches and to share inspectional information. FDA shared the results of 57 inspections with the other two participants and received results from 17 of their inspections. A conclusion of the two-year program was that the project contributed substantially to a better understanding of regional approaches to inspection and the building of mutual confidence. Because of the success of the pilot program the three regulatory bodies want to consider going forward to develop and implement a common policy related to re-inspection of shared sites in other countries.

Observations noted in the following review of documents represent serious quality deficiencies identified in the cited API facilities. These deficiencies are presented in four categories of Facility, Laboratory, Manufacturing, and Quality Unit to assist the reader. While only summaries of the citations are listed here, the source document is referenced with each citation for those readers interested in a more detailed evaluation.

Facility:

Issues identified by the FDA involved design and maintenance of the facilities.

Design

Facilities not designed to minimize potential contamination[39]

Failure to have API facilities of appropriate design and construction suitable for intended use[51]

Failure to have appropriate facility and controls in place to prevent cross-contamination[55]

Maintenance

Failure to properly maintain manufacturing buildings[36]

Failed maintenance procedures to prevent contamination[35]

Laboratory:

Issues identified by the FDA involved laboratory management, sampling plans, testing procedures, stability monitoring, laboratory investigations, and evaluation of contract laboratories.

Laboratory management

Failure to calibrate laboratory balance over the range of intended use[37]

Failure to include complete data derived from all tests conducted[40]

Failure to reanalyse 5 API lots originally analysed by a general DNA method not validated for the API when a specific, validated, and approved DNA method was available[42]

Failure to ensure approved test procedures are followed[44]

Failure to have complete and reliable laboratory control records derived from all tests[44]

Failure to appropriately qualify analytical instruments used for Broth Disk Elution testing[50]

Failure to have personnel qualified by education, training, experience, or a combination[50]

Failure to have adequate controls to prevent manipulation of raw data during routine testing[51]

Sampling plans

Failure to ensure sampling plans are scientifically sound[40]

Failure to have raw material and component sampling plans that represent the batch[40]

Lack of adequate written procedure for homogeneous raw material sampling[53]

Testing procedures

Failure to validate analytical methods used to test APIs[38]

Lack scientifically sound test procedures to ensure APIs meet specifications[39]

Failure to have system suitability for chromatographic test methods[40]

Failure to appropriately validate methods and to assess impact of method modification[40]

Failure to validate analytical methods used for potency testing[42]

Failure to validate electronic formulae used to calculate results[45]

Failure to use current USP methods to test APIs[45]

Failure to appropriately validate analytical method[50]

Stability monitoring

Failure to have stability methods validated to be stability indicating[40]

Failure to have adequate stability testing program to monitor APIs[45]

Failure to appropriately store stability samples under controlled conditions[45]

Laboratory investigations

Failure to conduct appropriate laboratory out-of-specification (OOS) investigation[37]

OOS investigation did not include all the available data[38]

Failure to document OOS results and to include other lots associated with the OOS[40]

Retested OOS lot without determining root cause and used the passing retest result to invalidate the original OOS without determining lab error[40]

Failure to thoroughly investigate failure of batch to meet specifications[43]

Failure to investigate and document OOS results for API and just retested and released batch[44]

Failure to investigate OOS impurities[47]

Failure to conduct adequate investigation concerning OOS for LOD[47]

Failure to have OOS procedure[50]

Evaluation of contract laboratories

Failure to properly evaluate contract laboratory[47]

Failure to adequately evaluate and qualify contract laboratory[54]

Manufacturing:

Issues identified by the FDA involved production management, equipment cleaning, process validation, production activities, and deviations/investigations.

Production management

Failure to have adequate number of personnel qualified by education, training, or combination to ensure APIs are manufactured in accord with GMPs[45]

Failed to ensure calibration and maintenance of critical equipment and instruments[45]

Failed to review and approve blank production records printed by a third party[45]

Failure to have equipment of appropriate design for intended manufacturing use[48]

Failure to follow procedures for evaluating critical material suppliers[48]

Failure to have documents for change control, OOS, etc. related to API manufacture[49]

Failure to have procedures or practices to prevent cross-contamination[52]

Failed to adequately evaluate suppliers of raw ingredients[54]

Equipment not of appropriate design and not qualified to ensure it does not alter API[54]

Equipment cleaning

Failure to document cleaning of major equipment[36]

Failure to document equipment cleaning records[37]

Incomplete cleaning validation for non-dedicated manufacturing equipment[38]

Failure to ensure cleaning procedures are validated for "big bags" used in manufacturing[40]

Failure to ensure cleaning records are maintained[40]

Equipment and utensils not cleaned at appropriate intervals[53]

Written procedures for non-dedicated equipment cleaning was not completed[53]

Equipment cleaning and use logs were not reviewed by supervisor as required by SOP[53]

Process validation

Master Production Records do not include complete production instructions[39]

Failure to ensure new and modified equipment are qualified and suitable for intended use[39]

Failure to identify and validate critical parameters and in-process attributes for APIs[40]

Failure to validate manufacturing process[47]

Failure of quality system to have validated manufacturing processes[49]

Failure to prepare, review, and approve documents for manufacture of APIs[52]

Production activities

Failure to record quality-related activities at the time they are performed[37]

Personnel fail to wear suitable clothing to protect API from contamination[39]

Failure of Operator to document activities at time they are performed[45]

Failure to document manufacturing operations at time they are performed[52]

Deviations/investigations

Failure to investigate production deviations[35]

Failure to report production deviations and investigate critical production deviations[39]

Quality:

Issues identified by the FDA involved quality management, product release/rejection, other quality related activities, and complaint handling.

Quality management

Quality Unit did not exercise its responsibility to ensure APIs met specifications and their production complied with GMPs[38]

Failure to maintain revision histories[39]

Changes to product, process, quality controls, equipment, and facilities not reported to FDA[43]

Failed to discover data altering practices and adequate GMP documenting practices[45]

Failure of Quality Unit to exercise its responsibility to ensure APIs meet specifications and production complies with GMPs[46]

Failure of Quality Unit to exercise responsibility to ensure APIs meet specifications and production complies with GMPs[51]

Firm is neither registered nor has it listed every API in commercial distribution[52]

Product release/rejection

Released quarantined API[35]

Failure to reject APIs contaminated with foreign material[39]

Released 36 batches with a test method not validated for its intended use[41]

Failure to ensure that all released lots and lots in inventory meet revised specification[42]

Issued outdated Certificate of Analysis that did not include the revised purity specification required by Biological License Application[42]

Failure to control, process, analyze, and approve or reject raw materials and finished APIs[52]

Released product from non-validated processes and no stability testing schedules[53]

First batch released under concurrent validation with 19 redline changes to process[53]

Other quality related activities

Failure to control issuance of lab records to ensure integrity of lab raw data[40]

Failure to adequately investigate critical deviations or failure of batch to meet specifications[41]

Inadequate investigation of critical deviations or failure of batch to meet specifications[42]

Failure to have adequate product quality reviews[44]

Failure of Quality Unit to review and approve all appropriate quality-related documents[44]

Failure of Quality Unit to ensure APIs are tested and results reported[46]

Quality Unit failed to investigate critical deviations or failure of a batch to meet specifications[47]

Quality Unit failed to adequately review batch records[47]

Quality Unit failed to initiate change control and assess impact after changes were made to Master Batch Record[47]

Quality Unit failed to follow procedures for annual product review[47]

Quality Unit failed to ensure materials are appropriately tested and results accurately reported[50]

Failure to have test records readily available during retention period at establishment[50]

Failure of Quality Unit to ensure materials are tested and results reported[51]

Complaint handling

Failure to quarantine returned APIs[36]

Failure to investigate all quality related complaints[48]

Failed to adequately investigate product quality complaint and did not extend to other lots[54]

11.6   Conclusion

API manufacturing represents a diverse array of processes within pharmaceutical production. From small molecule chemical entity to supremely complex biological moieties, these active components often require months, if not years, to synthesize or manufacture. The complexity of the applied processes, the often huge amounts of raw materials and intermediates required, and the global distribution of API manufacturers pose specific challenges to the inspectorates, but also to the companies striving to comply with the pertinent regulations.

The ever increasing risk from falsified medicines has put APIs in the regulators' spotlight again, as its manufacture and distribution forms a critical component of the global supply chain for drugs. This has led to an increase in inspectoral oversight and audits by the customers, i.e. the drug product manufacturers. Each of these have their own interpretation of the regulations and their own specific agenda, which makes it essential for API manufacturers to have a well-defined, scientifically sound and defendable quality management system in place that assures Good Manufacturing Practices are applied within the validated manufacturing steps as required by the regional regulatory authorities.

This chapter should be read in conjunction with other parts of this book as the authors attempted to omit duplication and repetition of information presented elsewhere as much as possible.

Appendix A: Checklist to aid the pharmaceutical manufacturer in preparing for the inspection of API facilities (Annex 2 of APIC Auditing Guide, with kind permission by Cefic (European Chemical Industry Council))

Auditing Guide

Annex 2 – Aide Mémoire

Company :	Auditor(s) :
Location, Country :	Date of Audit:

General Remark

Chapters 1 to 19 of this Aide Mémoire refer to the appropriate chapters of ICH Q7 *(Good Manufacturing Practice Guide for Active Pharmaceutical Ingredients)*. Chapter 20 relates to aspects of Quality Management Systems according to ISO 9001 or ICH Q10 Pharmaceutical Quality System.

Reference ICH Q7	Topics/Issue	Applicability		Complaint			Kind of Documentation**	Commentary	Question posed
		Yes	No	Yes	tbi*	No			
1	**Introduction**								
1.1	**Scope**								
	Has the company designated the point at which the production of the API begins? Can a rationale be provided for this decision? Has the decision been discussed with the respective authority? Are the quality critical steps identified?								
2	**Quality Management**								
2.1	**Principles**								
2.10	A Certified Quality Management System (e.g. ISO 9001) is implemented? (if yes, see chapter 20)								
2.11	Is there a quality policy? How is it brought to the attention of the employees? Is there a Quality Manual or equivalent documentation that describes in detail how the Quality System is implemented? How does Management review effectiveness of quality system								
2.12	Is the Quality Unit (QA/QC) independent of production?								
2.13	Is there an authorized person(s) for the release of IM and APIs? Who is the person(s)?								
2.14	Are all deviations documented and explained? Are critical deviations investigated in a timely manner? Is there a written procedure for handling investigations (6.53)? Average days for completion?								
2.15	How is it ensured that materials are not released or used before completion of evaluation by the QU? If not done by QU: Is an appropriate system in place?								
2.16	How is management notified of serious GMP deficiencies, quality related complaints and/or product defects? Average time needed for information?								

*tbi = to be implemented   ** Procedure, SOP, OI, memo, notes (personal), Q-manual

485

Reference ICH Q7	Topics/Issue	Applicability Yes	Applicability No	Complaint Yes	Complaint tbi*	Complaint No	Kind of Documentation**	Commentary	Question posed
2.2	**Responsibilities of the QU**								
2.20	Are there procedures that ensure that QU reviews and approves all quality related documents?								
2.21	Non-transferable responsibilities of QU: – release/rejection of APIs and IM (to be sold) – establish system to release/reject materials and labels – review of critical process steps batch records – ensure critical deviations are investigated – approving specifications and master instructions – approving all quality related documents – ensuring conduction of internal audits – approving contract manufacturers – approving changes with quality impact – approving validation documents – ensure complaints are resolved – ensuring calibration system is functioning according to procedure executed – ensuring that stability data is generated and reviewed – performing product quality reviews								
2.3	**Responsibilities for Production Activities**								
	– procedure for preparing, reviewing and approving instructions – reviewing batch production records – ensure all deviations and investigations are handled – cleaning of facilities – calibrations performed – validation documents generated – evaluation of proposed changes – ensure that facilities and equipment are qualified								
2.4	**Internal Audits**								
2.40	Are regular audits performed? Is there an audit schedule? Is the schedule followed?								

* tbi – to be implemented   ** Procedure, SOP, QI, memo, notes (report), Q-manual

486

Reference ICH Q7	Topics/Issue	Applicability		Complaint			Kind of Documentation**	Commentary	Question posed
		Yes	No	Yes	tbi*	No			
2.41	Are audit findings and corrective actions documented? Procedure to notify management of audit findings? Are corrective actions completed within agreed time (are there significant delays?)								
2.5	**Product Quality Review**								
2.50	Are regular Product Quality Reviews conducted for all products? Frequency (dedicated, campaign)? Content (at least): – review of critical IPC and API test results – review of all batches failed – review of all critical deviations – review of process changes and impact on quality – review of changes to analytical methods – review of results of ongoing stability programmes – review of returns, complaints, recalls – review of adequacy of corrective actions defined in previous review								
2.51	Evaluation and assessment for need of additional corrective actions to address recurring issues and/or need for process or cleaning revalidation								
3	**Personnel**								
3.1	**Personnel Qualifications**								
3.10	Adequate number of personnel? Qualification of personnel sufficient at different levels?								
3.11	Are responsibilities of all personnel engaged in manufacture in APIs in writing available? Are responsibilities periodically reviewed to ensure they are current?								
3.12	Is regular training conducted? Are records of training maintained? Is effectiveness of training evaluated? How?								

* tbi = to be implemented   ** Procedure, SOP, OI, memo, notes (personal), Q-manual

Reference ICH Q7	Topics/Issue	Applicability Yes	Applicability No	Complaint Yes	Complaint tbi*	Complaint No	Kind of Documentation**	Commentary	Question posed
3.2	**Personnel Hygiene**								
3.20	Do personnel wear clean clothing suitable for activity? Additional protective apparel where necessary (e.g. Final Product Packing Rooms)?								
3.21	How is it ensured that personnel have no direct contact with IM and APIs?								
3.22	How is it ensured that no smoking, drinking, chewing and storage of food takes place ?								
3.23	How are personnel with infectious diseases or open lesions identified? Is there a procedure in place that these persons have no product contact?								
3.3	**Consultants**								
	Are consultants used to advise on any GMP related activities? Is there an assessment of consultant's education, training and experience?								
4	**Buildings and Facilities**								
4.1	**Design and Construction**								
4.10	Can cleaning and maintenance be easily performed based on design of equipment and layout of facility? Have production and warehouse facilities been designed to prevent contamination or cross contamination? If not, how is contamination prevented?								
4.11	Is there adequate space for placement of equipment to prevent mix-up or contamination?								
4.12	Outdoor equipment raises concerns for contamination?								
4.13	Does flow of materials and personnel raise concerns for contamination?								

Reference ICH Q7	Topics/Issue	Applicability		Complaint			Kind of Documentation**	Commentary	Question posed
		Yes	No	Yes	tbi*	No			
4.14	Defined areas or control systems in place for the following activities: – receipt, identification, sampling of incoming materials – quarantine before release/reject – Sampling of intermediates or API's – holding of rejected materials before further disposition? – Packaging and labeling operations?								
4.15	Washing facilities and toilets available for personnel?								
4.16	Laboratory areas separated from production?								
4.2	**Utilities**								
4.20	All utilities that could impact on product quality are identified and qualified? Are the utilities monitored and actions taken when alert limits are exceeded?								
4.21	Adequate ventilation, air filtration and exhaust systems in place? Are these systems designed and operated to prevent contamination?								
4.22	Control of re-circulated air sufficient to avoid contamination?								
4.23	Permanently installed pipework appropriately identified? Is pipework maintained and located in such a way as to prevent contamination?								
4.24	Are drains designed to prevent back-siphonage or microbiological contamination in areas where product is exposed?								
4.3	**Water**								
4.30	Water demonstrated to be suitable for intended use?								
4.31	Is Process water meeting drinking water quality as a minimum standard? Is additional water treatment system in place? Is quality of all grades of process water monitored at points of								

* tbi = to be implemented   ** Procedure, SOP, OI, memo, notes (personal), Q-manual

Reference ICH Q7	Topics/Issue	Applicability		Complaint			Kind of Documentation**	Commentary	Question posed
		Yes	No	Yes	tbi*	No			
	use for physical/chemical attributes, total microbial counts, objectionable organisms? Are actions taken when limits are exceeded?								
4.32	Tighter specifications needed to ensure quality? What are the specifications?								
4.33	Validation of treatment of (higher) water treatment?								
4.34	If claims are made for sterile or parenteral use: Monitor microbial counts, objectionable microorganisms and endotoxins								
4.4	**Containment**								
4.40	For highly sensitizing materials are dedicated production areas (facilities, air systems, equipment) in use?								
4.41	Dedicated production area for high pharmacological activity								
4.42	Are there measures to prevent cross-contamination from personnel, materials etc. for example moving from one production area to another?								
4.43	Production of highly toxic, non-pharmaceutical products, for example pesticides excluded from pharmaceutical production facilities?								
4.5	**Lighting**								
	Adequate lighting for e.g. cleaning and maintenance								
4.6	**Sewage and Refuse**								
4.60	Sewage to be removed timely								
4.7	**Sanitation and Maintenance**								
4.70	Buildings to be kept properly maintained, repaired and cleaned								
4.71	Written procedures for cleaning for equipment and facilities in place								
4.72	Procedures for pest control in place?								

Reference ICH Q7	Topics/Issue	Applicability		Compliant			Kind of Documentation**	Commentary	Question posed
		Yes	No	Yes	tbi*	No			
5	**Process Equipment**								
5.1	**Design and Construction**								
5.10	Equipment suitably located, easy to clean and maintain?								
5.11	Equipment surfaces do not alter product quality								
5.12	Equipment only used within the qualified operation range?								
5.13	Major equipment and permanently installed pipework identified								
5.14	Lubricants not in contact with IM and APIs? Otherwise food grade lubricants used?								
5.15	Precautions (measures) taken where equipment is opened to prevent contamination? For example addition of seeds or sampling								
5.16	Are current engineering drawings available for equipment, installations and utility systems?								
5.2	**Equipment Maintenance and Cleaning**								
5.20	Preventive maintenance programme in place? Schedule followed?								
5.21	Written procedures for the cleaning of equipment in place? Do the procedures give sufficient detail to enable operators to clean each type of equipment in an effective and reproducible manner?								
5.22	Are equipment and utensils, such as sampling devices cleaned, stored and where appropriate sanitized or sterilized to prevent contamination or carry-over of a material that would affect the quality of the IM or API?								
5.23	Continuous production or dedicated production facilities: is equipment/facility cleaned at appropriate intervals to prevent build-up or carry over of contaminants for example degradants or objectionable levels of micro-organisms? Is the cleaning frequency justified and documented?								

tbi = to be implemented   ** *Procedure, SOP, OI, memo, notes (personal), Q-manual*

Reference ICH Q7	Topics/Issue	Applicability		Complaint			Kind of Documentation**	Commentary	Question posed
		Yes	No	Yes	tbi*	No			
5.24	Is equipment cleaned between production of different products?								
5.25	For multi-purpose equipment is the Maximum Acceptable Carry Over and other Acceptance criteria for residues justified and determined? Are the cleaning procedures validated?								
5.26	Equipment identified as to its content and cleanliness status?								
5.3	**Calibration**								
5.30	Instruments critical for IM and/or API quality are calibrated? How is critical defined? Written procedure in place? Schedule followed?								
5.31	Calibration done with standards that are traceable to certified standards?								
5.32	Records of calibration maintained?								
5.33	Calibration status of instruments known? How (label, electronic)?								
5.34	How is it ensured that instruments out of calibration are not used?								
5.35	If instruments have been shown out of calibration, are deviation investigations performed to determine if this fact has an influence on the release of the IM/API?								
5.4	**Computerised Systems**								
5.40	Are GMP related computer systems validated?								
5.41	IQ, OQ for Hard- and Software available to demonstrate suitability of computer hardware/software to perform task?								
5.42	Retrospective validation for existing systems if not validated at time of installation?								
5.43	What controls are in place to prevent unauthorized access? What controls are in place to prevent changes to data?								

Reference ICH Q7	Topics/Issue	Applicability		Complaint			Kind of Documentation**	Commentary	Question posed
		Yes	No	Yes	tbi*	No			
	What controls are in place to prevent omissions in data? Is there an audit trail / documents available where changes to data are recorded, who made the change, when the change was made and of the previous entry?								
5.44	Written procedures for the operation and maintenance of computerized systems available?								
5.45	Is the manual entry of critical data checked by additional means (second operator or system itself)?								
5.46	Are all quality related incidents and deviations relating to computerized systems investigated according to defined procedures investigated?								
5.47	Changes to the computerized system are made according to a defined procedure?								
5.48	How is data protected in cases of system breakdowns? Back-up system provided? Is Recovery from back-ups tested periodically?								
6.	**Documentation and Records**								
6.1	**Documentation System and Specifications**								
6.10	Is there a written procedure in place describing preparation, review, approval and distribution of all quality related documents?								
6.11	How is revision, superseding and withdrawal of documents controlled? Is a revision history maintained?								
6.12	Procedure in place for retaining all appropriate documents? Retention period specified?								
6.13	Retention period for APIs with expiry date: 1 year after expiry (min.) Retention period for APIs with retest date: 3 years after complete distribution (min.)								

* tbi = to be implemented   ** Procedure, SOP, OI, memo, notes (personal), Q-manual

Reference ICH Q7	Topics/Issue	Applicability		Complaint			Kind of Documentation**	Commentary	Question posed
		Yes	No	Yes	tbi*	No			
6.14	Are corrected entries in documents dated and signed? Original entry still readable?								
6.15	Are documents promptly retrievable (copies or electronic means acceptable)?								
6.16	Are specifications for all materials, IM and APIs established?								
6.17	Are electronic signatures authenticated and secure?								
6.2	**Equipment Cleaning and Use Records**								
6.20	Are there records for the major equipment used , cleaning and maintenance showing the following – date – time – product and batch number of each batch – person who performed cleaning – person who performed maintenance								
6.3	**Records of Raw Materials, IM, API Labeling and Packaging Materials**								
6.30	Records of each delivery should contain: – name of manufacturer/supplier – identity and quantity – supplier control or identification number – number allocated on receipt – date of receipt – acceptable condition of received goods assessed – result of tests and conclusion derived from this – trace of use – review of labels and packaging materials showing conformity with specifications – final decision release or reject								
6.31	Are master labels maintained?								

Reference ICH Q7	Topics/Issue	Applicability		Compliant			Kind of Documentation**	Commentary	Question posed
		Yes	No	Yes	tbi*	No			
6.4	**Master Production Instructions**								
6.40	Are Master Production Instructions for each IM/API – prepared – dated – signed – independently checked by QU								
6.41	Do Master Production Instructions contain the following: – name of product including document reference code – complete list of raw materials – accurate statement of quantities needed or calculation of quantity – production location and major equipment to be used – detailed production instructions including sequences, ranges of parameters, sampling instructions, IPC, time limits, expected yield – instructions for storage								
6.5	**Batch Production Records**								
6.50	Are Batch Production Records checked before issuance for correct version?								
6.51	Are the records showing an unique batch number (not for continuous production)?								
6.52	The batch record should contain the following: – date(s) and times (if appropriate) – identity of major equipment – identification of materials used – actual results – sampling performed – signatures of the person(s) performing the operation – IPC/laboratory test results – actual yield, if appropriate – description of packaging and labels used								

* tbi = to be implemented   ** Procedure, SOP, OI, memo, notes (personal), Q-manual

495

Reference ICH Q7	Topics/Issue	Applicability		Complaint			Kind of Documentation**	Commentary	Question posed
		Yes	No	Yes	tbi*	No			
	– deviation/investigation – results of release testing								
6.6	**Laboratory Control Records**								
6.60	Laboratory records should contain the following: – description of sample including name, batch number or code, date when sample was taken, quantity – reference to test method – cross reference to preparation of reference standards, reagents and/or standard solutions – complete record of all raw data – record of all calculations – statement of test result if they comply with specifications – signature and date of person(s) performing the testing – signature of second person demonstrating review for accuracy, completeness								
6.61	Other records to be maintained: – modification to test method – calibration of laboratory instruments – stability testing performed – OOS investigations								
6.7	**Batch Production Record Review**								
6.70	Is a written procedure for the handling of batch (laboratory) record review available?								
6.71	Are batch (laboratory) records of critical steps reviewed by the QU? Are they reviewed before the release of the API?								
6.72	Are all deviations, investigations and OOS reviewed as part of the batch record review?								
6.73	Is the QU releasing all IM that are shipped outside the control of the company?								

Reference ICH Q7	Topics/Issue	Applicability		Compliant			Kind of Documentation**	Commentary	Question posed
		Yes	No	Yes	tbi*	No			
7	**Materials Management**								
7.1	**General Controls**								
7.10	Are written procedures available for handling of receipt, identification, quarantine, storage, sampling, testing, approval or rejection of materials?								
7.11	System to evaluate suppliers of critical materials in place? Evaluation must show that supplier can consistently provide material meeting specifications (7.31)								
7.12	Materials purchased against agreed specifications? Purchased from an approved (by QU) supplier?								
7.13	If supplier is not the manufacturer, is the original manufacturer known?								
7.14	Change of source/supplier handled according to Change Control procedures (chap. 13)?								
7.2	**Receipt and Quarantine**								
7.20	Upon receipt materials visually examined for – correct labeling – container damage – broken seals – tampering or contamination Are materials held under quarantine until released for use? How is this done?								
7.21	Incoming materials are released before mixed with existing stocks? Are procedures in place to prevent discharging materials wrongly?								
7.22	If deliveries are made in non-dedicated tankers which assurance is provided to demonstrate no contamination (one or more of the following): – certificate of cleaning								

* tbi = to be implemented   ** Procedure, SOP, OI, memo, notes (personal), Q-manual

Reference ICH Q7	Topics/Issue	Applicability		Complaint			Kind of Documentation**	Commentary	Question posed
		Yes	No	Yes	tbi*	No			
	– testing for trace impurities – audit of the supplier								
7.23	Is each delivery of materials identified (code or batch number)? Is there a system in place to identify the status of each batch?								
7.3	**Sampling and Testing of Incoming Production Materials**								
7.30	Is at least one test conducted to verify the identity of incoming materials? If suppliers Certificate of Analysis is used instead of testing a system for evaluation must be in place.								
7.31	(see also 7.11) Are 3 full analyses conducted before reducing testing? Is a full analysis performed at appropriate intervals and compared with the suppliers certificate of analysis?								
7.32	How is it demonstrated that samples taken from the material are representative? Are sampling methods described with at least – number of containers to be sampled – which part of the container – amount of sample to be taken								
7.33	Is sampling done at defined locations preventing contamination?								
7.34	Are containers from which samples are taken identified?								
7.4	**Storage**								
7.40	Is material stored in a manner to prevent degradation and contamination?								
7.41	Are fiber drums, bags and boxes stored off the floor? Is stored material suitably spaced to permit cleaning and inspection?								
7.42	Do materials met their respective storage conditions? Is the FIFO principle followed?								

Reference ICH Q7	Topics/Issue	Applicability		Compliant			Kind of Documentation**	Commentary	Question posed
		Yes	No	Yes	tbi*	No			
7.43	In case materials is stored outdoors: – do labels remain legible – are the containers cleaned before opening – is it described in a procedure								
7.44	How are rejected materials held under a quarantine system?								
8	**Production and In Process Controls**								
8.1	**Production Operations**								
8.10	Are weighing and measuring devices of suitable accuracy for their intended use? Are the devices periodically calibrated with certified references?								
8.11	Do containers with subdivided material contain the following information: – name of material – code or control number – weight, if applicable – retest date, if applicable								
8.12	How are critical weighing, measuring or subdividing operation witnessed? Is an equivalent control used? If so what?								
8.13	Are all other critical operations witnessed or subjected to equivalent control?								
8.14	Are actual yields compared with expected yields at designated steps in production?								
8.15	How is the processing status of major units of equipment indicated?								
8.2	**Time Limits**								
8.20	Are all specified time limits of the operating instructions met?								
8.21	How are storage conditions for IM held for further processing determined?								

* tbi = to be implemented   ** Procedure, SOP, OI, memo, notes (personal), Q-manual

Reference ICH Q7	Topics/Issue	Applicability		Complaint			Kind of Documentation**	Commentary	Question posed
		Yes	No	Yes	tbi*	No			
8.3	**In-process Sampling and Controls**								
8.30	Are IPC established to monitor the progress and control the performance of the processing steps?								
8.31	Are critical IPC approved by the QU?								
8.32	How is the qualification (training) of the production personnel documented, if they perform the IPC?								
8.33	Are sampling methods for IPC described in writing?								
8.34	Does in-process sampling not cause the contamination of sample and/or product?								
8.4	**Blending of Batches of IM or APIs**								
8.40	Are OOS batches blended with other batches meeting the specifications? Are all batches individually tested prior to blending? And do they all meet specification?								
8.41	Is the blending process adequately documented and the blended batch tested for conformance to specifications?								
8.42	Does the batch record of the blended batch allow traceability back to the individual batches?								
8.43	Are blending operations validated if physical attributes of the API resulting from this step are known to be critical?								
8.44	How is it demonstrated that the blended batch does not affect stability?								
8.45	Is the expiry/retest date based on the oldest batch in the blend?								
8.5	**Contamination Control**								
8.50	How is it ensured that carryovers (e.g. degradants) into successive batches of the same IM/API do not affect the impurity profile of the API?								
8.51	What measures are taken in production to prevent contamination of IM/API?								

Reference ICH Q7	Topics/Issue	Applicability		Complaint			Kind of Documentation**	Commentary	Question posed
		Yes	No	Yes	tbi*	No			
8.52	What specific precautions are taken to avoid contamination of the API after purification?								
9	**Packaging and Identification Labelling of APIs and IM**								
9.1	**General**								
9.10	Are written procedures available describing – receipt – identification – quarantine – sampling – examination/testing – release of packaging materials and labels ?								
9.11	Are specifications for all packaging materials and labels established? Are suppliers of primary packaging materials in contact with the product qualified?								
9.12	Are records of each delivery of packaging materials and labels kept?								
9.2	**Packaging Materials**								
9.20	Can containers/packaging material used provide adequate protection against deterioration or contamination during transportation?								
9.21	Are containers cleaned so that they are suitable for their intended use?								
9.22	Are written procedures for cleaning in place for re-used containers? Are all previous labels removed or defaced?								
9.3	**Label Issuance and Control**								
9.30	Is access to label storage area limited to authorized personnel?								

*tbi = to be implemented   ** Procedure, SOP, OI, memo, notes (personal), Q-manual

501

Reference ICH Q7	Topics/Issue	Applicability		Complaint			Kind of Documentation**	Commentary	Question posed
		Yes	No	Yes	tbi*	No			
9.31	Are procedures in place to reconcile the quantities of labels issued and used? Are discrepancies investigated and approved by the QU?								
9.32	Are labels bearing batch numbers not used being destroyed? How is it documented?								
9.33	Are all out-dated and obsolete labels destroyed?								
9.34	Are printing devices checked that the imprint conforms to the print specified in batch record? Is an examination done to check if the correct label is on the packed IM/API? (9.45)								
9.35	Is a representative label included in the batch record?								
9.4	**Packaging and Labelling Operations**								
9.40	Are written procedures in place ensuring that correct packaging materials and labels are used?								
9.41	Is physical or spatial separation of labels done when multiple labeling operations are done at the same time?								
9.42	Labels should indicate the following information (at least): – name of product – identifying code and batch number – storage conditions, when such information is critical to assure quality								
9.43	If the IM/API is transferred outside of the control of the manufacturer the label as well as requirements of 9.42 contain: – name and address of manufacturer – quantity – special transport conditions, if applicable – special storage conditions, if applicable (10.22) – legal requirements, if applicable								

Reference ICH Q7	Topics/Issue	Applicability		Complaint			Kind of Documentation**	Commentary	Question posed
		Yes	No	Yes	tbi*	No			
	For APIs with expiry date: date to be included on label and certificate of analysis. For APIs with retest date: date to be included on label and/or certificate of analysis								
9.44	Are packaging and labeling facilities inspected before use to ensure that all materials not needed are removed? Is this inspection documented?								
9.45	Are seals and other security measures used that will alert the recipient that the material may have been altered? Please specify.								
10	**Storage and Distribution**								
10.1	**Warehousing Procedures**								
10.10	Are facilities for the storage of materials available supporting the claimed storage conditions (e.g, temperature, humidity)? Are records of the storage conditions kept?								
10.11	Are separate storage areas provided for quarantined, rejected, returned or recalled products? Or is an alternative system used? If so, how is it designed and qualified?								
10.2	**Distribution Procedures**								
10.20	How is it ensured that APIs/IM are not distributed outside of the company before the release of the QU?								
10.21	How are transportation conditions ensured so that the quality of the product will not be adversely affected?								
10.22	How does the manufacturer ensure that the transporter knows and follows the appropriate transport and storage conditions?								
10.23	Are there systems in place to easily permit a recall? Has its effectiveness been demonstrated?								

* tbi = to be implemented   ** Procedure, SOP, OI, memo, notes (personal), Q-manual

Reference ICH Q7	Topics/Issue	Applicability		Complaint			Kind of Documentation**	Commentary	Question posed
		Yes	No	Yes	tbi*	No			
11	**Laboratory Controls**								
11.1	**General Controls**								
11.10	Are adequate laboratory facilities available?								
11.11	Are all sampling plans and testing procedures reviewed and approved by the QU?								
11.12	Do the specifications set for the APIs include a control of the impurities? If the API has a specification for microbiological purity and/or endotoxins what appropriate action limits have been established?								
11.13	Are all OOS results investigated? Is resampling after OOS described in a procedure?								
11.14	Are written procedures in place for preparation of reagents and standard solutions?								
11.15	Are primary reference standards stored under appropriate conditions? Is the source of the primary standard documented?								
11.16	If the primary standard is not obtained from an officially recognized source, is appropriate testing conducted to fully establish the identity and purity of the primary standard?								
11.17	Are procedures in place to prepare, identify, test store and approve secondary reference standards? Is the suitability of the secondary standard determined prior to use by comparing it against the primary standard? Are secondary reference standards periodically re-qualified?								
11.2	**Testing of Intermediates and APIs**								
11.20	Is there an impurity profile established for every API?								
11.21	Is the impurity profile compared at appropriate intervals against the impurity profile in the regulatory submission or against historical data?								

Reference ICH Q7	Topics/Issue	Applicability		Complaint			Kind of Documentation**	Commentary	Question posed
		Yes	No	Yes	tbi*	No			
11.3	**Validation of Analytical Procedures**								
	See section 12.8								
11.4	**Certificates of Analysis**								
11.40	Are authentic Certificates of Analysis issued for each batch of IM/API?								
11.41	Information on the Certificate of Analysis: – name of IM/API – batch number and code number? – date of release – expiry date, if applicable – retest date, if desired-								
11.42	On the Certificate of Analysis, are all tests performed listed, together with acceptance limits and numerical results obtained?								
11.43	Certificates of Analysis should be – dated – signed by authorized personnel of the QU – show name, address and telephone number of manufacturer If Certificate of Analysis is issued by agents (chap. 18) the name, address and telephone number of the agents must be shown.								
11.44	If Certificate of Analysis is issued by agents the name, address and telephone number of the laboratory that performed the tests must be shown. It also should contain a reference to the original manufacturer and to the original Certificate of Analysis.								
11.5	**Stability Monitoring of APIs**								
11.50	Is an on-going stability testing programme conducted? Do the results of the stability programme justify storage conditions and expiry/retest dates (see also 11.61)?								

* tbi = to be implemented   ** Procedure, SOP, OI, memo, notes (personal), Q-manual

Reference ICH Q7	Topics/Issue	Applicability		Complaint			Kind of Documentation**	Commentary	Question posed
		Yes	No	Yes	tbi*	No			
11.51	Are the test methods used in stability validated and stability indicating?								
11.52	Are the stability samples stored in containers of the same material as the market containers?								
11.53	Are the first three commercial production batches placed on stability?								
11.54	Thereafter, is at least one batch per year added to the stability monitoring programme? Are annually tests performed to confirm stability?								
11.55	For APIs with less than 1 year stability: Is testing performed monthly for the first three months and at three month intervals after that?								
11.6	**Expiry and Retest Dating**								
11.60	Is an expiry/retest date assigned when the APIs are transferred outside of the control of the company?								
11.7	**Reserve/Retention Samples**								
11.70	Are reserve samples stored for 1 year after expiry date or 3 years after distribution (whatever is longer)? For APIs are reserve samples stored for at least 3 years after complete distribution?								
11.71	Are reserve samples stored in same packaging system or more protective than the marketed? Is the amount of sample sufficient to conduct at least 2 full compendial or internal specification analyses?								
12	**Validation**								
12.1	**Validation Policy**								
12.10	Is the company's overall validation policy documented? (Could be combined with 2.12)								

Reference ICH Q7	Topics/Issue	Applicability Yes	Applicability No	Complaint Yes	Complaint tbi*	Complaint No	Kind of Documentation**	Commentary	Question posed
12.11	Are all critical parameters defined during the development (or from historical data)? Are the operating ranges defined?								
12.12	Are all critical operation steps validated?								
12.2	**Validation Documentation**								
12.20	Is a validation protocol established? Is it approved by the QU?								
12.21	Is the following specified in the validation protocol: – critical process steps – acceptance criteria – type of validation – number of process runs?								
12.22	Is a validation report prepared summarising the results obtained, including recommendation of changes to correct deficiencies?								
12.23	Are variations from the validation protocol documented and justified?								
12.3	**Qualification**								
12.30	Is there policy or procedure for Qualification/Validation?. Is Validation Master Plan available? Is appropriate qualification (DQ, IQ, OQ, PQ) conducted for critical equipment and ancillary systems? Is operational qualification completed before Performance Qualification / process validation activities?								
12.4	**Approaches to Process Validation**								
12.40	Is process validation (PV) conducted before commercial distribution of API batches?								
12.42	Prospective Validation should normally be performed. What is the justification of performing other types? Is the validation completed before commercial distribution of the drug product?								

* tbi = to be implemented   ** Procedure, SOP, OI, memo, notes (personal), Q-manual

Reference ICH Q7	Topics/Issue	Applicability		Complaint			Kind of Documentation**	Commentary	Question posed
		Yes	No	Yes	tbi*	No			
12.43	If retrospective validation is conducted for well established processes, are the following requirements met: – critical process parameters have been identified – appropriate in-process criteria have been established – no significant process failures have occurred – impurity profiles have been established for the existing API								
12.44	Are batches selected for retrospective validation representative for all batches made during the review period?								
12.5	**Process Validation Programme**								
12.50	Are at least 3 consecutive successful production batches made for prospective and concurrent validation? For retrospective validation are 10 to 30 consecutive batches examined? If fewer batches are examined, what is the justification for it?								
12.51	Has process validation confirmed that the process can be reproducibly controlled within critical operating parameters and the impurity profile is within the specified limits?								
12.6	**Periodic Review of Validated Systems**								
12.60	Are systems and processes periodically evaluated to verify that they are still operating in a valid manner (e.g. through product quality review)?								
12.7	**Cleaning Validation**								
12.70	Are cleaning procedures validated? If not, is there a justification? Is cleaning validation directed to situations where contamination or carryover poses the greatest risk?								
12.71	If various APIs/IM are produced in the same equipment and the same cleaning process is used, is a representative API/IM selected for cleaning validation (on the basis of solubility, difficulty to clean and calculation of residue limits based on potency, toxicity and stability)?								

Reference ICH Q7	Topics/Issue	Applicability Yes	No	Complaint Yes	tbi*	No	Kind of Documentation**	Commentary	Question posed
12.72	Does the cleaning validation protocol include – equipment to be cleaned – procedures – materials used – acceptable cleaning levels – parameters to be monitored – analytical methods – type of samples (swab, rinse) – how samples are collected and labeled								
12.73	Does the type of sampling detect insoluble and soluble residues? Is the sampling method capable to quantitatively measure levels of remaining residues?								
12.74	Are the analytical methods sensitive enough to detect residues or contaminates? How are residue limits established (on minimum known pharmacological, toxicological or physiological activity or the most deleterious component)?								
12.75	If claims on microbiological and/or endotoxin specifications are made, does the cleaning validation take this into account?								
12.76	Are the cleaning procedures monitored at appropriate intervals to ensure their effectiveness?								
12.8	**Validation of Analytical Methods**								
12.80	Are the analytical methods developed by the company validated? How are Pharmacopoeial methods qualified?								
12.81	How is the degree of analytical validation (e.g. for different steps of production) justified?								
12.82	Is the analytical equipment qualified?								

* tbi = to be implemented   ** Procedure, SOP, OI, memo, notes (personal), Q-manual

509

Reference ICH Q7	Topics/Issue	Applicability		Complaint			Kind of Documentation**	Commentary	Question posed
		Yes	No	Yes	tbi*	No			
12.83	Are records of modified validated analytical methods maintained? Is the reason for the modification documented?								
13	**Change Control**								
13.10	Is a formal change control system in place capable of evaluating all changes?								
13.11	Written procedures should be provided for the identification, documentation, review and approval of changes.								
13.12	Are all changes impacting the quality of the API/IM approved by the QU?								
13.13	Are changes classified (e.g. major, minor)? If not, how is the impact on the quality of the API being evaluated? How is level of testing, validation, documentation determined (scientific judgement)?								
13.14	How is it ensured that after a change all affected documents are revised?								
13.15	Are the first batches evaluated after the change has been implemented?								
13.16	If critical changes have been made, has the impact on expiry/retest dates and process validation been evaluated?								
13.17	Are medicinal product manufacturers notified about changes that could impact the API quality (especially physical attributes)?								
14	**Rejection and Re-Use of Materials**								
14.1	**Rejection**								
14.10	Are IM/APIs failing to meet specifications identified? How?								
14.2	**Reprocessing**								
	Are all steps where reprocessing is conducted part of the filing documents?								

Reference ICH Q7	Topics/Issue	Applicability		Complant			Kind of Documentation**	Commentary	Question posed
		Yes	No	Yes	tbi*	No			
14.3	**Reworking**								
14.30	Is an investigation performed before a decision is taken to rework a batch?								
14.31	Have reworked batches been subjected – to appropriate evaluation – stability testing – to show equivalency to original process? Is concurrent validation performed if more than one batch is affected? Is a report issued if only one batch is affected?								
14.32	Is the impurity profile of the reworked batch compared with the one of the established process? If routine analytical methods are inadequate, are additional methods used?								
14.4	**Recovery of Materials and Solvents**								
14.40	Do procedures exist for the recovery of materials? Do the recovered materials meet specifications for their intended use?								
14.41	Do recovered solvents used in different processes meet appropriate standards?								
14.42	Are recovered solvents been tested for suitability before being combined with fresh or approved solvents?								
14.5	**Returns**								
14.50	Are returned APIs/IM identified and quarantined?								
14.51	Are returned materials evaluated on their quality before re-use?								
14.52	Are records of returned goods available containing – name and address of the consignee – API/IM, batch number and quantity – Reason of return – Use or disposal of API/IM								

* tbi = to be implemented   ** Procedure, SOP, OI, memo, notes (personal), Q-manual

Reference ICH Q7	Topics/Issue	Applicability		Complaint			Kind of Documentation**	Commentary	Question posed
		Yes	No	Yes	tbi*	No			
15	**Complaints and Recalls**								
15.10	Is a written procedure available describing the handling of complaints?								
15.11	Do the complaint records include the following: – name and address of complaint – name and phone number of person submitting the complaint – complaint nature (including name and batch number of API) – date complaint is received – action taken (including person taking the action) – any follow-up, if applicable – response provided to the originator of complaint including date of response – final decision on API-								
15.12	Are the records of complaints retained? For how long? Are trends or recurring complaints evaluated?								
15.13	Is there a recall procedure in place?								
15.14	Does the recall procedure specify – who should be involved – how the recall is initiated – who should be informed – how recalled material is treated								
16	**Contract Manufacturers (including Laboratories)**								
16.10	Is it ensured that all contract manufacturers engaged comply with the GMP requirements of ICH Q7?								
16.11	How is the contract manufacturer evaluated for GMP compliance?								
16.12	Is there a written contract (agreement) with the contract manufacturer? Are the GMP responsibilities defined in detail?								
16.13	Does the contract permit to audit the contract manufacturer?								

Reference ICH Q7	Topics/Issue	Applicability		Complaint			Kind of Documentation**	Commentary	Question posed
		Yes	No	Yes	tbi*	No			
16.14	Is subcontracting by the contract manufacturer excluded? If not, how is it ensured that the contract giver is involved in prior evaluation of the subcontractor?								
16.15	Are all records kept at the contract manufacturers site? How is it ensured that these are readily available?								
16.16	Does the contract manufacturer have a change control system? How is it ensured that the contract giver is informed about all intended changes of the contract manufacturer to the process? Does the contract giver approve all changes?								
17	**Agents, Brokers, Traders, Distributors, Repackers, and Relabellers (Agent)**								
17.1	**Applicability**								
17.10	Is this not the original manufacturer of the API? (Then this section applies.)								
17.11	Does the Agent comply with the GMP requirements as defined in ICH Q7?								
17.2	**Traceability of Distributed APIs and IM**								
17.20	Is the following information retained: – identity of original manufacturer – address of original manufacturer – purchase orders – transportation documentation – receipt documents – name or designation of API – manufacturers batch number – distribution records – authentic certificate of analysis, including those of the original manufacturer – expiry/retest date								
17.3	**Quality Management**								
17.30	Has the Agent established a system of managing quality as defined in section 2?								

* tbi = to be implemented   ** Procedure, SOP, OI, memo, notes (personal), Q-manual

Reference ICH Q7	Topics/Issue	Applicability		Complaint			Kind of Documentation**	Commentary	Question posed
		Yes	No	Yes	tbi*	No			
17.4	**Repackaging. Relabeling and Holding of APIs and IM**								
17.40	How does the Agent ensure that during repackaging, relabeling and holding of APIs/IM no mix-ups and loss of identity and purity of the API/IM occurs? Are these operations conducted under conditions described in Q7?								
17.41	Is repackaging done under conditions to avoid contamination?								
17.5	**Stability**								
17.50	Are stability studies conducted if the API is repacked in a different type of container?								
17.6	**Transfer of Documentation**								
17.60	Does the Agent transfer all quality and regulatory information from the original manufacturer to the customer?								
17.61	Does the agent provide the name of the original manufacturer and the batch number to the customer?								
17.62	Is the specific guidance for Certificates of Analysis described in section 11.4 followed?								
17.7	**Handling of Complaints and Recalls**								
17.70	Does the Agent maintain records of all complaints and recalls that were brought to their attention?								
17.71	Does the Agent review the complaint together with the original manufacturer for determining further action?								
17.72	Do the records of the Agents include responses from the original manufacturer to a complaint?								
17.8	**Handling of Returns**								
17.80	Are returns to the Agent handled in the way described in section 14,52? Is documentation maintained of returned APIs/IM?								

Reference ICH Q7	Topics/Issue	Applicability		Compliant			Kind of Documentation**	Commentary	Question posed
		Yes	No	Yes	tbi*	No			
18	**Specific Guidance for APIs Manufactured by Cell Culture/Fermentation**								
18.1	**General**								
18.10	Are the GMP principles of the other sections complied with?								
18.11	What measures are taken for Biotech processes to ensure that raw materials (media, buffer components) are no source of microbiological contamination? If applicable, is the bioburden, viral contamination and/or endotoxins controlled at appropriate stages of production?								
18.12	Which controls are performed for steps prior to this guide, e.g. cell banking?								
18.13	Which equipment and environmental controls are used to minimize contamination? Are adequate acceptance criteria for quality and frequency's for monitoring set at the various steps of production?								
18.14	Are the following controls taken into account: – maintenance of WCB – proper inoculation and expansion of the culture – control of critical operating parameters – monitoring the process for cell growth, viability and productivity – harvesting and purification procedures – monitoring of bioburden – viral safety concerns (ICH Q5a)								
18.15	Is removal of media components, host cell proteins process and product related impurities and contamination demonstrated?								
18.2	**Cell Bank Maintenance and Record Keeping**								
18.20	Is the access to the cell banks limited to authorized personnel?								
18.21	Do the storage conditions of the cell banks ensure that viability is maintained and contamination prevented?								
18.22	Are records of the use of vials from the cell banks and storage conditions maintained?								

* tbi = to be implemented   ** Procedure, SOP, OI, memo, notes (personal), Q-manual

Reference ICH Q7	Topics/Issue	Applicability		Complaint				Kind of Documentation**	Commentary	Question posed
		Yes	No	Yes	tbi*	No				
18.23	Are cell banks monitored periodically for suitability for use?									
18.24	For handling of cell banks check ICH Q5a									
18.3	**Cell Culture / Fermentation**									
18.30	Are closed and contained systems used when aseptic additions are needed? If open vessels are used which measures and controls are used to minimise risk of contamination?									
18.31	If use of open equipment can cause microbial contamination which environmental controls are done?									
18.32	Are personnel handling the cultures appropriately gowned?									
18.33	Are critical operating parameters including cell growth, viability and productivity monitored?									
18.34	Is cell culture equipment cleaned after use?									
18.35	Is culture media sterilized before use?									
18.36	Are procedures in place to detect contamination and to determine necessary action? Is the impact of the contamination evaluated?									
18.37	Are records of contamination maintained?									
18.38	Is multi-purpose equipment sufficiently tested to minimize contamination?									
18.4	**Harvesting, Isolation and Purification**									
18.40	Are harvesting steps performed in equipment and areas designed to minimize risk of contamination?									
18.41	Can the harvesting and purification procedures remove or inactivate organisms in a way that the API is recovered with consistent quality?									
18.42	Is all equipment cleaned properly after use?									
18.43	Is purification performed under controlled environmental conditions if open systems are used?									

Reference ICH Q7	Topics/Issue	Applicability		Compliant			Kind of Documentation**	Commentary	Question posed
		Yes	No	Yes	tbi*	No			
18.44	If equipment is used for multiple products additional controls and testing is to be conducted.								
18.5	**Viral Removal/Inactivation steps**								
18.50	See ICH Q5a								
18.51	Are viral removal and inactivation steps performed within their validated parameters?								
18.52	Are appropriate precautions being taken to prevent viral contamination from pre-viral to post-viral removal/inactivation? Do open processing take place in areas that are separate from other processing activities and have separate air handling units?								
18.53	If equipment is used for different purification steps is it appropriately cleaned?								
19	**APIs for Use in Clinical Trials**								
19.1	**General**								
19.11	Are the controls used consistent with the stage of development? Are procedures flexible enough to provide changes as knowledge of the process increases? If APIs are intended to be used for clinical trials, is the API produced in suitable facilities with appropriate controls to ensure the quality?								
19.2	**Quality**								
19.20	Is there an appropriate GMP concept in place? Is a procedure for approval of batches in place?								
19.21	Is there an independent QU in place?								
19.22	Are raw materials, packaging materials IM and APIs tested?								
19.23	Are process and quality problems evaluated?								

* tbi = to be implemented   ** Procedure, SOP, OI, memo, notes (personal), Q-manual

517

Reference ICH Q7	Topics/Issue	Applicability		Complaint			Kind of Documentation**	Commentary	Question posed
		Yes	No	Yes	tbi*	No			
19.24	Does the labeling of APIs for use in clinical trials indicate the material as being for investigational use?								
19.3	**Equipment and Facilities**								
19.30	Is it ensured that during all phases of clinical development the equipment is qualified, instruments calibrated, clean and suitable for it intended use?								
19.31	Are materials handled in a way to minimize contamination?								
19.4	**Control of Raw Materials**								
19.40	Are raw materials evaluated or tested?								
19.5	**Production**								
19.50	Is the production of APIs for use in clinical trials documented appropriately according to the stage of production? Do these documents include information about materials used, equipment, processing and scientific observations?								
19.6	**Validation**								
19.60	Is the equipment used qualified and the instruments calibrated? (Validation is not expected!)								
19.61	If batches are produced for commercial use, then section 12 is to be applied.								
19.7	**Changes**								
19.70	Are all changes adequately recorded?								
19.8	**Laboratory Controls**								
19.80	Are analytical methods used scientifically sound? (No analytical validation required)								
19.81	Is a system to retain reserve samples in place?								
19.9	**Documentation**								
19.90	Is a system in place to document the information gained during the development?								

Reference ICH Q7	Topics/Issue	Applicability		Complaint			Kind of Documentation**	Commentary	Question posed
		Yes	No	Yes	tbi*	No			
19.91	Is the development of analytical methods appropriately documented?								
19.92	Is it ensured that all information is retained for an appropriate length of time?								
20	**Quality Management System**								
20.1	**Quality Issues**								
20.10	Has the organization established, documented, implemented and maintained a quality management system in accordance with the requirements of ISO 9000:2000?								
20.11	Is the effectiveness of the quality management system continually improved?								
20.12	Does the organization manage these processes in accordance with the requirements of ISO 9000:2000?								
20.13	Does the quality management system documentation include: – Documented statement of a quality policy and quality objectives – Quality Manual – Documented procedures required by ISO 9000:2000 – Documents needed by the organization to ensure the effective planning, operation and control of its processes – Records required by ISO 9000:2000								
20.14	Are documents required for the quality management system controlled?								
20.15	Has a documented procedure been established identifying the following controls needed: – Approval of documents for adequacy prior to issue – Review, update as necessary and re-approval of documents – Ensure that changes and the current revision status of documents are identified – Ensure that relevant versions of applicable documents are available at points of use								

*tbi = to be implemented   ** Procedure, SOP, OI, memo, notes (personal), Q-manual

Reference ICH Q7	Topics/Issue	Applicability		Complaint			Kind of Documentation**	Commentary	Question posed
		Yes	No	Yes	tbi*	No			
	– Ensure that documents remain legible and readily identifiable – Ensure that documents of external origin are identified and their distribution controlled – Preventing the unintended use of obsolete documents, and to apply suitable identification to them if they are retained								
20.16	Have records been established and maintained to provide evidence of conformity to requirements and of the effective operation of the quality management system?								
20.17	Has a documented procedure been established to define the following controls needed – Identification – Storage – Retrieval – Protection – Retention time – Disposition								
20.2	**Management Responsibility**								
20.20	Has top management provided evidence of its commitment to the development and implementation of the quality management system and for the continual improvement of its effectiveness?								
20.21	Has top management ensured that customer requirements are determined and met with the aim of enhancing customer satisfaction?								
20.22	Has top management established a quality policy?								
20.23	Has top management ensured that quality objectives are established at relevant functions and levels within the organization?								
20.24	Has top management ensured that responsibilities, authorities are defined and communicated within the organization?								

Reference ICH Q7	Topics/Issue	Applicability		Complaint			Kind of Documentation**	Commentary	Question posed
		Yes	No	Yes	tbi*	No			
20.25	Has top management appointed member(s) of management who have responsibility and authority for quality management?								
20.26	Has top management ensured that appropriate communication processes have been established within the organization?								
20.27	Does the top management review the quality management system, at planned intervals, to ensure its continuing suitability, adequacy and effectiveness?								
20.28	Do the outputs from the management review include the decisions and actions?								
20.3	**Resource Management**								
20.30	Have the resources for quality management been determined and provided?								
20.31	Is competency for personnel who perform work affecting product quality based on appropriate education, training, skills, and experience?								
20.32	Has the organization determined the necessary competency for personnel performing work affecting product quality?								
20.33	Does the organization identify, provide, and maintain the facilities including: Buildings, Workspace and associated utilities, Process Equipment, hardware and software, Supporting services?								
20.34	Has the environment needed to achieve conformity of product requirements been determined and managed?								
20.4	**Product Realisation**								
20.400	Is planning of the organization's product realization consistent with the requirements of the other processes of the quality management system?								

* tbi = to be implemented   ** Procedure, SOP, OI, memo, notes (personal), Q-manual

521

Reference ICH Q7	Topics/Issue	Applicability Yes	Applicability No	Complaint Yes	Complaint tbi*	Complaint No	Kind of Documentation**	Commentary	Question posed
20.401	Has the organization determined requirements specified by the customer, including the requirements for delivery and post-delivery activities?								
20.402	Prior to the commitment to the customer (e.g. submission of tenders, acceptance of contracts or orders or acceptance of change orders) are requirements adequately reviewed?								
20.403	Has the organization determined and implemented effective arrangements for communicating with customers?								
20.404	Are inputs relating to product requirements defined, documented and maintained as a record?								
20.405	Are outputs of the design and development provided in a form that enables verification against the design and development inputs?								
20.406	Are systematic reviews performed in accordance with planned arrangements at suitable stages of the design and development?								
20.407	Is design and development verification performed in accordance with planned arrangements to ensure that the design outputs have met the design and development input requirements?								
20.408	Is design and development validation performed in accordance with planned arrangements?								
20.409	Are design and/or development changes identified and recorded?								
20.410	Are the purchasing processes controlled to ensure purchased product (or service) conforms to requirements?								
20.411	Does purchasing information describe the product to be purchased?								
20.412	Have the inspection or other activities necessary for ensuring that purchased product meets specified purchase requirements been established and implemented?								

Reference ICH Q7	Topics/Issue	Applicability Yes	Applicability No	Complaint Yes	Complaint tbi*	Complaint No	Kind of Documentation**	Commentary	Question posed
20.413	Are the production and service provision planned and carried out under controlled conditions?								
20.414	Have processes where deficiencies may become apparent only after the product is in use or the service has been delivered been validated?								
20.415	Is the product identified by suitable means throughout product realization?								
20.416	Does the organization exercise care with customer property while it is under the organization's control or being used by the organization?								
20.417	Is conformity of product preserved during internal processing and delivery to the intended destination?								
20.418	Has the organization determined the monitoring and measurement to be undertaken and the monitoring and measurement devices needed to provide evidence of conformity of product to determined requirements?								
20.5	**Measurement, Analysis and Improvement**								
20.50	Have the monitoring, measurement, analysis and improvement processes been planned, and implemented?								
20.51	Is information relating to customer perception monitored by the organization as to whether customer requirements have been met?								
20.52	Are internal audits conducted at planned intervals to determine whether the quality management system: – Conforms to planned arrangements, requirements of ISO 9001 and the quality management system – Is effectively implemented and maintained								
20.53	Are suitable methods applied for monitoring and where applicable, measurement of the quality management system processes necessary to meet customer requirements?								

* tbi = to be implemented   ** Procedure, SOP, OI, memo, notes (personal), Q-manual

Reference ICH Q7	Topics/Issue	Applicability Yes	No	Complaint Yes	tbi*	No	Kind of Documentation**	Commentary	Question posed
20.54	Are product characteristics monitored and measured to verify that product requirements are met?								
20.55	Is nonconforming product identified and controlled to prevent unintended use or delivery?								
20.56	Is appropriate data determined, collected and analysed to demonstrate the suitability and effectiveness of the quality management system and to evaluate where continual improvement of the effectiveness of the quality management system can be made?								
20.57	Does the organization continually improve the effectiveness of the quality management system?								
20.58	Are corrective actions taken to eliminate the cause of nonconformities and to prevent recurrence?								
20.59	Has the organization determined actions to eliminate the causes of potential nonconformities in order to prevent occurrence?								

This template has been developed by APIC/CEFIC. You may use the templates for your internal auditing purpose but for the purpose of a Third Party Audit, please note that only APIC Certified Auditors are authorised to perform an official APIC Audit that is coordinated by the API Compliance Institute. While efforts have been made to assure the accuracy APIC/CEFIC can not be held liable for any errors or omissions. You are not allowed to delete this disclaimer when using this template.

Appendix B: FDA Compliance Program Guidance Manual 7356.002F (Active Pharmaceutical Ingredient (API) Process Inspection)

CHAPTER 56 – DRUG QUALITY ASSURANCE

FIELD REPORTING REQUIREMENTS

Forward a copy of each *Establishment Inspection Report* (EIR) for inspections classified as Official Action Indicated (OAI) due to current good manufacturing practice (CGMP) deficiencies as part of any regulatory action recommendation submitted to HFD-300. For all **domestic** inspections that result in a recommendation to CDER for regulatory action based on current good manufacturing practice deficiencies, forward the recommendation and any associated documentation to CDER and other offices as specified in the Regulatory Procedures Manual. Warning Letter recommendations, with a copy of the proposed Warning Letter and supporting documents, should be sent to CDER's Division of Manufacturing and Product Quality (DMPQ), Case Management Branch (HFD-326) via the e-mail address CDER Drug WL. All other regulatory action recommendations and associated documentation based on domestic inspection are to be sent by mail to CDER's Office of Compliance (HFD-300). For all **foreign** inspections, regardless of inspection outcome, forward by mail the complete and original EIR, including exhibits, to DMPQ, Investigations & Pre-Approval Compliance Branch, Foreign Inspection Team (FIT), HFD-325. See Part III, *Special Instructions for Foreign Drug Inspections*.

This program provides guidance in evaluating compliance with CGMP and providing comprehensive regulatory coverage of all aspects of production and distribution of active pharmaceuticals ingredients (APIs) to assure that such products comply with Section 501(a)(2)(B) of the Federal Food, Drug, and Cosmetic Act (the Act). As soon as the District becomes aware of any significant inspectional, analytical, or other information developed under this program that may affect the agency's new drug and abbreviated new drug approval decisions with respect to a firm, the District should report the information immediately according to current FACTS and EES procedures. This includes promptly filing and deleting OAI notifications.

Districts are to use this revised compliance program for CGMP inspections of API facilities.

Table of Contents

PART I – BACKGROUND

General

APIs are subject to the adulteration provisions of Section 501(a)(2)(B) of the Act, which requires all drugs to be manufactured in conformance with CGMP. No distinction is made between an API and a finished pharmaceutical in the Act and the failure of either to comply with CGMP constitutes a violation of the Act. FDA has not promulgated CGMP regulations specifically for APIs or drug components (as we have for finished pharmaceuticals). Thus, the use of "CGMP" in this document refers to the requirements of the Act rather than the requirements of 21 CFR Parts 210 and 211 regulations for finished pharmaceuticals.

FDA has long recognized that the CGMP requirements in the good manufacturing practice regulations for finished pharmaceuticals (21 CFR Parts 210 and 211) are valid and applicable in concept to Active Pharmaceutical Ingredient (API) manufacturing. These concepts include, among others, building quality into the drug by using suitable equipment and employing appropriately qualified and trained personnel, establishing adequate written procedures and controls designed to assure manufacturing processes and controls are valid, establishing a system of in-process material and final drug tests, and ensuring stability of drugs for their intended period of use. In 2001, FDA adopted an internationally harmonized guidance to industry on API CGMPs in conjunction with regulatory partners in the International Conference on Harmonization of Technical Requirements for Registration of Pharmaceuticals for Human Use (ICH). This guidance is ICH Q7A, *Good Manufacturing Practice Guidance for Active Pharmaceutical Ingredients*. ICH Q7A represents the Food and Drug Administration's (FDA's) current thinking on CGMPs for API's. Thus, API and related manufacturing and testing facilities that follow this guidance generally will be considered to comply with the statutory CGMP requirement. However, alternate approaches may be used if such approaches satisfy the requirements of Section 501(a)(2)(B) of the Act as long as the approach ensure that the API meets its purported or represented purity, identity, and quality characteristics.

The term "active pharmaceutical ingredient" (API) is used in this Program consistent with the meaning of this term as defined in ICH Q7A. An active pharmaceutical ingredient is defined in ICH Q7A as "any substance or mixture of substances intended to be used in the manufacture of a drug product and that, when used in the production of a drug, becomes an active ingredient in the drug product. Such substances are intended to furnish pharmacological activity or other direct effect in the diagnosis, cure, mitigation, treatment or prevention of disease or to affect the structure and function of the body." Currently, other terms are also used by FDA and industry to mean an API. "Drug substance" and "bulk pharmaceutical chemical" (BPC) are terms commonly used to mean API and, for BPC, inactive ingredients. The use of these terms to describe active ingredients may be considered equivalent to the term used here, API.

FDA expects API manufacturers to apply CGMPs to the API process beginning with the use of starting materials, and to validate critical process steps that impact the quality and purity of the final API. Controls over material quality are expected to increase as the process approaches the final API. The level of control needed is highly dependent on the manufacturing process and increases throughout the process as it proceeds from early intermediate steps to final isolation and purification steps. The appropriate level of control depends on the risk or criticality associated with each specific process step.

ICH Q7A contains general guidance to industry on the extent and application of CGMP for manufacturing APIs under an appropriate system for managing quality. It is also intended to help ensure that APIs meet the quality and purity characteristics that they purport or are represented to possess. ICH Q7A is to be used as a guideline for inspecting API manufacturers and related facilities. If an investigator believes that a particular practice conforming to this guidance is believed to be deficient, the investigator or district should consult with CDER DMPQ before making an observation that is in conflict with ICH Q7A. A firm may also use alternate approaches to those described in ICH Q7A.

API manufacturers must register and APIs in commercial distribution must be listed under section 510(g) of the Act unless exempted under 21 CFR 207.10. Foreign drug manufacturers are also required to register and list all drugs imported or offered for import into the United States. Refer to 21 CFR 207.40 for additional information on establishment registration and drug listing requirements for foreign drug facilities.

The inspection guidance in this program is structured for the efficient use of resources planned for routine surveillance coverage of API manufacturing facilities, recognizing that in-depth coverage of all systems and all processes is not feasible for all firms on a biennial basis. It also provides for follow-up compliance coverage as needed.

Scope of APIs Covered by this Program

An API process is a related series of operations which result in the preparation of an active pharmaceutical ingredient. Major operations or steps in an API process may include multi-step chemical synthesis and fermentation, purification, crystallization, drying, milling, packing, labeling, and testing.

Some drugs processed similarly to an API may in fact be bulk finished product and subject to the requirements of 21 CFR Parts 210 and 211. If the drug material will not undergo further processing or compounding after its synthesis/fermentation/extraction, but is merely repackaged into market containers, it is a bulk finished product. However, investigators should use this program as guidance when covering the synthesis/fermentation processes that result in such APIs rather than the program for dosage forms (CP 7356.002).

This program does not cover all vaccines, whole cells, whole blood and plasma, blood and plasma derivatives (plasma fractionation), and gene therapy APIs as these drugs are regulated under the jurisdiction of the Center for Biologics Evaluation and Research.

The following APIs are to be inspected using CP7256.002M, Inspections of Licensed Biological Therapeutic Drug Products:

- biotechnology-derived APIs, including those expressed from mammalian or bacterial cell cultures

- polypeptides

Neither this Compliance Program nor ICH Q7A will provide guidance on the sterilization and aseptic processing of sterile APIs (see Q7A Section 1.3). Investigators are to use the finished product regulations (21 CFR 210 and 211) as guidance and follow CP 7356.002A, Sterile Drug Process Inspections, when inspecting the sterile processing of APIs labeled as sterile. Investigators are also to use FDA guidance on aseptic processing, Sterile Drug Products Produced by Aseptic Processing — Current Good Manufacturing Practice, 2004, in evaluating aseptic processing conditions for sterile APIs.

PART II – IMPLEMENTATION

Objective

The primary objective of this compliance program is to provide comprehensive CGMP inspectional coverage of the domestic and foreign API industry in all profile classes (i.e., types of API manufacturing processes) to determine whether a manufacturer is operating in a state of control. An API manufacturer is considered to be operating in a state of control when it employs conditions and practices that assure compliance with the intent of Section 501(a)(2)(B) of the Act. A firm in a state of control produces APIs for which there is an adequate level of assurance of quality, identity and purity.

A firm is not in a sufficient state of control if any one system, as defined in this program, is found to be significantly non-compliant with CGMPs, such that the quality, identity and purity of the API resulting from that system cannot be adequately assured. Documented CGMP deficiencies provide the evidence for concluding that a system is not operating in a state of control. See Part V, Regulatory/Administrative Strategy, for a discussion of compliance actions based on inspection findings demonstrating that a system(s) is not in a state of control.

Profile classes generalize inspection coverage from a small number of specific APIs to all APIs in that class. This program establishes a systems approach to further generalize inspection coverage from a small number of profile classes to an overall evaluation of the firm. This allows for pre-approval program inspections to focus on the specific issues related to a given application and improves the review process by providing timely and efficient support for application decisions.

Inspection of API manufacturers should be conducted and reported using the system definitions and organization in this compliance program. Focusing on systems, rather than just profile classes, will increase efficiency in conducting inspections because the systems are often applicable to multiple profile classes. An inspection under this program is profileable and will result in a determination of acceptability/non-acceptability for all API profile classes. Inspection coverage should be representative of all API profile classes manufactured by the firm. All other profile classes should be covered under the main program CP 7356.002, or related program circular, as appropriate.

Program Management Instructions

The Field will conduct API manufacturing inspections and maintain profiles or other monitoring systems with the goal that each API firm will receive biennial inspectional coverage. CDER will also identify firms for inspection coverage under this program to fulfill CDER and agency annual performance goals and as part of an initiative to ensure risk-based prioritization of inspection coverage.

Unless specifically directed by CDER, the District Office is responsible for determining the frequency and depth of coverage given to each API firm consistent with this compliance program's instructions. CGMP inspectional coverage under this program shall be sufficient to assess the state of compliance for each firm.

An inspection under this program is defined as audit coverage of 2 or more systems (the "systems" are defined below in this section and are consistent with the main program, 7356.002), with mandatory coverage of the Quality System. Inspecting at least two systems (i.e., the Quality System and one other system) will provide the basis for an overall CGMP decision.

Coverage of a system should be sufficiently detailed, with specific examples selected, so that the system inspection outcome reflects the state of control in that system for every profile class. If a particular representative system is adequate, it should be adequate for all profile classes manufactured by the firm.

If an API selected for inspection coverage is associated with a unique processing or control function in a system not chosen for coverage you may cover the unique function for that API. In doing so, you need not give full coverage to that system. For example, if an API chosen for coverage uses high purity water alone in its manufacture, you may inspect the water purification system without having to give full inspection coverage of the Materials System.

In some circumstances, it may not be possible to generalize certain deficiencies in a system to all API profile classes. If so, the unaffected profile classes may be considered acceptable if found otherwise acceptable.

Selecting unique functions within a system will be at the discretion of the investigator. Any given inspection need not cover every system.

Complete inspection of one system may necessitate further follow up of some aspects of another system to fully document the findings. However, this coverage does not constitute nor require complete coverage of the other system.

A general scheme of systems for auditing the manufacture of API consists of the following:

1. **Quality System** assures overall compliance with CGMPs and internal procedures and specifications.
2. **Facilities and Equipment System** includes activities which provide an appropriate physical environment and resources used in the production of APIs.
3. **Materials System** includes measures and activities to control starting materials, intermediates, and containers. It includes validation of computerized and inventory control processes, storage, and distribution controls.
4. **Production System** includes measures and activities to control the manufacture of APIs, including in-process sampling and testing, and process validation.
5. **Packaging and Labeling System** includes measures and activities that control the packaging and labeling of intermediates and APIs.
6. **Laboratory Control System** includes measures and activities related to laboratory procedures, testing, analytical methods development and methods validation or verification, and the stability program.

Detailed inspection coverage guidance under these systems is given in **Appendix A** of this program.

Inspection Planning

This program is intended to provide for a risk-based inspection strategy. Inspection depth should therefore reflect appropriate risks associated with a particular firm's operations, such as the firm's compliance history, the technology employed, the labeled and purported characteristics, and the intended use in the finished product, if known, of the APIs.

When a system is inspected, the inspection of that system may be considered applicable to all API products which use it. Investigators should select an adequate number and type of APIs to accomplish coverage of the system. APIs selected for coverage should be representative of the firm's overall abilities in manufacturing within CGMPs. (A profile classification scheme is used to categorize APIs by the nature of their processing, as described below.)

Profile class codes or APIs selected for coverage are to be representative of all APIs processed at the firm being inspected. Profile class codes may also be grouped by similarity, such that coverage of one profile class is sufficient to demonstrate CGMP conditions for another profile class. For example, inspecting a CSS API could amount to surrogate coverage of CSN. Similarly, inspecting a CBI could amount to surrogate coverage of other profile classes, such as CFN, CFS, and perhaps CEX.

The public health significance of certain CGMP deviations may be lower when the API is intended for a dosage form that has no dosage limitation, such as in products like calamine lotion or some OTC medicated shampoos. Such APIs should be given inspection coverage of reduced depth and intensity.

Profile Classes

The inspection findings will be used as the basis for updating all profile classes in the profile screen of the FACTS EIR coversheet that is used to record profile/class determinations. **Normally, an inspection under this system approach will result in all profile classes being updated.** Effective with this program circular are a list of profile class codes that are used to report the processes covered during API inspections. These are:

PROFILE CLASS	FULL DESCRIPTION
CSN	Non Sterile API by Chemical Synthesis
CSS	Sterile API by Chemical Synthesis
CFN	Non Sterile API by Fermentation
CFS	Sterile API by Fermentation
CEX	Plant/Animal Extraction API
CTL	Control Testing Laboratory
CTX	Testing Laboratory plus Manufacturer
CRU	Crude Bulk Not Elsewhere Classified (CRU of bulk intermediates, and contract micronizers)

Types of Inspections

There are two basic types of inspections: surveillance and compliance. Surveillance inspections are conducted on a routine basis to satisfy FDA's responsibilities to inspect drug manufacturing facilities. Compliance inspections are conducted in response to violative surveillance inspections and when a need arises to inspect a facility for-cause.

This program follows the approach in the main compliance program, 7356.002. There are two alternate approaches to inspecting a facility to satisfy FDA inspection obligations; these are termed "Full Inspection" and "Abbreviated Inspection." These are described in Part III, Inspectional, of this program.

PART III – INSPECTIONAL

Inspections of API manufacturers, whether foreign or domestic, should be conducted by experienced investigators with education and/or training particularly in fermentation (see also 7356.002M for additional inspection guidance) and chemical synthesis manufacturing methods. Use of chemists and/or microbiologists during API inspections is recommended, particularly for evaluating laboratory operations (e.g., analytical methods evaluation, analytical data, lab procedures and instrumentation), analytical review of methods used to establish impurity profiles, fermentation manufacturing processes, and complex multi-step chemical synthesis processes.

Investigators conducting API inspections must understand the basic differences between the processes used for the production of APIs and those used for finished dosage forms. APIs are usually produced by chemical synthesis or by cell culture and extraction. Thus, the production of APIs typically involves significant changes of starting materials or intermediates by various chemical, physical, and biological processing steps. The ultimate objective in API processing generally is to achieve a pure compound of certain identity, whereas the ultimate objective of finished dosage form manufacturing generally is to achieve the uniform distribution of an API among many dosing units designed to deliver a precise amount of API to a specific area of the body.

Since manufacturers of APIs are often referenced in many drug applications, each inspection should cover representative APIs when covering the systems selected (e.g., if inspecting the Production System for a site making an API by fermentation and another by synthesis, the inspection should include physical inspection and audit a sampling of records for both types of processing). This strategy, together with the classification of all profile classes upon completion of the inspection, will maximize the use of agency resources and avoid repeated visits to the same manufacturing site to cover different API profile classes referenced in subsequent applications. Any inspection of an API manufacturer should be recorded as a CGMP qualifying inspection.

Inspections should cover any specific APIs referenced in the assignment and any other representative APIs not inspected in the last two years. For foreign API firms, investigators should cover only APIs intended to be marketed or already marketed in the United States.

APIs selected for coverage should include those that are referenced in drug applications, are therapeutically significant, are intended for use in parenteral drug products, are difficult to manufacture, or are documented as having past compliance problems. However, this does not preclude the selection of less therapeutically significant APIs to evaluate specific APIs (or profile classes) not previously given in-depth coverage at the facility.

Investigators conducting API inspections should understand the general inspection strategy set forth in this program. Recognizing that API firms vary greatly in size, diversity of operations, and quality assurance systems, investigators should carefully plan their inspectional strategy at each firm. Further guidance on preparing an inspection strategy appears later.

Investigators should also review the firm's rationale for the point at which CGMPs begin, which is expected to vary by type of process (e.g., synthetic, fermentation, extraction, purification).

For an API inspection that is initiated by a pre-approval assignment, CP 7346.832, Pre-Approval Inspections/Investigations, inspection time should be reported under the appropriate program assignment codes referenced in both compliance programs based on the actual time spent in each program.

Inspection Approaches

This program provides two surveillance inspectional options: Full Inspection Option and Abbreviated Inspection Option. Either option may satisfy the biennial inspection requirement.

Full Inspection Option

The Full Inspection Option is a surveillance or compliance inspection which is meant to provide a broad and in-depth evaluation of the firm's conformity to CGMPs. The Full Inspection Option is an inspection of at least four of the six systems as listed in Part II and Appendix A of this program, one of which must be the Quality System.

A Full Inspection is appropriate:

a. For an initial FDA inspection of a facility, or after a significant change in management or organizational procedures, such as might occur after a change in ownership.

b. For a firm with a history of non-compliance or a recidivist firm whose ability to comply is short-lived. To determine if the firm meets this criterion, the District should utilize all information at its disposal, such as current and past inspection findings, results of sample analyses, complaints, recalls, and compliance actions.

c. To evaluate if important changes have occurred in the firm's state of control by comparing current operations against the EIR for the previous Full Inspection (e.g., by conducting a Full

Inspection at every fourth inspection cycle.) In addition to changes in management or ownership, the following types of changes are typical of those that warrant the Full Inspection Option:

    i.   New potential for cross-contamination arising through changes in processing or type of APIs using that equipment.

    ii.   Use of new technology requiring new expertise, significant equipment changes and/or additions, or new facilities.

d.   When District management or CDER specifically requests this option.

e.   To follow up on a Warning Letter or other regulatory action.

Abbreviated Inspection Option

The Abbreviated Inspection Option is a surveillance or compliance inspection which is meant to provide an efficient update evaluation of the firm's conformity to CGMPs. A satisfactory Abbreviated Inspection will provide documentation for continuing a firm in an acceptable CGMP compliance status. The Abbreviated Inspection Option is an inspection audit of at least two systems but not more than three systems, one of which must be the Quality System. During the course of an Abbreviated Inspection, verification of Quality System activities may require limited coverage in other systems.

An Abbreviated Inspection is appropriate when the Full Inspection Option is not warranted, including:

a.   To maintain surveillance over a historically compliant firm's activities and to provide input to the firm on maintaining and improving the CGMP level of assurance of quality of its APIs.

b.   When an intended Full Inspection finds objectionable conditions as listed in Part V of this program in one or more systems (a minimum of two systems must be completed) and District management and, as necessary, CDER Office of Compliance, concurs with reducing inspection coverage in order to expedite the issuance of a Warning Letter to correct violations.

Compliance Inspections

Compliance Inspections are inspections done "for-cause" and to evaluate or verify corrective actions after a regulatory action has been taken. The coverage given in compliance inspections must be related to the areas found deficient and subjected to corrective actions.

In addition, coverage must be given to other systems because a determination must be made on the overall compliance status of the firm after the corrective actions are taken. The firm is expected to address all of its operations in its corrective action plan after a previously violative inspection, not just the deficiencies noted in the FDA-483. The Full Inspection Option should be used for a compliance inspection, especially if the Abbreviated Inspection Option was used during the violative inspection.

Compliance Inspections include "For-Cause Inspections." For-Cause Inspections are for the purpose of investigating a specific problem that has come to the attention of the agency and may not result in the coverage of systems as described in this program. The problem may be identified by a complaint, recall, or other indicator of defective API or poorly controlled process. Coverage of these problems may be assigned under other compliance programs or PACs; however, expansion of the coverage to a CGMP inspection is to be reported under this program. For-Cause Inspections may be assigned under this program as the need arises.

Selecting Systems for Coverage

A complete description of each system and the areas for coverage are in **Appendix A** of this program. The selection of the system(s) for coverage and the relative depth or intensity of audit coverage should take into consideration the relative significance of a particular system for the firm's specific operating conditions, history of previous coverage, and history of CGMP compliance. It is expected that a Full Inspection will not be conducted every two years at most firms. Districts should select different systems for inspection coverage as a cycle of Abbreviated Inspections is carried out to build comprehensive information on the firm's total manufacturing activities over time.

Preparing the Inspection Strategy

This guidance is in addition to that given in the Investigations Operations Manual (IOM).

1. Select two or more, as appropriate, systems for inspection coverage as guided by this program (see Inspection Approaches, above). **Appendix A** contains a detailed description of the inspection coverage to be given each system when selected for inspection.

2. Select significant APIs for inspection coverage, if not specified in the assignment. Significant APIs are those which utilize all the systems in the firm very broadly and/or use special manufacturing features, e.g., complex chemical synthesis, highly sensitizing material, material of an infectious nature, or a new chemical entity made under an approved drug application. Review the firm's FACTS listing, Drug Master Files (DMF) or A/NDA files.

3. If a CDER product or CGMP/regulatory reviewer (compliance officer) is assigned to participate as a member of the inspection team, the lead investigator is to brief them on the intended inspection strategy and explain their supporting role and responsibilities for the inspection. The lead investigator should consult the reviewer on any specific A/NDA Chemistry, Manufacturing and Controls issues (whether pre-market or post-market) to be covered during the inspection.

4. Review the impurity profile for each API process to be covered during the inspection and compare these to the impurity profiles submitted in the application or DMF, if filed. (Investigators and Chemists should be particularly familiar with USP <1086> **Impurities in Official Articles**.) If the impurity profile has not been filed to CDER, review the guidance on establishing impurity profiles in ICH Q3A and Q3C.

5. Review any compendia monographs for the APIs to be inspected to verify conformity, as appropriate.

6. Before or during the inspection, determine if the firm has made process changes by comparing current operations against the EIR for the previous inspection. Also compare the current operations with those filed in the DMF or the drug application to determine whether the firm is complying with commitments made to the agency. (See also CP 7346.832 for conducting a pre-approval inspection of an API.) The following changes are typical of those that would warrant extensive coverage during the inspection:

    a. New potential for cross-contamination arising through changes in API processes or product-type lines, to include processing numerous APIs of varying toxicity in common equipment and/or facilities.

    b. Use of new technology requiring new expertise, significantly new equipment or new facilities.

    c. Changes in starting materials, intermediates, equipment, facilities, support systems, processing steps, packaging materials, or computer software, particularly those that are not referenced in the DMF or application.

7. For foreign firms, DFI will assist investigators in obtaining file information from the appropriate CDER reviewing division or compliance unit. Investigators may also request background information about the site assigned for inspection directly from the US Agent before the initiation of the inspection.

Special Inspection Reporting Instructions:

Investigators should describe in the EIR their inspection coverage and findings in sufficient detail for further agency evaluation of the firm's state of control and conformance to CGMPs. ICH Q7A may be used as a guideline in describing coverage and any findings and deficiencies observed. However, do not reference specific ICH Q7A sections in the FDA 483 observations or in the EIR. The FDA 483, if issued, is to be organized into sections for each of the systems covered. In addition to the IOM format and information reporting requirements, all EIRs of API manufacturers must include:

1. A list of APIs manufactured (or categories of drugs, if many) along with the general manufacturing process for each (e.g., chemical synthesis, fermentation, extraction of botanical material).

2. For foreign API manufacturers, the names, titles, complete mailing address, telephone and fax number of the firm's U.S. Agent.

3. For foreign API manufacturers, a report of all APIs imported into the United States in the last two years, their consignees, and an estimate of the frequency and quantity of shipments to these consignees.

4. A description of each of the systems selected for coverage, (i.e., areas, processes, and operations), what was covered, who was interviewed, and what manufacturing activities were taking place during the inspection.

5. An explanation of the choice of APIs selected for coverage.

6. Any significant changes to a firm's packaging, labeling, product line, or processes, particularly those changes not properly filed, submitted, or reported in a DMF or A/NDA.

Special Instructions for Foreign Drug Inspections:

The Division of Field Investigations (DFI) schedules foreign inspections, makes travel arrangements for inspection teams, and resolves logistical problems. CDER's Office of Compliance, Foreign Inspection Team (FIT) receives and reviews all foreign establishment inspection reports, receives and reviews all foreign firms' responses to an FDA 483, and handles all correspondence regarding inspection outcomes with foreign firms. CDER/FIT maintains the complete file for each foreign drug facility.

Investigators should instruct management at foreign firms to submit their original written response to an FDA 483 directly to CDER's Office of Compliance, with a copy to the investigator. The original response with appropriate documentation should be submitted to the following address:

> Food and Drug Administration
> Foreign Inspection Team, HFD-325
> Division of Manufacturing and Product Quality
> Center for Drug Evaluation and Research
> 11919 Rockville Pike
> Rockville, Maryland 20852-2784
> USA

Investigators and analysts are to submit their written comments to a foreign firm's response to their issued FDA 483 directly to CDER's Foreign Inspection Team (FIT) as soon as possible. After appropriate district office review and endorsement, all foreign establishment inspection reports will be promptly forwarded to FIT for review and final classification.

FIT will draft and coordinate the issuance of Warning Letters, Untitled Letters, and other correspondence to foreign firms. FIT will also recommend automatic detention of foreign firms/APIs, make recommendations to review units, and request follow-up inspections, as appropriate.

PART IV – ANALYTICAL

API samples collected by the investigator for the purpose of evaluating quality are to be submitted to the appropriate servicing laboratory. A list of each analyzing laboratory for API testing is maintained in Compliance Program Guidance 7356.002 and 7346.832. However, it should be noted that physical API samples are not required to support regulatory or administrative action against a violative firm or drug.

Forensic Chemistry Center (FCC) will request profile (also called "forensic" and "fingerprint") samples of both foreign and domestic source APIs directly from the manufacturer. Investigators are to collect API samples for profile analysis only upon specific request for collection from FCC. Such requests will be made through DFI. If an investigator is instructed to collect a profile sample, FCC will provide specific instructions as to method and amount of collection and shipping. FCC contact information is in Part VI, *Program Contacts*.

Prior to each foreign API site inspection, DFI will provide FCC with the inspection dates, the investigator's name, firm's name, address, telephone number, fax number, FEI number, any related product and application numbers, and the name of the contact person. FCC will then directly request a sample from the firm as needed. FCC may contact the investigator to request their collection of any specific information. The inspection dates will provide FCC information so they can access FACTS to obtain the EIR coversheet.

FCC is responsible for API profile sample collection and analysis and will provide periodic reports of such analysis and assist CDER in evaluating this program's effectiveness.

PART V – REGULATORY/ADMINISTRATIVE STRATEGY

An inspection report that documents that one or more systems is out of control should be classified OAI. Districts may recommend the issuance of a Warning Letter in accordance with the RPM. Normally, the issuance of a Warning Letter or the taking of other regulatory or administrative action should result in a classification of all profile classes as unacceptable. A CDER disapproval of a recommendation for Warning Letter or other regulatory action should result in a classification of all profile classes as acceptable.

A Warning Letter with a CGMP charge (i.e., 501(a)(2)(B) adulteration) involving a domestic API manufacturer requires CDER review and concurrence before issuance. See and follow FDA **Regulatory Procedures Manual** procedures for clearing Warning Letters and Untitled Letters.

A recommendation for regulatory action for API CGMP deficiencies is to cite the statute (501(a)(2)(B) or United States Code, 21 USC 351(a)(2)(B)) and not the finished pharmaceutical regulations at 21 CFR 210 and 211. A recommendation should also not cite to ICH Q7A, but may use ICH Q7A as a guideline in describing the deficiencies observed. Any regulatory action based upon CGMP noncompliance for APIs should demonstrate how the observed deviations could or did result in actual or potential defects or risk to contamination. In evaluating whether to recommend regulatory or administrative action, consider the critical attributes of the API, its therapeutic significance, and its intended use in finished drug product manufacturing.

Evidence that supports a significant deficiency or pattern of deficiencies within a system may demonstrate the failure of a system. A failure of a system puts all drugs at risk and is to be promptly corrected. The following lists the deficiencies that should result in a recommendation for regulatory action to CDER; other deficiencies may also warrant regulatory action:

1.   Contamination of APIs with filth, objectionable microorganisms, toxic chemicals, or significant amounts of other types of chemicals, or a reasonable potential for such contamination because of a finding of a demonstrated route of contamination. (Facilities and Equipment System; Production System)

2. Failure to show that API batches conform to established specifications, such as NDA, USP, customer specifications, and label claims. See also Compliance Policy Guide (CPG) 7132.05. (Quality System)

3. Failure to comply with commitments in drug applications, including DMFs, which should be accurate and current with respect to all required information, such as manufacturing process, impurity profiles (if filed), and other specifications or procedures associated with the manufacture of the API. (Quality System)

4. Distribution of an API that does not conform to established specifications. (Quality System)

5. Deliberate blending of API batches to dilute or hide filth or other noxious contaminants, or blending to disguise a critical quality defect in an attempt to obtain a batch that meets its specifications. (Production System)

6. Failure to demonstrate that water, including validation of the process water purification system, and any other solvents used in the final step of the API process are chemically and microbiologically suitable for their intended use and does not adversely alter the quality of the API. (Materials System)

7. Lack of adequate validation of critical steps in the API process, particularly concerning final separation and purification of the API, or when there is evidence that an API process is not adequately controlled. Lack of adequate control may be indicated by repeated batch failures or wide variation in final yields as compared to process average over time. See also the revised CPG 7132c.08, Process Validation Requirements for Drug Products and Active Pharmaceutical Ingredients Subject to Pre-Market Approval. (Quality System; Production System)

8. Implementation of retrospective process validation for an existing API process when the process has changed significantly, when the firm lacks impurity profile data, or when there is evidence of repeated batch failures due to process variability. (Quality System; Production System)

9. Failure to establish an impurity profile for each API process. FDA expects manufactures to establish complete impurity profiles for each API as part of the process validation effort. This includes collecting data on (1) actual and potential organic impurities that may arise during synthesis, purification, and storage of the API; (2) inorganic impurities that may derive from the API process; and (3) organic and inorganic solvents used during the manufacturing process that are known to carry over to the API. Impurity profile testing of each batch or after a specified number of batches may detect new impurities that may appear because of a deliberate or non-deliberate change in the API manufacturing process. (Laboratory Control System)

10. Failure to show that a reprocessed batch complies with all established standards, specifications, and characteristics. (Quality System; Laboratory Control System)

11. Failure to test for residues of organic/inorganic solvents used during manufacturing that may carryover to the API using analytical procedures with appropriate levels of sensitivity. (Laboratory Control System)

12. Failure to have a formal process change control system in place to evaluate changes in starting materials, facilities, support systems, equipment, processing steps, and packaging materials that may affect the quality of APIs. (All systems)

13. Failure to maintain batch and quality control records. (Quality System)

14. Incomplete stability studies to establish API stability for the intended period of use, and/or failure to conduct forced degradation studies on APIs to isolate, identify and quantify potential degradants that may arise during storage. (Laboratory Control System)

15. Use of laboratory test methods that are inadequate or have not been validated; or, the use of an inadequately qualified or untraceable reference standard. (Laboratory Control System)

16. Packaging and labeling in such a way that introduces a significant risk of mislabeling. (Packaging and Labeling System)

PART VI – REFERENCES, ATTACHMENTS, AND PROGRAM CONTACTS

References

1. ICH Guidance for Industry, Q7A *Good Manufacturing Practice Guidance for Active Pharmaceutical Ingredients*, August 2001 [http://www.fda.gov/cder/guidance/4286fnl.htm]

2. FDA *Guideline for Submitting Supporting Documentation in Drug Applications for the Manufacture of Drug Substances*, February 1987 [http://www.fda.gov/cder/guidance/drugsub.pdf]

3. Drug Manufacturing Inspections Compliance Program 7356.002, and related programs [http://www.fda.gov/ora/cpgm/#drugs]

4. Process Validation Requirements for Drug Products and Active Pharmaceutical Ingredients Subject to Pre-Market Approval, Compliance Policy Guide 490.100 (7132c.08), March 12, 2004 [http://www.fda.gov/ora/compliance_ref/cpg/cpgdrg/cpg490-100.html]

5. Performance of Tests for Compendial Requirements on Compendial Products, Compliance Policy Guide 420.400 (7132.05), October 1, 1980 [http://www.fda.gov/ora/compliance_ref/cpg/cpgdrg/cpg420-400.html]

6. The United States Pharmacopoeia/National Formulary (USP/NF) (available on-line through WebLERN)

7. FDA Regulatory Procedures Manual, [http://www.fda.gov/ora/compliance_ref/rpm/default.htm]

8. ICH Q3A *Impurities in New Drug Substances* [Word] or [PDF] (Issued 2/10/2003, Posted 2/10/2003)

9. ICH Q3C Impurities: Residual Solvents or Adobe Acrobat version (Issued 12/24/1997, Posted 12/30/1997); Q3C Tables and List [Word] or [PDF] (Posted 11/12/2003); Appendix 4, Appendix 5, and Appendix 6 (Appendices were issued with the Q3C draft guidance documents)

Program Contacts

Center for Drug Evaluation and Research (CDER)

For questions relating to product quality, the application of CGMPs, and validation of active pharmaceutical ingredient processes contact:

> Division of Manufacturing and Product Quality, Office of Compliance
> Telephone: (301) 827-9005 or (301) 827-9012
> Fax: (301) 827-8909

CDER maintains a list of contacts to help answer CGMP regulatory, technical, and program questions. This subject contact list is at http://www.fda.gov/cder/dmpq/subjcontacts.htm.

Office of Regulatory Affairs (ORA)

For questions on profile sampling of APIs:

> Food and Drug Administration
> Forensic Chemistry Center (HFR-MA500)
> Bulk Drug Group
> 6751 Steger Dr.
> Cincinnati, Ohio 45237-3097
> Telephone: (513) 679-2700, extension 185 or 181
> Fax: (513) 679-2761

Office of Regional Operations (ORO)

Division of Field Investigations
ORO/DFI (HFC-130)
Telephone: (301) 827-5653
Fax: (301) 443-3757
Division of Field Science
ORO/DFS (HFC-140)
Telephone: (301) 827-1223

Office of Enforcement (OE)

Division of Compliance Information & Quality Assurance Staff
OE/MPQAS (HFC-240)
Telephone: (240) 632-6820

PART VII – CENTER RESPONSIBILITIES

Center responsibilities are as described in Drug Manufacturing Inspections Compliance Program Guidance 7356.002 and Pre-Approval Inspection/Investigations Compliance Program Guidance 7346.832.

APPENDIX A: Description of Each System and Areas of Coverage

QUALITY SYSTEM

Assessment of the Quality System has two phases. The first phase is to evaluate whether the Quality Unit has fulfilled the responsibility to review and approve all procedures related to production, quality control, and quality assurance and assure the procedures are adequate for their intended use. This also includes the associated recordkeeping systems. The second phase is to assess the data collected to identify quality problems and may link to other major systems for inspectional coverage.

For each of the following bulleted items, the firm should have written and approved procedures and documentation resulting therefrom. The firm's adherence to written procedures should be verified through observation whenever possible. These areas are not limited to the final API's, but may also include starting materials and intermediates. These areas may indicate deficiencies not only in this system but also in other systems that would warrant expansion of coverage. All areas under this system should be covered; however the actual depth of coverage may vary from the planned inspection strategy depending upon inspectional findings.

- Adequacy of staffing to ensure fulfillment of quality unit duties

- Periodic quality reviews as described in ICH Q7A Section 2.5, *Product Quality Review*; inspection audit coverage should include API types that are representative of manufacturing at this site; inspection audit should also examine some batch and data records associated with each API quality review to verify that firm's review was sufficiently complete; and, audit should confirm that firm has identified any trends and has corrected or mitigated sources of unacceptable variation.

- Complaint reviews (quality and medical): documented; evaluated; investigated in a timely manner; includes corrective action where appropriate. Determine whether pattern of complaints and records of internal rejection or reprocessing/reworking of API batches warrant expanding the inspection.

- Discrepancy and failure investigations related to manufacturing and testing: documented; evaluated; critical deviations investigated in a timely manner and expanded to include any related APIs and material; includes corrective action where appropriate.
- Change Control (including "process improvements"): documented; evaluated; approved; need for revalidation assessed.
- Returns/Salvages: assessment; investigation expanded where warranted; final disposition.
- Rejects: investigation expanded where warranted; corrective action where appropriate.
- System to release raw materials.
- Batches manufactured since last inspection to evaluate any rejections or conversions (i.e., from drug to non-drug use) due to processing problems.
- Reprocessing and/or reworking events are properly approved and evaluated for impact on material quality.
- Recalls (including any attempt to recover distributed API not meeting its specifications or purported quality), determine cause and corrective actions taken.
- Stability Failures: investigation expanded where warranted; disposition. Determine if stability data supports API retest or expiry dates and storage conditions.
- Validation: Status of validation/revalidation activities (e.g., computer, manufacturing process, laboratory methods), such as reviews and approvals of validation protocols and reports.
- Training/qualification of employees in quality control unit functions.

ICH Q7A references for **Quality System**:

- Section 2, *Quality Management*
- Section 13, *Change Control*
- Section 14, *Rejection and Reuse of Materials*
- Section 15, *Complaints and Recalls*
- Section 16, *Contract Manufacturers (including laboratories)*.

FACILITIES AND EQUIPMENT SYSTEM

For each of the following, the firm should have written and approved procedures and documentation resulting therefrom. The firm's adherence to written procedures should be verified through observation whenever possible. These areas may indicate deficiencies not only in this system but also in other systems that would warrant expansion of coverage. When this system is selected for coverage in addition to the Quality System, all areas listed below should be covered; however, the actual depth of coverage may vary from the planned inspection strategy depending upon inspectional findings.

1. Facilities
   - Cleaning and maintenance.
   - Facility layout, flow of materials and personnel for prevention of cross-contamination, including from processing of non-drug materials.
   - Dedicated areas or containment controls for highly sensitizing materials (e.g., penicillin, beta-lactams, steroids, hormones, and cytotoxics).
   - Utilities such as steam, gas, compressed air, heating, ventilation, and air conditioning should be qualified and appropriately monitored (note: this system includes only those utilities whose output is not intended to be incorporated into the API, such as water used in cooling/heating jacketed vessels).
   - Lighting, sewage and refuse disposal, washing and toilet facilities.

- Control system for implementing changes in the building.
- Sanitation of the building including use of rodenticides, fungicides, insecticides, cleaning and sanitizing agents.
- Training and qualification of personnel.

2. Process Equipment

- Equipment installation, operational, performance qualification where appropriate.
- Appropriate design, adequate size and suitably located for its intended use.
- Equipment surfaces should not be reactive, additive, or absorptive of materials under process so as to alter their quality.
- Equipment (e.g., reactors, storage containers) and permanently installed processing lines should be appropriately identified.
- Substances associated with the operation of equipment (e.g., lubricants, heating fluids or coolants) should not come into contact with starting materials, intermediates, final APIs, and containers.
- Cleaning procedures and cleaning validation and sanitization studies should be reviewed to verify that residues, microbial, and, when appropriate, endotoxin contamination are removed to below scientifically appropriate levels.
- Calibrations using standards traceable to certified standards, preferably NIST, USP, or counterpart recognized national government standard-setting authority.
- Equipment qualification, calibration and maintenance, including computer qualification/validation and security.
- Control system for implementing changes in the equipment.
- Documentation of any discrepancy (a critical discrepancy investigation is covered under the Quality System).
- Training and qualification of personnel.

ICH Q7A references for **Facilities and Equipment System**:

- Section 4, *Buildings and Facilities*
- Section 5, *Process Equipment*
- Section 6, *Documentation and Records*

MATERIALS SYSTEM

For each of the following, the firm should have written and approved procedures and documentation resulting therefrom. The firm's adherence to written procedures should be verified through observation whenever possible. These areas are not limited to the final API, but may also incorporate starting materials and intermediates. These areas may indicate deficiencies not only in this system but also in other systems that would warrant expansion of coverage. When this system is selected for coverage in addition to the Quality System, all areas listed below should be covered; however, the actual depth of coverage may vary from the planned inspection strategy depending upon inspectional findings.

- Training/qualification of personnel.
- Identification of starting materials, containers.
- Storage conditions.

- Holding of all material and APIs, including reprocessed material, under quarantine until tested or examined and released.

- Representative samples are collected, tested or examined using appropriate means and against appropriate specifications.

- A system for evaluating the suppliers of critical materials.

- Rejection of any starting material, intermediate, or container not meeting acceptance requirement.

- Appropriate retesting/reexamination of starting materials, intermediates, or containers.

- First-in / first-out use of materials and containers.

- Quarantine and timely disposition of rejected materials.

- Suitability of process water used in the manufacture of API, including as appropriate the water system design, maintenance, validation and operation.

- Suitability of process gas used in the manufacture of API (e.g., gas use to sparge a reactor), including as appropriate the gas system design, maintenance, validation and operation.

- Containers and closures should not be additive, reactive, or absorptive.

- Control system for implementing changes.

- Qualification/validation and security of computerized or automated process.

- Finished API distribution records by batch.

- Documentation of any discrepancy (a critical discrepancy investigation is covered under the Quality System).

ICH Q7A references for **Materials System**:

- Section 7, *Materials Management*
- Section 10, *Storage and Distribution*
- Section 4.3, *Water*
- Section 6, *Documentation and Records*

PRODUCTION SYSTEM

For each of the following, the firm should have written and approved procedures and documentation resulting therefrom. The firm's adherence to written procedures should be verified through observation whenever possible. These areas are not limited to the final API, but may also incorporate starting materials and intermediates. These areas may indicate deficiencies not only in this system but also in other systems that would warrant expansion of coverage. When this system is selected for coverage in addition to the Quality System, all areas listed below should be covered; however, the actual depth of coverage may vary from the planned inspection strategy depending upon inspectional findings.

- Training/qualification of personnel.
- Establishment, adherence, and documented performance of approved manufacturing procedures.
- Control system for implementing changes to process.
- Controls over critical activities and operations.
- Documentation and investigation of critical deviations.
- Actual yields compared with expected yields at designated steps.

- Where appropriate established time limits for completion of phases of production.
- Appropriate identification of major equipment used in production of intermediates and API.
- Justification and consistency of intermediate specifications and API specification.
- Implementation and documentation of process controls, testing, and examinations (e.g., pH, temperature, purity, actual yields, clarity).
- In-process sampling should be conducted using procedures designed to prevent contamination of the sampled material.
- Recovery (e.g., from mother liquor or filtrates) of reactants; approved procedures and recovered materials meet specifications suitable for their intended use.
- Solvents can be recovered and reused in the same processes or in different processes provided that solvents meet appropriate standards before reuse or commingling.
- API micronization on multi-use equipment and the precautions taken by the firm to prevent or minimize the potential for cross-contamination.
- Process validation, including validation and security of computerized or automated process
- Master batch production and control records.
- Batch production and control records.
- Documentation of any discrepancy (a critical discrepancy investigation is covered under the Quality System).

ICH Q7A references for **Production System**:

- Section 6, *Documentation and Records*
- Section 8, *Production and In-Process Controls*
- Section 12, *Validation*
- Section 18, *Specific Guidance for APIs Manufactured by Cell Culture / Fermentation*

See also 7356.0002M for additional inspection guidance on fermentation, extraction, and purification processes.

PACKAGING AND LABELING SYSTEM

For each of the following, the firm should have written and approved procedures and documentation resulting therefrom. The firm's adherence to written procedures should be verified through observation whenever possible. These areas are not limited to the final API, but may also incorporate starting materials and intermediates. These areas may indicate deficiencies not only in this system but also in other systems that would warrant expansion of coverage. When this system is selected for coverage in addition to the Quality System, all areas listed below should be covered; however, the actual depth of coverage may vary from the planned inspection strategy depending upon inspectional findings.

- Training/qualification of personnel.
- Acceptance operations for packaging and labeling materials.
- Control system for implementing changes in packaging and labeling operations
- Adequate storage for labels and labeling, both approved and returned after issued.
- Control of labels which are similar in size, shape, and color for different APIs.
- Adequate packaging records that will include specimens of all labels used.
- Control of issuance of labeling, examination of issued labels and reconciliation of used labels.

- Examination of the labeled finished APIs.
- Adequate inspection (proofing) of incoming labeling.
- Use of lot numbers, destruction of excess labeling bearing lot/control numbers.
- Adequate separation and controls when labeling more than one batch at a time.
- Adequate expiration or retest dates on the label.
- Validation of packaging and labeling operations including validation and security of computerized process.
- Documentation of any discrepancy (a critical discrepancy investigation is covered under the Quality System).

ICH Q7A references for **Packaging and Labeling System**:
- Section 9, *Packaging and Identification Labeling of APIs and Intermediates*
- Section 17, *Agents, Brokers, Traders, Distributors, Repackers, and Relabellers* (applies to the handling of APIs after original site of manufacture and before receipt by the dosage manufacturer)

LABORATORY CONTROL SYSTEM

For each of the following, the firm should have written and approved procedures and documentation resulting therefrom. The firm's adherence to written procedures should be verified through observation whenever possible. These areas are not limited to the final API, but may also incorporate starting materials and intermediates. These areas may indicate deficiencies not only in this system but also in other systems that would warrant expansion of coverage. When this system is selected for coverage in addition to the Quality System, all areas listed below should be covered; however, the actual depth of coverage may vary from the planned inspection strategy depending upon inspectional findings.

- Training/qualification of personnel.
- Adequacy of staffing for laboratory operations.
- Adequacy of equipment and facility for intended use.
- Calibration and maintenance programs for analytical instruments and equipment.
- Validation and security of computerized or automated processes.
- Reference standards; source, purity and assay, and tests to establish equivalency to current official reference standards as appropriate.
- System suitability checks on chromatographic systems.
- Specifications, standards, and representative sampling plans.
- Validation/verification of analytical methods.
- Required testing is performed on the correct samples and by the approved or filed methods or equivalent methods.
- Documentation of any discrepancy (a critical discrepancy investigation is covered under the Quality System).
- Complete analytical records from all tests and summaries of results.
- Quality and retention of raw data (e.g., chromatograms and spectra).
- Correlation of result summaries to raw data; presence and disposition of unused data.
- Adherence to an adequate Out of Specification (OOS) procedure which includes timely completion of the investigation.

- Test methods for establishing a complete impurity profile for each API process (note: impurity profiles are often process-related).

- Adequate reserve samples; documentation of reserve samples examination.

- Stability testing program, including demonstration of stability indicating capability of the test methods.

ICH Q7A references for **Laboratory System**:

- Section 11, *Laboratory Controls*

- Section 6, *Documentation and Records*

- Section 12, *Validation*

ICH Q7A Sections 3, *Personnel*, and 6, *Documentation and Records*, apply to all systems. Section 19, *APIs for Use in Clinical Trials*, applies to APIs intended for the production of dosages solely for use in a clinical trial.

The organization and personnel, including appropriate qualifications and training, employed in any given system, will be evaluated as part of that system's operation. Production, control, or distribution records are required to maintain CGMPs and those selected for review should be included for inspection audit within the context of each of the above systems. Inspection of contract companies should be within the system for which the intermediate or API, or service is contracted and also include evaluation of their Quality System.

References

1. www.ich.org

2. Sulfanilamide Disaster, www.fda.gov/AboutFDA/WhatWeDo/History/ProductRegulation/SulfanilamideDisaster/default.htm

3. http://ecfr.gpoaccess.gov/cgi/t/text/text-idx?c=ecfr&tpl=/ecfrbrowse/Title21/21cfr58_main_02.tpl

4. FDA Compliance Program Guidance Manual Program 7356.002, Active Pharmaceutical Ingredient (API) Process Inspection Part A, www.fda.gov/downloads/ICECI/ComplianceManuals/ComplianceProgramManual/ucm125404.pdf

5. Guide to Inspection of Bulk Pharmaceutical Chemicals, US Food and Drug Administration, Revised Edition, May 1994 - no longer listed on the FDA website

6. www.fda.gov/downloads/Drugs/GuidanceComplianceRegulatoryInformation/Guidances/ucm070289.pdf

7. Directive 2001/83/EC of the European Parliament and of the Council of 6 November 2001 on the community code relating to medicinal products for human use, http://ec.europa.eu/health/documents/eudralex/vol-1/index_en.htm

8. http://ec.europa.eu/health/documents/eudralex/index_en.htm

9. http://ec.europa.eu/health/files/eudralex/vol-4/2007_09_gmp_part2_en.pdf

10. EMEA Guidance on the occasions when it is appropriate for Competent Authorities to conduct inspections at the premises of Manufacturers of Active Substances used as starting materials, Doc. Ref. EMEA/INS/GMP/50288/2005, 2 March 2005

11. Good manufacturing practices for Active ingredient manufacturers. European Federation of Pharmaceutical Industries and Associations, European Chemical Industry Council, April, 1996, www.apic.cefic.org

12. PhRMA Guidelines for the Production, Packing, Repacking, or Holding of Drug Substances. Part I, Pharmaceutical Technology 1995; 19(12): 22-32

13. PhRMA Guidelines for the Production, Packing, Repacking, or Holding of Drug Substances. Part II, Pharmaceutical Technology 1996; 20(1): 50-63

14. GMP Guide for API (draft). Pharmaceutical Inspection Cooperation Scheme, 1998

15. Good Manufacturing Practice for Active Pharmaceutical Ingredients. European Medicines Agency, CPMP/ICH/4106/00, November 2000

16. Good Manufacturing Practice for Active Pharmaceutical Ingredients. U.S. Department of Health and Human Services, Food and Drug Administration, Federal Register 2001; 66 (186), 49028-9

17. Good Manufacturing Practice for Active Pharmaceutical Ingredients. Ministry of Health, Labor, and Welfare November 2001, PMSB/ELD Notification No. 1200

18. Q11 Development and Manufacture of Drug Substances. International Conference on the Harmonization of Technical Requirements for Registration of Pharmaceuticals for Human Use, Draft Step 2 Document, 28 April 2011

19. Q8 (R2) Pharmaceutical Development. International Conference on the Harmonization of Technical Requirements for Registration of Pharmaceuticals for Human Use, Revision 2, 2009

20. Q9 Quality Risk Management. International Conference on the Harmonization of Technical Requirements for Registration of Pharmaceuticals for Human Use, 2006

21. Q10 Pharmaceutical Quality System. International Conference on the Harmonization of Technical Requirements for Registration of Pharmaceuticals for Human Use, 2009

22. http://ec.europa.eu/health/files/gmp/2012_01_20_gmp_cp_en.pdf

23. PE 009-9 (Part II) PIC/S GMP Guide (Part II: Basic Requirements For Active Pharmaceutical Ingredients), http://tinyurl.com/7ocfpl4

24. This document is no longer available from the internet - contact WHO

25. http://whqlibdoc.who.int/trs/WHO_TRS_961_eng.pdf

26. Inspektion von Qualifizierung und Validierung in pharmazeutischer Herstellung und Qualitätskontrolle , Aide mémoire 07121105, www.zlg.de/download/AM/QS/07121105.pdf

27. http://tinyurl.com/73dhseq

28. www.nihs.go.jp/mhlw/yakuji/yakuji-e_20110502-02.pdf

29. www.apic.cefic.org

30. www.efpia.org

31. www.gmp-compliance.org/eca_news_1857_6295,6301,6302,6303,6468.html

32. www.ispe.org

33. www.rx-360.org

34. www.rx-360.org/LinkClick.aspx?fileticket=aJXiedOMBNw%3D&tabid=209

35. Warning Letter: 320-12-004, Akzo Nobel Chemicals S.A. de CV, US FDA, 16-Nov-2011

36. Warning Letter: Ampac Fine Chemicals LLC., US FDA, 25-Jun-2010

37. Warning Letter: CMS #150412, ChemPacific Corporation, US FDA, 07-Apr-2011

38. Warning Letter: 320-11-014, Dr. Reddy's Laboratories, Ltd., US FDA, 03-Jun-2011

39. Warning Letter: 320-10-005, Jilin Shulin Synthetic Pharmaceutical Co. Ltd., US FDA, 13-May-2010

40. Warning Letter: 320-10-009, Kyowa Hakko Kogyo Co. Ltd., US FDA, 29-Sep-2010

41. Warning Letter: SJN-2010-03, Lilly del Caribe, Inc., US FDA, 05-Feb-2010

42. Warning Letter: NEW-27-11W, Lonza Biologics, Inc., US FDA, 01-Sep-2011

43. Warning Letter: 320-12-06, Merck KGaA, US FDA, 15-Dec-2011

44. Warning Letter: 320-11-011, Moehs Iberica, S.L., US FDA, 14-Apr-2011

45. Warning Letter: Nanjing Pharmaceutical Factory Co. Ltd., US FDA, 22-Jun-2010

46. Warning Letter: 320-11-010, Ningbo Smart Pharmaceutical Co. Ltd., US FDA, 30-Mar-2011

47. Warning Letter: NWE-03-11W, PCI Synthesis, US FDA, 2-Dec-2011

48. Warning Letter: 11-09, Scientific Protein Laboratories, LLC, US FDA, 20-Jan-2011

49. Warning Letter: 320-11-06, Synbiotics Ltd., US FDA, 16-Dec-2010

50. Warning Letter: FLA-11-17, Toxin Technology, Inc., US FDA, 08-Feb-2011

51. Warning Letter: 320-10-01, Xian Libang Pharmaceutical Co. Ltd., US FDA, 28-Jan-2010

52. Warning Letter: 320-11-20, Yag Mag Labs Private Ltd., US FDA, 12-Sep-2011

53. FDA 483: FEI # 3005231248, Hospira Boulder, Inc., US FDA, 21-Jun-2011

54. FDA 483: FEI # 2111173, Scientific Protein Laboratories LLC, US FDA, 03-Sep-2010

55. Warning Letter: 320-11-19, Sichuan Pharmaceutical Co. Ltd., US FDA, 09-Sep-2011

CHAPTER 12

Optimizing your regulatory compliance

Mark Tucker
President of Mark Tucker LLC, USA

12.1   Introduction

Several chapters in this book describe the many aspects of regulatory inspections and a focused preparation plan. What I would like to do in this chapter is to move from focus on an individual inspection to implementation of a proactive, continuous improvement program that will result in a company that is inspection ready at all times.

It is often hard for people to imagine a proactive compliance model. Many people think about compliance as an audit program, where people trained in GMPs go to a site, audit the processes and procedures for a week and then make a compliance status determination. The site then responds to the findings, making commitments to improve. This model has very limited usefulness; what you get with an audit is a snap shot in time, and the time is always very limited. Thus, you have limited visibility into the real working of the operation.

In today's regulatory environment, this type of compliance program is out of date. As outlined in ICH Q10[1], there is an expectation that there is an integrated Quality organization that partners with Operations and Research and Development to deliver high quality products to patients, and that allows Senior Management to be aware of and involved in mitigating non-compliances.

12.2   Proactive compliance

What does a company that has incorporated a proactive compliance program look like? First, when entering the plant for the first time, it is impossible to tell who is in Quality and who is in Operations. Both organizations thoroughly understand the business and the regulatory environment they work in. Both are actively aware and involved in discrepancy analysis and remediation. They both work to produce high quality products by seeking out sources of variability and errors and engineering them out of the process and procedures. A firm with a proactive compliance program has developed a process, involving Operations, Quality, R&D and Compliance, to self-identify potential issues before there are any negative effects, assess the risk associated with the potential issues and prioritize remediation of the issues so that the most severe are corrected first. Finally, Senior Management is aware of any significant issues at the site, can assess and verify lower level decision making and ensure the correct prioritization and resources are available for remediation efforts.

12.3   The proactive risk identification/remediation system

What does a proactive compliance system need to be successful (**Figure 1**)? First and foremost, the proactive compliance system must have the support of the entire management team; this cannot be just a Quality initiative.

The system must involve all parts of the organization relevant to manufacturing a quality product. The system must be able to identify potential compliance and product quality issues, based on discrepancies and issues, but also before problems actually arise. The system must have an audit function that not only fulfills the traditional audit role (a statutory requirement of most health authorities), but is also able to engage credibly with different parts of the organization to provide input before work is performed. The system must be able to assess the risk associated with an issue, and ensure that the highest risk gaps are remediated first. The system must be able to provide meaningful reports and metrics to management to help ensure that resources are being allocated correctly to improve compliance and product quality. The system must maintain a trained group of professionals from all parts of the organization to identify and assess risks, develop reports and communicate with their management. Finally, the system must tie into the inspection management team charged with preparing for and hosting health authority inspections.

12.4   Senior management

As mentioned above, Senior Management must instill a culture of proactive compliance throughout the organization and invest resources in the proactive compliance system, particularly in the areas of proactive identification and remediation of issues.

This is often a weakness in organizations I assess. At first glance, it might seem very expensive for Operations to help maintain a system that reacts to concerns before they are problems and to be actively involved in inspection/discrepancy management. In a segregated organization where Quality and Operations are not partners, Operations often does not feel the budgetary pain of discrepancies/investigations/non compliances; these costs are handled within the Quality group. They also do not feel the pressure and scrutiny of Health Authority inspectors. A proactive model asks Operations to invest more in this process while the investment of Quality, and the organization as a whole, decreases. Silo-ed organizations that are unable to look at the big picture often fail to realize the full value of a proactive system.

Proactive Compliance

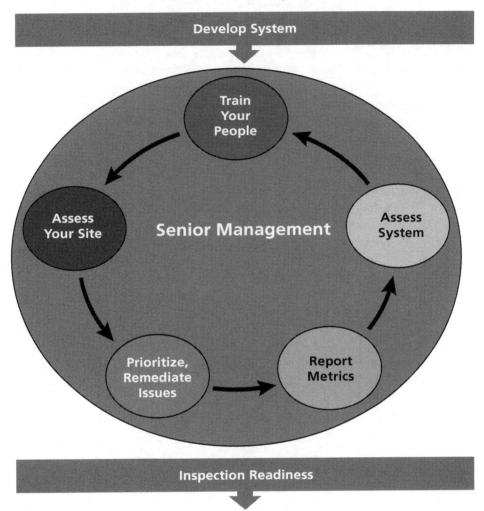

Figure.1: Proactive Compliance Model.

Senior Management must set clear goals around risk reduction at the site (or sites/corporation). In general, these goals should be multi-tiered. For example, a site or a company may have an overall goal for a 10% decrease in total risk. In addition, the goal may also require that any risk that is would be considered a major or critical finding during a Health Authority inspection have a remediation plan in effect.

Senior Management must also actively participate in the process, meeting at some frequency to evaluate their state of compliance. This Senior Management participation is invaluable for several reasons. First, when all parts of the organization participate, it demonstrates alignment and commitment of Senior Management to the process. Second, it allows Senior Management to be informed about issues they might not otherwise be aware of, and participate in the solving of issues outside their areas. Third, it raises the business acumen of the Quality leaders and the Compliance acumen of the Operational staff as they listen to the concerns of each area. Fourth, it is an excellent training ground for lower level managers attending and participating in the process both by being being coached on their decision making but also by exposure to issues and concerns outside their areas.

Finally, Senior Management must use the entire proactive compliance system, but particularly audits and audit findings, to improve operations, not to punish site personnel. The entire internal audit process is sabotaged if people, at any level, are penalized for audit findings. The goal is to have the entire company cooperating to discover errors and weaknesses for the betterment of the product and the company prior to an inspection or issues occurring. Senior management is crucial for setting the correct attitude; a "gotcha!" mindset, which is often a senior management default setting, can kill even the best proactive program.

12.5   Training

Once your issues identification, risk ranking and reporting processes are outlined (see below), it is imperative, particularly for multi-site organizations, to develop, maintain and implement a robust training system for those participating in the process. For the system to be effective, it must be iterative, and for an iterative process to work, it must be consistent between iterations.

Because of this, it is important that the process, and particularly the risk identification and risk ranking processes, be as well defined and consistent as possible. That is only possible if the teams remain constant (as much as possible, the same people participate each time) and the training system is effective and global in nature.

12.6   Issue identification

Issue identification is the most critical part of the process and relies on a team of people with high operational and GMP knowledge. All operational and quality areas should be represented and assessed.

The criteria for what is an issue in each area must be clearly defined, and is dependent on the final purpose of the list (**Figure 2**). If this list is destined to be used only for inspection preparedness, you may only want to include issues that would be deemed critical or major by a health authority. If this list is to be used as a quality tool outlining all potential GMP gaps, the criteria will necessarily be much broader. For example, are all discrepancies going to be included on the list, or will only those discrepancies deemed to have a potential product impact be included? In QC, will all Out Of Trend results (OOTs) be included, or only results reported as Out Of Specification (OOS)?

Regardless of how the list will be used, it is imperative that the criteria for inclusion be consistently applied from site to site and from iteration to iteration within a site.

12.7   Assessment and prioritization

The most common format I have seen for risk assessment and prioritization is a modified FMEA (Failure Mode and Effect Analysis, **Figure 3**). As with the inclusion criteria described above, the scoring criteria must be very well defined, clear and understood by all members of the team. The assessing team must again be very knowledgeable of both the operations being discussed and GMPs.

It is also necessary that the scoring system be able to separate issues in a fashion that underscores the significant from the less significant, and allows a clear indication of improvement as remediation efforts occur (**Figure 4**).

12.8   Remediation

The system will only be as effective in improving your operations as the efforts to close gaps are. Therefore, it is critical that real root causes be identified and actions taken to fix the underlying disease, not the symptoms. The remediation effort should be based on the prioritization efforts described above, with the highest scoring issues remediated first. Individual action items should be assigned, due dates recorded and the responsible person identified for each action (**Figure 3**).

Following remediation efforts, the residual risk should be evaluated and any additional follow ups (if necessary) re-prioritized through the risk assessment and prioritization process.

#	Risk Input	Include the issue in your risk assessment if it meets one of the following criteria:	Frequency for Assessing this input
1	Regulatory commitments	• Unmet or past-due post-market commitments related to GMPs.	Quarterly
2	QC Testing	• Adverse trend in metrics and assay trending • Use of invalidated method of lack of compliance with validated method • Incomplete or missing documentation related to methods transfer • Incomplete or inadequate sample shipping validation • Insufficient qualification of equipment/instrumentation, including manufacturing test instruments used for in-process testing • Inadequate assay performance, e.g. susceptibility to error or increase in retest rate • Incomplete documentation or inadequate document control • Issues with inventory accuracy and sample chain of custody • Insufficient documentation of rationales for sampling and testing plan • Changes not appropriately filed with health authority agency • Gaps in analyst qualification and training	Monthly
3	Batch Disposition/ List	• Multiple lots rejected or terminated for the same reason • A lot (or multiple lots) that were shipped to the wrong destination • Multiple lots impacted by the same investigation • Any lot associated with 3 or more investigations • Any re-processed lots • Any lots in quarantine for >3 months due to Quality issues	Monthly
4	Audit Findings (internal & third-party)	• Critical and major audit findings • Critical and major self-inspection findings • Minor audit observations, recommendations, and comments should be assessed and added to the log if they meet additional input criteria • Overdue audits per schedule and SOP	Within one month of receipt of completed audit report (including final response)
5	Annual Product Reviews/Product Quality Reviews	• APRs/PQRs not completed on time • Inadequate response to Action items • Action Items not identified appropriately • Adverse trend in metrics	Annually per product

#	Risk Input	Include the issue in your risk assessment if it meets one of the following criteria:	Frequency for Assessing this input
6	Final Vial Inspection (Trends)	• Adverse trend in metrics • Changes to action limits (especially if limits became less stringent) • Inadequate documentation of rationale for defect categories, acceptance criteria, and limits • Any lots undergoing multiple re-inspections • Lack of identification of particulate material and acceptable investigation & root-cause analysis • For novel defects, lack of identification or lack of incorporation into current defect library • Inadequate inspection process and inspector qualification process	Monthly

7	Product Complaints	• Adverse trend in metrics
		  – > 3 similar complaints on the same or multiple lots
		  – Adverse trends in complaints without adequate follow-up
		  – < 95% adherence to the SOP timelines for closure
		• Any product complaint that is confirmed by a production event or investigation
		• Inadequate investigations pertaining to complaints
			Quarterly

8	Contaminations	• Any Mycoplasma contaminations • Any viral contaminations (eg., MVM) • Any other unusual contaminations (e.g., yeast, mold "uncommon" isolates) • Any consecutive contaminations on same vessel class (e.g., 20L, 12K)	Quarterly
9	Change Control	• Adverse trend in change control metrics • Any change that requires a regulatory submission as a result of: – Incorrect assessment of changes per GMP/procedural requirements (includes lack of partner notification)	Quarterly

Figure 2. Some examples of potential risk assessment input values. It is very important to set and understand the boundaries you will use to populate your "risk log", and at what frequency you expect updates.

Input	#	Issue (Risk)	Risk Category	Compliance score	Readiness score	HA Focus score	Initial RPN score	Action Plan	Owner
Investigations	86_12	Description: Contamination of culture in Cell Culture during the product campaign. Investigation into first three events was robust and did not consider all possible routes of contamination. Root cause was not identified until 4th contamination event of similar nature.	Environmental Monitoring					Develop a defense package to present to HA.	B. Williams 4/11/10
								Combine the 4 investigations into one Out of Trend investigation.	V. Ziegler 3/15/10
								Review Contamination trending procedures and open CAPAs if necessary to improve to prevent this event from recurring.	K. Franks 3/15/10
		Root cause of issue determined to be incorrect set up. Sample port connection was not aseptically connected and allowed for bacterial ingress. CAPAs opened and completed that clarified SOP. Training of staff on new procedure completed.	Quality Systems	8	6	8	384	Retrain discrepancy reviewers on trending process and what constitutes a trend.	S. Smith 3/1/10
								Analyze and improve investigator training to ensure robust investigations in future. Evaluate investigation closure requirements to determine if we have the right people approving investigations.	M. Jones 8/1/10

Figure 3. FMEA risk log example.

Non-Compliance/Severity		Readiness		Agency Focus (Risk)	
Criteria	Rank	Criteria	Rank	Criteria	Rank
Critical Compliance Risk • Repeat inspection observation (within the network) • Violates previous inspection • Past-due and/or incomplete commitment • "Critical" internal audit observation • Causes released product to be at risk • Subject to regulatory or market action • Systemic break-down of GMP systems	10	**Not Prepared** • Site is **not prepared to address issue** during an inspection • Action plan **has not been developed** commitment	10	**Highly Detectable** • **Highly likely** to be detected during an inspection due to repeated occurrence • **Highly likely to** be detected on tour (standard tour route) or from a list of rejected lots, investigations, or Annual Product Review • Would be a follow up to previous regulatory/inspection commitment • May be current agency focus such as failure investigations, BPDRs, Quality Unit oversight, or aseptic processing	10
Major Compliance Risk • Violation of GMPs • "Major" internal audit observation • Quality system elements exist, but are not consistently or effectively applied • Does not comply with submission • Lack of procedures or consistent lack of adherence to procedures • Any investigation escalated to senior management • Likely to result in a major regulatory observation	8	**Not Prepared** • Site is **not prepared to address issue** during an inspection • Action plan **developed but action has not been initiated**	8	**Very Likely to be Detected** • **Very likely** to be detected during an inspection • **Possibly detected** on tour or from other standard requests provided during inspections • May be current agency focus such as AEs and complaints, record , keeping facilities and equipment, vendor qualification, raw materials, or validation	8

Non-Compliance/Severity		Readiness		Agency Focus (Risk)	
Moderate Compliance Risk • Minor departure from GMPs • Inadequate procedures • "Minor" internal audit observation • Likely to result in a minor regulatory observation • Batch rejection	6	**Somewhat Prepared** • Action plan has been developed • Work in progress • Open action items scheduled to be complete > 1 month	6	**Likely to be Detected** • Likely to be detected during an inspection • Possibly detected during documentation review • May be associated with an agency focus such as shipping and distribution, analytical methods, specifications and testing, stability and laboratory controls, environmental monitoring	6
Low Compliance Risk • Inconsistent cross-site practices • Unlikely to result in a regulatory observation • Recommendation from audit report	4	**Somewhat Prepared** • Open action items scheduled to be complete < 1 month	4	**Somewhat Likely to be Detected** • Somewhat likely to be detected during an inspection • May be categorized as an issue associated with electronic issues, cleaning or training record	4
No Compliance Risk • Compliant with policies, procedures and GMPs	2	**Well Prepared** Action plan is 100% complete Site is prepared to address the issue during an inspection	2	**Not likely to be Detected** Not at all likely to be detected during an inspection	2

Figure 4. An example of an inspection risk scoring criteria tool.

12.9   Metrics and reporting

As part of the Quality System, Senior Management, including Quality, Compliance and Operations leaders, should review the status of their sites and corporations in meeting current GMPs. The metrics and method for reporting issues should allow Senior Management to meet their obligations described above.

In general, I have found it useful to develop a multi-level reporting mechanism, designed with the granularity required for different levels of the organization. For example, the highest levels of the organization may want to only know the top risks, action plans for those risks, and anticipated completion dates, and any missed targets. However, at the function level, they may want to have access to every risk in their area, with the remediation plans, action dates and responsible people talking about any issues and concerns with the agreed plans.

Regardless of the level of metrics and reports being developed, it is important to ensure they are useful to the organization and are measuring the right things (see section on Continual Improvement below).

Top risks, remediation plans and due dates.
Issues, concerns and roadblocks to agreed remediations.
Number of findings per audit at site versus issues already self identified, with trending.
Number of issues that would result in critical or major findings, with trending.
Total "risk" at site, with trending.

Table 1. Examples of possible metric categories.

12.10   Auditing

Preferably, the auditing function is independent of both Quality and Operations. The auditing program must, by regulation/law, have the capability to perform audits of sites/areas on some regularly scheduled frequency, issue findings and track finding remediations to closure. This auditing function is well known and should come as a surprise to no one.

However, in today's environment, the auditing organization needs to be much more than this. The audit program should be able to access new construction plans, draft master validation plans/draft validation packages, and commissioning plans, for example, to provide input before the work is performed. This can help save on costly reworks/remediations and /or being found non-compliant by regulatory agencies, where the timeline to respond and implement change moves out of your company's control.

The auditors have to be knowledgeable about the business as well as the regulations, and help the organization remain current in its knowledge of regulatory expectations. Auditors need to participate in the risk assessment of gaps identified in the proactive compliance system to ensure the compliance aspect of risk is not overlooked or diluted.

In addition, the audit program must develop a mechanism to incorporate the self identification and remediation of issues by the broader organization into the audit program. For example, the audit group may publish a section in their report that details potential issues the site had identified and prioritized appropriately. These issues may not be a finding in the audit report because the appropriate actions have already been identified and reviewed. Incorporating the self-identified issues helps the broader organization see the audit function as a partner and not an outsider, because they are not just identifying negatives, but also stressing the positives at a site. It also provides visibility to Senior Management regarding which sites are or are not doing a good job of self identification/remediation.

12.11 Continual improvement

As part of the metrics reported to Management, you will want to incorporate a review of lessons learned following each inspection or audit. In addition, part of the Senior Management review should include effectiveness measures of the process. The purpose of the review is to understand where you may have internal blind spots or lack of awareness of current Health Authority expectations, and to improve your ability to identify, assess and remediate risks.

Where there any surprises during the inspection/audit?
If so, in what area?
What are the possible causes? Disagreement with inspector/auditor interpretation of regs/procedures/standards? Lack of awareness of current policy/regulations? Process oversight?
Did we prioritize the issues correctly? Did the inspectors focus on the areas we thought they would?
Did the issues we prepared for get reviewed? Was the preparation on target? Were we over prepared? Under prepared?

Table 2. Examples of things to consider following every inspection/audit to improve the process

12.12   Conclusion

By implementing a program as described above, you will be well on your way to establishing a robust, compliant operation that will continually improve by reducing errors and variability. This in turn will lead to higher quality products with a lower cost of quality, allowing you to compete more effectively with your competitors. In addition, as you continually improve, you will ensure patients get the high quality medicines they need and deserve, and Health Authorities will be able to trust your organization to do the right thing.

12.13   References

1   ICH Q10: Guideline on pharmaceuical quality system. Geneva, International Conference on Harmonisation of Technical Requirements for Registration of Pharmaceuticals for Human Use, 2008.